STUDENT SOLUTIONS MANUAL
to accompany

CHEMISTRY
PRINCIPLES AND REACTIONS
Fourth Edition

MASTERTON · HURLEY

CASSANDRA T. EAGLE
DAVID G. FARRAR
Both of Appalacian State University, NC

Harcourt College Publishers

Fort Worth   Philadelphia   San Diego   New York   Orlando   Austin
San Antonio   Toronto   Montreal   London   Sydney   Tokyo

Requests for permission to make copies of any part of the work should be mailed to the following address: Permissions Department, Harcourt, Inc., 6277 Sea Harbor Drive, Orlando, Florida 32887-6777.

Printed in the United States of America

ISBN 0-03-026917-2

012 202 765432

We dedicate this *Student Solutions Manual* to our daughter, Amanda.

May you continue to ask "why?" and search for your own solutions as you grow.

Love,

Mom and Dad

Acknowledgements

The authors thank all the wonderful babysitters who played with and nurtured our two-year-old daughter, Amanda, while we worked on this manuscript:

Elizabeth Blanchard

Becky Connerly

Lindsey Harkins

Stacey Hembree

Michelle McCrain

Marie Miller

Candice Taylor

Lori Valentine

Holly Walden

To the Student

Thank you for purchasing this *Student Solutions Manual*. It is our hope that it helps you increase your understanding of Introductory Chemistry. The goal of this Manual is to help you understand how to do the problem, not to be a substitute for logic. We caution you against memorizing how to do a problem or learning "tricks" for solving a problem which are based on simplifications that may not always hold true. There are many possible solutions to some problems. If you have an alternate way of solving a problem, please verify that method with your instructor.

Table of Contents

Chapter 1: Matter and Measurements

2. Refer to Section 1.1.

    a. Platinum is an **element**. It is found on the periodic table.

    b. Table salt is a **compound**. It is not an element because it is not found in the periodic table. It is not a mixture because it can not be broken down into more than one ingredient.

    c. Soy sauce is a **mixture**. We can ascertain this by reading the list of ingredients.

    d. Sugar is a **compound**. It is not an element because it is not found in the periodic table. It is not a mixture because it can not be broken down into more than one ingredient.

4. Refer to Section 1.1.

    a. Wine is a homogeneous mixture, the mixture is the same throughout. This is called a **solution**.

    b. Gasoline is a **solution** of a variety of organic compounds.

    c. Chocolate chip cookie batter is a **heterogeneous mixture**. Not every spoonful has the same number of chips.

6. Refer to Section 1.1.

    a. The components of a car's gas emission are separated by gas-liquid **chromatography**, which is discussed in the text.

    b. Carbon and water can be separated by **filtration**. Water is liquid while carbon (a common form is charcoal) is a solid. When the mixture is passed through the filter, the carbon is trapped by the filter paper while the water passes through.

8. Refer to Section 1.1, Table 1 and the Table of Atomic Masses in the front of your text.

    a. sodium   Na

    b. nitrogen   N

c. nickel   Ni

d. lead   Pb

10. Refer to Section 1.1 and Table 1.1.

a. Si   silicon
b. S   sulfur
c. Fe   iron
d. Zn   zinc

12. Refer to Section 1.2.

Answers refer to quantities and units used in the United States.

a. The contents of aspirin tablets are given in units of mass, generally milligrams (mg).

b.   A bottle of fruit juice is measured in volume using fluid ounces and milliliters (mL).

c. A thermostat uses units of temperature, degrees Fahrenheit (°F).

14. Refer to Section 1.2, Example 1.4 and Table 1.2.

Convert one of the numbers to the units of the other. Once the numbers are expressed in common units, they can be compared directly.

a. $303 \text{ m} \times \dfrac{1 \text{ km}}{1000 \text{ m}} = 0.303 \text{ km}$, thus: $303 \text{ m} < 303 \times 10^3 \text{ km}$.

b. $500 \text{ g} \times \dfrac{1 \text{ kg}}{1000 \text{ g}} = 0.500 \text{ kg}$, thus: $500 \text{ g} = 0.500 \text{ kg}$.

d. $1.50 \text{ cm}^3 \times \dfrac{(1 \times 10^{-2} \text{ m})^3}{1 \text{ cm}^3} \times \dfrac{1 \text{ nm}^3}{(1 \times 10^{-9} \text{ m})^3} = 1.50 \times 10^{21} \text{ nm}^3$,

thus: $1.50 \text{ cm}^3 > 1.50 \times 10^3 \text{ nm}^3$.

16. Refer to Section 1.2 and Example 1.1.

Convert from degrees Fahrenheit to Celsius, and then from Celsius to Kelvin.

$t_{°F} = 1.8t_{°C} + 32$
$T_K = t_{°C} + 273.15$

$350°F = 1.8\ t_{°C} + 32°F$
$350°F - 32°F = 1.8\ t_{°C}$
$t_{°C} = 318°F\ /\ 1.8 = 177°C$

> Note that 1.8 is an exact number, and thus does not limit the number of significant figures, resulting in three (3) significant figures in the answer.

$T_K = t_{°C} + 273.15 = 177 + 273.15 = 450\ K.$

> Note that 177 shows the greater uncertainty (meaning that it has fewer digits past the decimal than 273.15), thus the answer must also have no digits after the decimal.

18. Refer to Section 1.2 and Example 1.1 and Problem 16 (above).

Convert from Kelvin to Celsius, and then from Celsius to degrees Fahrenheit.

$T_K = t_{°C} + 273.15$
$t_{°F} = 1.8t_{°C} + 32$

$308\ K = t_{°C} + 273.15$
$t_{°C} = 308\ K + 273.15 = 35°C$

$t_{°F} = (1.8)(35°C) + 32°F = 63 + 32 = 95°F.$

20. Refer to Section 1.2 and Example 1.2.

a. 12.7040   Six (6) significant figures; zeros after the decimal are significant.

b. 200.0   Four (4) significant figures; zeros after the decimal are significant.

c. 276.2   Four (4) significant figures.

d. 4.00×10^3 Three (3) significant figures; zeros after the decimal are significant.

e. 100   This number is ambiguous; it could be 1, 2 or 3 significant figures.

$$\text{volume} \;=\; \pi(r)^2 h \;=\; (3.1416)\,(2.500\ \text{cm})^2\,(1.20\text{cm})$$

$$=\; (3.1416)(6.250\ \text{cm}^2)(1.20\ \text{cm}) \;=\; 23.6\ \text{cm}^3$$

Note that since the height has the least number of significant figures, 3, the answer can only have three significant figures.

a. $\dfrac{2.63\ \text{g}}{4.982\ \text{cm}^3} = 0.528\ \text{g}/\text{cm}^3$

The numerator has the lesser number of significant figures (s.f.), 3, and thus limits the number of s.f. in the answer to 3 (recall that leading zeros are **not** significant).

b. $\dfrac{13.54\ \text{mi}}{5.00\ \text{hr}} = 2.71\ \text{mi}/\text{hr}$

The denominator has the lesser number of significant figures, 3, and thus limits the number of s.f. in the answer to 3.

c. $13.2\ \text{g} + 1468\ \text{g} + 0.04\ \text{g} = 1481\ \text{g}$

The second number shows the greatest uncertainty (meaning that it has the least number of digits past the decimal), thus the answer must also have no digits after the decimal.

d. $\dfrac{2\ \text{g} + 0.127\ \text{g} + 459\ \text{g}}{6.2\ \text{cm}^3 - 0.567\ \text{cm}^3} = \dfrac{461\ \text{g}}{5.6\ \text{cm}^3} = 82\ \text{g}/\text{cm}^3$

In the numerator, the first number shows the greatest uncertainty (with no digits past the decimal), thus the sum must also have no digits after the decimal. In the denominator, the first number shows the greatest uncertainty, thus the sum must also have only one digit after the decimal. These sums result in the numerator having 3 significant figures (s.f.) while the denominator has only 2. Thus the final answer can only have 2 s.f.'s.

Use the IRS rules for rounding: if the number being dropped is five (5) or greater, round up, otherwise round down.

a. 7.49 g

b. 298.69 cm

c. 1 x 10^1 lb (Use scientific notation to avoid the ambiguity 10 would give.)

d. 12.1 oz

28. *Refer to Section 1.2.*

a. 4020.6 mL = 4.0206 x 10^3 mL

b. 1.006 g (This is already in proper scientific notation.)

c. 100.1°C = 1.001 x 10^2°C

30. *Refer to Section 1.2.*

a. Exact. One foot equals 12 inches by definition.

b. Exact. One cannot have a fraction of a geranium, it may be broken, sickly or stunted, but it is still a geranium.

c. Mileage is **not** exact, it can vary from day to day with driving conditions.

32. *Refer to Section 1.2, Tables 1.2 and 1.3 and Example 1.4.*

a. $22.3\,\text{mL} \times \dfrac{1\,\text{L}}{1000\,\text{mL}} = 0.0223\,\text{L} = 2.23 \times 10^{-2}\,\text{L}$

   3 significant figures because the initial volume is expressed in three, and the conversion factor is an exact number.

b. $22.3\,\text{mL} \times \dfrac{1\,\text{cm}^3}{1\,\text{mL}} \times \dfrac{(1\,\text{in})^3}{(2.54\,\text{cm})^3} = 1.36\,\text{in}^3$

   3 significant figures because the initial volume and the conversion factors are expressed in three.

c. $22.3\,\text{mL} \times \dfrac{1 \times 10^{-3}\,\text{L}}{1\,\text{mL}} \times \dfrac{1.057\,\text{qt}}{1\,\text{L}} = 0.0236\,\text{qt}$

   3 significant figures because the initial volume is expressed in three while the metric to english conversion factor has 4 s.f. and the metric conversion factor is an exact number.

34. Refer to Section 1.2, Tables 1.2 and 1.3, Example 1.4 and Problem 32 (above).

a. $1 \text{ na. mi. } \times \dfrac{6076.12 \text{ ft}}{1 \text{ na. mi.}} \times \dfrac{1 \text{ mi}}{5280 \text{ ft}} = 1.15078 \text{ mi.}$

6 significant figures, the distance is given as an exact number, thus the number of s.f.'s is constrained by the conversion factors, and 5280 ft/mi. is an exact number..

b. $1 \text{ na. mi. } \times \dfrac{6076.12 \text{ ft}}{1 \text{ na. mi.}} \times \dfrac{12 \text{ in}}{1 \text{ ft}} \times \dfrac{1 \text{ m}}{39.37 \text{ in}} = 1852 \text{ m.}$

4 significant figures, the distance is given as an exact number, as is the 12 in/ft conversion factor, thus the number of s.f.'s is constrained by the other two conversion factors.

c. $22 \text{ knots } \times \dfrac{1 \text{ na. mi./hr}}{1 \text{ knot}} \times \dfrac{1.151 \text{ mi}}{1 \text{ na mi.}} = 25 \text{ mi./hr} = 25 \text{ mph}$

2 significant figures, the first conversion factor is an exact number, while the second (from part a above) has 4 s.f.'s, thus the answer is constrained by the 2 s.f.'s given in the initial number.

36. Refer to Section 1.2, Table 1.3 and Problem 34 (above).

First calculate the area in sq. ft occupied by one hectare. Then convert from acres to hectares to sq. ft.

One hectare is defined by a square 100 m on a side, thus:
1 hectare = 100 m x 100 m = 10,000 m²

$1 \text{ hectare} = 10{,}000 \text{ m}^2 \times \left(\dfrac{39.37 \text{ in}}{1 \text{ m}} \right)^2 \times \left(\dfrac{1 \text{ ft}}{12 \text{ in}} \right)^2 = 1.076 \times 10^5 \text{ ft}^2$

$1 \text{ acre} \times \dfrac{1 \text{ hectare}}{2.47 \text{ acres}} \times \dfrac{1.076 \times 10^5 \text{ ft}^2}{1 \text{ hectare}} = 4.36 \times 10^4 \text{ ft}^2$

38. Refer to Section 1.2, Table 1.3 and Example 1.5.

Convert 1 pint to liters, then calculate the percentage.

$1 \text{ pt} \times \dfrac{4 \text{ qt}}{8 \text{ pt}} \times \dfrac{1 \text{ L}}{1.057 \text{ qt}} = 0.4730 \text{ L}$

$$\% = \frac{part}{total} \times 100\% = \frac{0.4730\,L}{6.0\,L} \times 100\% = 7.9\%$$

---

40. Refer to Section 1.2 and Table 1.3.

Volume = area × height. Convert the area to units of cm² and the height to cm and calculate volume. Then convert from cm³ (mL) to liters.

$$area = 3.02 \times 10^6\,mi^2 \times \left(\frac{1.609\,km}{1\,mi}\right)^2 \times \left(\frac{1 \times 10^5\,cm}{1\,km}\right)^2 = 7.82 \times 10^{16}\,cm^2$$

$$height = 2\,in \times \frac{2.54\,cm}{1\,in} = 5.08\,cm$$

$$volume = 7.82 \times 10^{16}\,cm^2 \times 5.08\,cm = 3.97 \times 10^{17}\,cm^3 = 3.97 \times 10^{17}\,mL$$

$$3.97 \times 10^{17}\,mL \times \frac{1\,L}{1000\,mL} = 3.97 \times 10^{14}\,L$$

---

42. Refer to Section 1.2 and Table 1.3.

Calculate the amount of silver (in ounces) in one dollar and the value of that silver. Is it more or less than one dollar? Note that 90.0% means 90.0 grams per 100 grams.

$$1\,silver\,dollar \times \frac{27.0\,g}{1\,silver\,dollar} \times \frac{90.0\,g}{100\,g} \times \frac{0.03527\,oz}{1\,g} = 0.857\,oz.\,silver$$

$$0.857\,oz\,silver \times \frac{\$4.32}{1\,oz} = \$3.70$$

Thus the silver in one silver dollar is worth more than the dollar value of the coin.

---

44. Refer to Sections 1.2 and 1.3 and Example 1.6..

Convert the volume to milliliters and calculate density (density = mass (g)/ volume (mL)).

$$\frac{252\,g}{3/4\,cup} \times \frac{1\,cup}{225\,mL} = 1.49\,g/mL$$

46. Refer to Section 1.3 and Problem 44.

Calculate the volume of the object from the change in the volume of the graduated cylinder, then calculate the density.

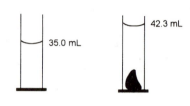

Volume of object = 42.3 mL – 35.0 mL = 7.3 mL

$$\text{density} = \frac{\text{mass (g)}}{\text{volume (mL)}} = \frac{11.33\,g}{7.3\,mL} = 1.6\,g/mL$$

48. Refer to Section 1.3, Table 1.3 and Problem 40 (above).

Using the mass and density, calculate the volume of the foil. You can then calculate the thickness since you have the length and width.

$$\text{volume} = 8.9\,g \times \frac{1\,cm^3}{2.70\,g} = 3.3\,cm^3$$

volume = length x width x thickness

$$3.3\,cm^3 = \left(12\,in \times \frac{2.54\,cm}{1\,in}\right) \times \left(11\,in \times \frac{2.54\,cm}{1\,in}\right) \times \text{thickness}$$

thickness = 0.0039 cm = 0.039 mm

50. Refer to Section 1.3.

Recall that 5.00% indicates that there are 5.00 g acetic acid per 100 g vinegar. Use density to calculate the mass of vinegar in 5.00 L, and then use the percentage to calculate how much of that vinegar is acetic acid.

$$5.00\,L\ \text{vingar} \times \frac{1000\,mL}{1\,L} \times \frac{1.01\,g\ \text{vinegar}}{1.00\,mL\ \text{vinegar}} \times \frac{5.00\,g\ \text{acetic acid}}{100\,g\ \text{vinegar}} = 253\,g\ \text{acetic acid}$$

52. Refer to Section 1.3 and Example 1.7.

Use the solubility of potassium chloride as a conversion factor.

a. at 30°C: $48.6\,g\ \text{water} \times \dfrac{37.0\,g\ \text{potasium chloride}}{100\,g\ \text{water}} = 18.0\,g\ \text{potassuim chloride}$

b. at 70°C: 52.0 g potasium chloride $\times \dfrac{100 \text{ g water}}{48.3 \text{ g potasium chloride}} = 108$ g water

c. For this part, calculate the amount of potassium chloride that will dissolve in 75 g water at each of the temperatures. Is it more than 30.0 g?

at 30°C: 75.0 g water $\times \dfrac{37.0 \text{ g potasium chloride}}{100 \text{ g water}} = 27.8$ g potassium chloride

at 70°C: 75.0 g water $\times \dfrac{48.3 \text{ g potasium chloride}}{100 \text{ g water}} = 36.2$ g potassium chloride

Thus, not all the potassium chloride would dissolve at 30°C, but would at 70°C.

---

54. Refer to Section 1. 3.

a. **physical.** Color is observed in the absence of a chemical reaction.

b. **physical.** The state (solid) is observed in the absence of a chemical reaction.

c. **physical.** Solubility is observed in the absence of a chemical reaction.

d. **chemical.** This is a chemical reaction.

---

56. Refer to Section 1.3 and Problem 48 (above).

Calculate the volumes using the densities of each.

Volume of lead $= 1 \text{ g} \times \dfrac{1 \text{ cm}^3}{11.34 \text{ g}} = 8.818 \times 10^{-2} \text{ cm}^3$

Volume of oxygen $= 1 \text{ g} \times \dfrac{1 \text{ cm}^3}{1.31 \times 10^{-3} \text{ g}} = 763 \text{ cm}^3$

One gram of oxygen takes up almost 10,000 times the volume of the lead. Gases will always have a much greater volume than equal amounts of a solid.

58. Refer to Sections 1.2 and 1.3 and Table 1.3.

$$108 \text{ carats} \times \frac{2.00 \times 10^2 \text{ mg}}{1 \text{ carat}} \times \frac{1 \text{ g}}{1000 \text{ mg}} \times \frac{1 \text{ lb}}{453.6 \text{ g}} = 4.76 \times 10^{-2} \text{ lbs}$$

$$108 \text{ carats} \times \frac{2.00 \times 10^2 \text{ mg}}{1 \text{ carat}} \times \frac{1 \text{ g}}{1000 \text{ mg}} \times \frac{1 \text{ cm}^3}{3.51 \text{ g}} \times \left(\frac{1 \text{ in}}{2.54 \text{ cm}}\right)^3 = 0.376 \text{ in}^3$$

60. Refer to Section 1.3 and Problem 48 (above).

Using the mass and density, calculate the volume, and from that the cylinder's radius.

$$153.2 \text{ g} \times \frac{1 \text{ cm}^3}{4.55 \text{ g}} = 33.7 \text{ cm}^3$$

$V = \pi r^2 h$

$33.7 \text{ cm}^3 = (3.14)(r^2)(7.75 \text{ cm})$

$r^2 = 1.38 \text{ cm}^2$

$r = 1.18 \text{ cm}$

diameter = $2r = 2.36$ cm

62. Refer to Sections 1.1 and 1.3.

a. Chemical properties involve a chemical change in the substance, while physical properties can be observed with no change in the identity of the substance.

b. Distillation is used to separate a homogeneous mixture and involves a phase change from liquid to gas and then back to liquid. Filtration is used to separate a the liquid from the solid in a heterogeneous mixture and does not involve a phase change.

c. The solute is the substance dissolved in the solvent.

64. Refer to Section 1.3 and Problem 56 (above).

The less dense substance will float on top of the more dense substance. Thus, from top to bottom, the substances will be: ethyl alcohol, lead and mercury.

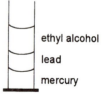

ethyl alcohol

lead

mercury

We want: $t_{°F} = 2(t_{°C})$. Substitute this into the equation for converting Fahrenheit to Celsius.

$t_{°F} = 1.8(t_{°C}) + 32$

$2(t_{°C}) = 1.8(t_{°C}) + 32$

$0.2(t_{°C}) = 32$

$t_{°C} = 160°C$

When $t_{°C} = 160°C$, $t_{°F} = 1.8(160) + 32 = 320°F$

67. Refer to Section 1.1 and Table 1.3.

Convert gallons to km^3. Then convert nm to km and solve for area (remember: volume = area x depth).

$$31.5 \text{ gal} \times \frac{4 \text{ qt}}{1 \text{ gal}} \times \frac{57.75 \text{ in}^3}{1 \text{ qt}} \times \frac{(1 \text{ m})^3}{(39.37 \text{ in})^3} \times \frac{(1 \text{ km})^3}{(1000 \text{ m})^3} = 1.19 \times 10^{-10} \text{ km}^3$$

$$100 \text{ nm} \times \frac{1 \text{ m}}{1 \times 10^9 \text{ nm}} \times \frac{1 \text{ km}}{1000 \text{ m}} = 1.0 \times 10^{-10} \text{ km}$$

$$\frac{1.19 \times 10^{-10} \text{ km}^3}{1.0 \times 10^{-10} \text{ km}} = 1.2 \text{ km}^2$$

68. Refer to Section 1.3 and Table 1.3

First calculate the volume of Al wire needed, then calculate the length. (Remember that diameter = 2r.)

$$12 \text{ g} \times \frac{1 \text{ cm}^3}{2.70 \text{ g}} = 4.4 \text{ cm}^3$$

$$r = \frac{0.200 \text{ in} \times 2.54 \text{ cm/in}}{2} = 0.254 \text{ cm}$$

$V = \pi r^2 h$

$4.4 \text{ cm}^3 = (3.14)(0.254 \text{ cm})^2(h)$

$h = 22 \text{ cm}$

Calculate the amount of air one breathes in a year and the amount of lead this corresponds to. Then determine the amount retained in one's lungs.

$$\frac{8.50 \times 10^3 \text{ L}}{1 \text{ day}} \times \frac{365 \text{ days}}{1 \text{ year}} = 3.10 \times 10^6 \text{ L air / year}$$

$$\frac{7.0 \times 10^{-6} \text{ g Pb}}{1 \text{ m}^3} \times \frac{1 \text{ m}^3}{1 \times 10^3 \text{ L}} \times \frac{3.10 \times 10^6 \text{ L}}{1 \text{ year}} = 2.17 \times 10^{-2} \text{ g Pb / year}$$

2.17×10^{-2} g / year $\times 0.75 \times 0.5 = 8.1 \times 10^{-3}$ g Pb / year   (retained by lungs)

Chapter 2: Atoms, Molecules and Ions

2. Refer to Section 2.1 and "Chemistry, the Human Side."

There is no detectable change in mass during the course of a chemical reaction.

Consider baking a pound cake using the old fashion recipe: 1 lb each of flour, sugar, butter and eggs. Your product (the resulting cake and steam released during baking) will weigh four pounds, equal to the four pounds of reactants (ingredients).

4. Refer to Section 2.1 and "Chemistry, the Human Side."

a. If the mass of the reactants equals the mass of the products, one is demonstrating the **law of conservation of mass**.

b. **None** of the laws are illustrated here.

c. Marble from different places in the world will have the same amounts of calcium due to the **law of constant composition**.

6. Refer to Section 2.2 and Figure 2.4"

Rutherford discovered the nucleus. His research group removed the electrons from helium atoms and then shot them at a gold foil target. He observed that some of those helium atoms bounced backwards. Rutherford deduced that this was because the helium atoms were colliding with the gold nuclei in the foil.

8. Refer to Section 2.2, Example 2.1 and the Periodic Table (inside front cover).

Se: Atomic number (Z) = 34 = 34 protons.
Mass number (A) = 46 neutrons + 34 protons = 80

$$^{80}_{34}\text{Se}$$

10. Refer to Section 2.2, Example 2.1 and the Periodic Table (inside front cover).

a. C-12: Mass number (A) = 12
Atomic number (Z) = 6 (from periodic table)
$$^{12}_{6}\text{C}$$

C-13:   Mass number (A) = 13
        Atomic number (Z) = 6 (from periodic table)
        $_6^{13}\text{C}$

b. i.   All atoms of the same element have the same atomic number and therefore must have the same number of protons.
    ii. Since the number of protons is invariant, if the mass changes it must be due to a change in the number of neutrons.
    iii. Since the number of electrons in a neutral molecule must equal the number of protons, and the number of protons is invariant, the number of electrons must also be the same.

---

12. Refer to Section 2.2 and Problems 10 (above) and 11.

Write out the nuclear symbols for the given elements, then consider the questions.

$_{20}^{40}\text{Ca}$ $_{20}^{41}\text{Ca}$ $_{19}^{41}\text{K}$ $_{18}^{41}\text{Ar}$

a. Ca-41, K-41 and Ar-41 are isobars since they have identical mass numbers.
    Ca-40 and Ca-41 are isotopes since they have identical atomic numbers.

b. Ca-40 and Ca-41 have the same number of protons.

c. Ca-41, K-41 and Ar-41 have identical mass numbers but different numbers of neutrons.

---

14. Refer to Section 2.2, Example 2.1 and the Periodic Table (inside front cover).

a. The number of protons = the atomic number (written above the chemical symbol) = 34.

b. Number of neutrons = mass number – atomic number.
    Number of neutrons = 75 – 34 = 41.

c. Number of electrons = number of protons (in a neutral atom) = 34.

d. Se^{2-} must have two more electrons than protons [(# of protons) – (# of electrons) = charge]. The number of protons equals the atomic number.
    # of neutrons = 41
    # of protons = 34
    # of electrons = 36

16. Refer to Section 2.2 and Examples 2.1 and 2.3.

Remember: Number of protons = atomic number
Number of neutrons = mass number – atomic number.
(# of electrons) = (# of protons) – charge

Nuclear Symbol	Atomic Charge	# of Protons	# of Neutrons	# of Electrons
$^{64}_{30}Zn$	0	30	34	30
$^{28}_{14}Si^{+4}$	+4	14	14	10
$^{32}_{16}S^{-2}$	-2	16	16	18

18. Refer to Section 2.2 and Example 2.1.

Number of protons = sum of the atomic numbers
Number of neutrons = (sum of the mass numbers) – (# of protons).
(# of electrons) = (# of protons) – charge

a. N_2: Two nitrogens, each with an atomic number of 7, therefore
the number of protons is: 2 x 7 = 14.
This molecule has no charge, thus the number of electrons = 14 – 0 = 14.

b. N_3^-: Three nitrogens, each with an atomic number of 7, therefore
the number of protons is: 3 x 7 = 21.
This molecule has a -1 charge, thus the number of electrons = 21 – (-1) = 22.

c. N_5^+: Five nitrogens, each with an atomic number of 7, therefore
the number of protons is: 5 x 7 = 35.
This molecule has a +1 charge, thus the number of electrons = 35 – 1 = 34.

d. N_5N_5: Ten nitrogens, each with an atomic number of 7, therefore
the number of protons is: 10 x 7 = 70.
This molecule has no charge, thus the number of electrons = 70 – 0 = 70.

20. Refer to Section 2.3, Figure 2.7 and the Table of Atomic Masses (inside front cover).

a.   S = sulfur
b.   Sc = scandium
c.   Se = selenium
d.   Si = silicon
e.   Sr = strontium

a. S is a non-metal (it is to the right of the metalloids).
b. Sc is a metal (it is to the left of the metalloids).
c. Se is a non-metal (it is to the right of the metalloids).
d. Si is a metalloid (as stated in the text).
e. Sr is a metal (it is to the left of the metalloids).

A period is a horizontal row, numbered from top to bottom. The first period consists of hydrogen and helium.

a. The second period is the row beginning with Li and ending with Ne. B is a metalloid. The two elements to the left of B are metals, the five to the right of B (C, N, O, F, and Ne) are non-metals.

b. The fourth period is the row beginning with K and ending with Kr. Ge and As are metalloids. The 13 elements to the left of the metalloids are metals, the three to the right (Se, Br, and Kr) are non-metals.

c. The sixth period is the row beginning with Cs and ending with Rn (including the lanthanide series). The 30 elements to the left of the thick black line are metals, the two to the right (At and Rn) are the non-metals.

Groups are arranged in vertical columns in the periodic table, numbered from left to right.

a. **Group 13** (B to Tl) has one metalloid (B) and no non-metals.

b. Groups 3-12 are the transition metals and groups 14-18 all have at least one non-metal. Likewise, group 1 has a non-metal (H). Thus only **groups 2 and 13** have neither non-metals nor transition metals.

c. Two groups have neither metals nor metalloids: **groups 17 and 18.**

28. *Refer to Section 2.4 and Example 2.2.*

Count up the total number of each element. Remember that a subscript after the parentheses indicates that everything within the parenthesis is multiplied by that number.

a. $(CH_3)_2NH$
    C: 2
    H: 2 x 3 + 1 = 7 H
    N: 1
    thus: C_2H_7N

b. $CH_3(CH_2)_2OH$
    C: 2 x 1 + 1 = 3
    H: 3 + 2 x 2 + 1 = 8
    O: 1
    thus: C_3H_8O

30. *Refer to Sections 2.4 and 2.6 and Table 2.4.*

a. H_2O

b. NH_3

c. N_2H_4 ·

d. SF_6 (the "hexa" prefix indicate that 6 fluorides are attached to the sulfur).

e. PCl_5 (the "penta" prefix indicate that 5 chlorides are attached to the phosphorus).

32. *Refer to Sections 2.4 and 2.6 and Example 2.7.*

a. Se_2Cl_2 diselenium dichloride

b. N_2O_4 dinitrogen tetraoxide

c. PH_3 phosphine (common name)

d. IF_7 iodine heptafluoride

e. SiC   silicon carbide

34. Refer to Section 2.5.

The sum of the charges must equal zero; if not, add extras of the needed ion to "balance" the charges.

K^+ and Cl^- give KCl $(+1 + (-1) = 0)$

K^+ and S^{-2} give K_2S $(2(+1) + (-2) = 0)$

K^+ and Ca^{+2} impossible combination since the charges can never add up to zero.

Ca^{+2} and Cl^- give $CaCl_2$ $(+2 + 2(-1) = 0)$

Ca^{+2} and S^{-2} give CaS $(+2 + (-2) = 0)$

36. Refer to Section 2.5, Tables 2.2 and 2.3, and Example 2.6.

The sum of the charges of the ions must equal zero. Increase the number of the deficient ion until the charges "balance."

a. Cobalt(II) acetate. The Roman numeral tells us that cobalt is a +2 ion (Co^{2+}). Acetate is a -1 ion ($C_2H_3O_2^-$). For the charges to add up to zero, we must have two acetates.
   $Co(C_2H_3O_2)_2$

b. Barium oxide. Barium in in group 2 and thus has a +2 charge. Oxygen is in group 16 and thus has −2 charge. These charges add up to zero.
   BaO

c. Aluminum sulfide. Aluminum is in group 13 and thus has a +3 charge (Al^{+3}), while sulfide has a −2 charge (S^{-2}). For these charges to add up to zero, we must have 2 aluminum and 3 sulfide ions.
   Al_2S_3

d. Potassium permanganate. Potassium, in group 1, has a +1 charge (K^+)and permanganate has a −1 charge (MnO_4^-). These charges add up to zero
   $KMnO_4$

e. Sodium hydrogen carbonate. Sodium is in group 1 and thus has a +1 charge (Na^+). Hydrogen carbonate ion is a -1 ion (HCO_3^-). These charges add up to zero.
   $NaHCO_3$

38. Refer to Section 2.5, Tables 2.2 and 2.3 and the Periodic Table.

Name the compounds by naming the cations (positive ions) followed by the anions (negative ions). For transition metals, indicate the charge of the metal with Roman numerals after the chemical symbol.

a. KNO_3 **potassium nitrate**

b. $(NH_4)_2SO_4$ **ammonium sulfate**

c. $Mg_3(PO_4)_2$ **magnesium phosphate**

d. $FeCl_3$ **iron(III) chloride**. Since each chloride has a –1 charge, the Fe must have a +3 charge for the charges to balance (add up to zero).

e. Cr_2O_7 **chromium(VII) oxide**. Since each oxygen has a –2 charge, each chromium must have a +7 charge.

40. Refer to Sections 2.5 and 2.6, Example 2.8 and Problem 38 (above).

a. $HCl(aq)$ The aq symbol indicates that the HCl is in solution and is therefore an acid.   **hydrochloric acid**

b. $HClO_3(aq)$ The aq symbol indicates that the $HClO_3$ is in solution and is therefore an acid.   **chloric acid**

c. **iron(III) sulfite**

d. **barium nitrite**

e. **sodium hypochlorite**

42. Refer to Sections 2.5 and 2.6, Tables 2.2 and 2.3 and Problem 38 (above).

	Name	Formula
a	sodium dichromate	$Na_2Cr_2O_7$
b	**bromine triiodide**	BrI_3
c	copper(II) hypochlorite	$Cu(ClO)_2$
d	**disulfur dichloride**	S_2Cl_2
e	potassium nitride	K_3N

a. Sodium has a +1 charge (Na^+) while dichromate has a –2 charge ($Cr_2O_7^{-2}$), thus there must be two sodiums to balance the charges.

b. This is a molecular compound since both elements are non-metals, so it is important to add the proper prefix to denote that there are three iodides.

c. Copper has a +2 charge (Cu^{+2}) while hypochlorite has a –1 charge (ClO^-), thus there must be two hypochlorites to balance the charges.

d. This is a molecular compound since both elements are non-metals, so it is important to add the proper prefixes to denote that there are two sulfurs and two chlorides.

e. Potassium has a +1 charge (K^+) while nitride has a –3 charge (N^{-3}), thus there must be three potassiums to balance the charges.

---

44. Refer to Sections 2.3, 2.5 and 2.6.

a. A_2B_3:   A has 7 protons, therefore $Z=7$ and A is nitrogen (N).
B is in group 16, period 2 and so must be oxygen (O).
N_2O_3 dinitrogen trioxide.

b. The major component of bones is Ca, which forms Ca^{2+}
The anion described is PO_4^{3-}.
$Ca_3(PO_4)_2$ calcium phosphate.

c. Zn forms the Zn^{2+} ion.
Sulfate is SO_4^{2-}. Loss of an oxygen gives the sulfite ion (SO_3^{2-}).
$ZnSO_3$ zinc sulfite.

---

46. This problem combines concepts from all sections of Chapter 2.

a. H_2 (molecule)   # of protons = sum of the atomic numbers = 2 x 1 = 2
# of electrons = # of protons – charge = 2 – 0 = 2
# of neutrons = sum of mass numbers - # of protons = 2 – 2 = 0

H^- (anion)   # of protons = atomic number = 1
# of electrons = # of protons – charge = 1 – (-1) = 2
# of neutrons = mass number - # of protons = 1 – 1 = 0

H^+ (cation)   # of protons = atomic number = 1
# of electrons = # of protons – charge = 1 – (+1) = 0
# of neutrons = mass number - # of protons = 1 – 1 = 0

b. A metal in group 2 with 12 protons ($Z = 12$) is magnesium (Mg). Metals form cations (positive ions). Being in group 2, Mg would form a +2 ion (Mg^{+2}) and react with H^- to give **MgH_2** (remember that charges must balance). **Magnesium hydride.**

c. H^+ combines with anions (negative ions) and would react with Cl^- to form HCl, an **acid**.

48. *Refer to Sections 2.5 and 2.6.*

a. "Compounds containing carbon atoms are molecular" is **usually true**. Carbon, being a non-metal can combine with other non-metals to form molecular compounds such as those mentioned in Section 2.4 (indeed, a whole field of chemistry, called organic chemistry, is dedicated to the study of such compounds). Carbon can, however, also form ionic compounds such as calcium carbide (used in old miners lamps).

b. "A molecule is made up of non-metal atoms" is **always true**, by definition.

c. "An ionic compound has at least one metal atom" is **usually true**. Most ionic compounds do contain a metal, but there are ionic compounds in which the cation (positive ion) is not a metal, such as in ammonium chloride, NH_4Cl.

50. *Refer to Section 2.1.*

There are 4 hydrogen *molecules* (H_2) and 7 chlorine *molecules* (Cl_2), giving 8 hydrogen *atoms* and 14 chlorine *atoms*. Since they react in a one to one ratio, giving HCl, after all the hydrogens react, there will still be 6 unreacted chlorine *atoms* (as 3 unreacted chlorine *molecules*) as shown pictorially below.

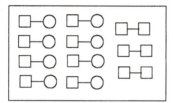

52. *Refer to Section 2.5 and Figure 2.11.*

The potassium ions (K^+) and chloride ions (Cl^-) would surround each other similar to the representation of NaCl in Figure 2.11.

54. *Refer to Section 2.2.*

This problem is best approached algebraically. We are given that the # of neutrons is 60% more than the number of protons. If we set # of protons = x, then

# of neutrons = $x + 0.6x$.

mass number (A) = (# of protons) + (# of neutrons) = 234

substituting gives: $A = (x) + (x + 0.6x) = 234$

$2.6x = 234$

$x = 90$. Thus, the element has 90 protons ($Z = 90$) and is thorium.

$^{234}_{90}$Th

56. *Refer to Section 2.1.*

Compound A: $\dfrac{\text{mass H}}{\text{mass C}} = \dfrac{2.39 \text{ g}}{28.5 \text{ g}} = 0.0839$.

Compound B: $\dfrac{\text{mass H}}{\text{mass C}} = \dfrac{11.6 \text{ g}}{34.7 \text{ g}} = 0.334$.

a. Compound C: $\dfrac{\text{mass H}}{\text{mass C}} = \dfrac{5.84 \text{ g}}{16.2 \text{ g}} = 0.360$.

$\dfrac{0.360}{0.0839} = 4.30$ This example does **not** follow the law of multiple proportions.

b. Compound C: $\dfrac{\text{mass H}}{\text{mass C}} = \dfrac{3.47 \text{ g}}{16.2 \text{ g}} = 0.214$.

$\dfrac{0.214}{0.0839} = 2.55$ This example does **not** follow the law of multiple proportions.

c. Compound C: $\dfrac{\text{mass H}}{\text{mass C}} = \dfrac{2.72 \text{ g}}{16.2 \text{ g}} = 0.168$.

$\dfrac{0.168}{0.0839} = 2.00$ This example **does** follow the law of multiple proportions since the H/C ratio in compound A is a whole number multiple of the H/C ratio in compound C.

a. ethane: $\dfrac{4.53 \text{ g H}}{18.0 \text{ g C}} = 0.252 \text{ g H} / \text{g C}$

ethene: $\dfrac{7.25 \text{ g H}}{43.20 \text{ g C}} = 0.168 \text{ g H} / \text{g C}$

While both these compounds have only hydrogen and carbon, they have different ratios of these elements.

b. $\dfrac{\text{ethane}}{\text{ethene}} = \dfrac{0.252 \text{ g H} / \text{g C}}{0.168 \text{ g H} / \text{g C}} = 1.5$

Thus ethane has 3 hydrogens for every 2 in ethene.

Reasonable formulae: ethane - CH_3 (actual formula is C_2H_6)
ethene - CH_2 (actual formula is C_2H_4)

Calculate the mass and volume of one Al-27, then calculate the density.

mass $= 13(\text{mass of a proton}) + 14(\text{mass of a neutron}) + 13(\text{mass of an electron})$
$= 13(1.6726 \times 10^{-24} \text{ g}) + 14(1.6749 \times 10^{-24} \text{ g}) + 13(9.1094 \times 10^{-28} \text{ g})$
$= 2.1744 \times 10^{-23} \text{ g} + 2.3449 \times 10^{-23} \text{ g} + 1.1842 \times 10^{-26} \text{ g}$
$= 4.5205 \times 10^{-23} \text{ g}$

volume $= \dfrac{4}{3} \pi r^3 = \dfrac{4}{3} \pi (1.43 \times 10^{-8} \text{ cm})^3 = 1.22 \times 10^{-23} \text{ cm}^3$

density $= \dfrac{\text{mass}}{\text{volume}} = \dfrac{4.5205 \times 10^{-23} \text{ g}}{1.22 \times 10^{-23} \text{ cm}^3} = 3.71 \text{ g/cm}^3$

The actual density of Al is 2.70 g/cm^3. There is quite a bit of free space around each aluminum atom, much as marbles in a jar have free space around them.

The difference between a Be atom and a Be^{2+} ion is two electrons.

mass of Be^{2+} = 1.4965×10^{-23}g - $2(9.1094 \times 10^{-28}$g$) = 1.4963 \times 10^{-23}$g

a. $200 \text{ breaths} \times \dfrac{500 \text{ mL}}{\text{breath}} \times \dfrac{2.5 \times 10^{19} \text{ molecules}}{\text{mL}} = 2.5 \times 10^{24} \text{ molecules}$

b. $\dfrac{2.5 \times 10^{24}}{1.1 \times 10^{44}} = 2.3 \times 10^{-20}$

c. $\text{breath} \times \dfrac{500 \text{ mL}}{\text{breath}} \times \dfrac{2.5 \times 10^{19} \text{ molecules}}{\text{mL}} \times 2.3 \times 10^{-20} = 2.9 \times 10^{2} \text{ molecules}$

Chapter 3: Mass Relations in Chemistry; Stoichiometry

> **2.** *Refer to Section 3.1 and the Periodic Table and Table of Atomic Masses (inside front cover).*

a. Ratio of $\dfrac{\text{bromine}}{\text{neon}} = \dfrac{79.90\,\text{amu}}{20.18\,\text{amu}} = 3.959$

b. Ratio of $\dfrac{\text{bromine}}{\text{calcium}} = \dfrac{79.90\,\text{amu}}{40.08\,\text{amu}} = 1.994$

c. Ratio of $\dfrac{\text{bromine}}{\text{helium}} = \dfrac{79.90\,\text{amu}}{4.003\,\text{amu}} = 19.96$

> **4.** *Refer to Section 3.1 and Example 3.1.*

Convert from percent to decimal by dividing by 100.
$0.04\% / 100\% = 4 \times 10^{-4}$
$0.20\% / 100\% = 2.0 \times 10^{-3}$
$99.76\% / 100\% = 9.976 \times 10^{-1}$

Atomic mass of oxygen
$= 16.00(9.976 \times 10^{-1}) + 17.00(4 \times 10^{-4}) + 18.00(2.0 \times 10^{-3})$
$= 15.69 + 0.007 + 0.036 = 16.00\,\text{amu}$

> **6.** *Refer of Section 3.1, Example 3.1 and the Table of Atomic Masses.*

If B-10 is 19.9% abundant, then B-11 is 80.1% (100-19.9) abundant.
The average mass (called simply atomic mass) of B = 10.811 amu (from the periodic table).
average mass = (fraction of B-10)(mass of B10) + (fraction of B-11)(mass of B-11)
set (mass of B-11) = x, and then solve for x.
$10.811\,\text{amu} = (0.199)(10.0129\,\text{amu}) + (0.801)x$
$10.811\,\text{amu} = 1.99\,\text{amu} + 0.801x$
$8.82 = 0.801x$
$x = 11.0\,\text{amu}$

Isotope	Mass	Abundance
1	49.94605 amu	x
2	51.94051 amu	$87.87 - x$
3	52.94065 amu	$y = 9.765$
4	53.93888 amu	2.365

The total abundance must equal 100, thus:
$87.87 + y + 2.365 = 100$, and
$y = 9.77$
$51.9961 = 49.94605(x) + 51.94051(0.8787 - x) + 52.94065(0.0977) + 53.93888(0.02365)$
$51.9961 = 49.94605x + 45.64013 - 51.94x + 5.17 + 1.27565$
$51.9961 = 52.09 - 1.99x$
$1.99x = 0.09$
$x = 0.05 = 5\%$

Abundance of isotope 1 = 5%
Abundance of isotope 2 = 83%
Abundance of isotope 3 = 9.77%

---

10. *Refer to Section 3.1 and Figures 3.1 and 3.2.*

a. Two, HCl-35 and HCl-37.

b. HCl-35: 35 + 1 = 36; and HCl-37: 37 + 1 = 38.

c.

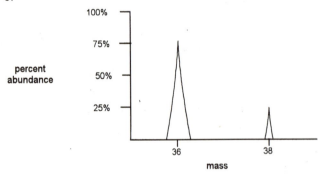

---

12. *Refer to Section 3.1 and Example 3.2.*

a. A mole of chocolate chips is Avogadro's number (6.022×10^{23}), regardless of the mass of one chip, just as a dozen is 12, regardless of the size.

b. 1×10^{-9} moles $\times \dfrac{6.022 \times 10^{23} \text{ chips}}{1 \text{ mol.}} \times \dfrac{1 \text{ cookie}}{15 \text{ chips}} = 4.015 \times 10^{13}$ cookies

(That's a lot of cookies!)

---

14. Refer to Section 3.1 and Example 3.2.

By definition, one mole of tungsten (W) = 6.022×10^{23} atoms, and one mole = 183.9 g

a. $\dfrac{1 \text{ mole}}{6.022 \times 10^{23} \text{ atoms}} \times \dfrac{183.9 \text{ g}}{1 \text{ mole}} = 3.054 \times 10^{-22}$ g/atom

b. $\dfrac{6.022 \times 10^{23} \text{ atoms}}{1 \text{ mol.}} \times \dfrac{1 \text{ mol.}}{183.9 \text{ g}} \times \dfrac{1 \times 10^{-12} \text{ g}}{1 \text{ pg}} = 3.275 \times 10^{9}$ atoms/pg

---

16. Refer to Sections 3.1 and 3.2.

a. $10.000 \text{ g Si} \times \dfrac{1 \text{ mole}}{27.977 \text{ g}} = 0.35744$ moles

b. $0.35744 \text{ mol.} \times \dfrac{6.022 \times 10^{23} \text{ atoms}}{1 \text{ mole}} = 2.152 \times 10^{23}$ atoms

c. $2.152 \times 10^{23} \text{ atoms} \times \dfrac{14 \text{ protons}}{1 \text{ atom}} = 3.013 \times 10^{24}$ protons

$2.152 \times 10^{23} \text{ atoms} \times \dfrac{14 \text{ neutrons}}{1 \text{ atom}} = 3.013 \times 10^{24}$ neutrons

$2.152 \times 10^{23} \text{ atoms} \times \dfrac{14 \text{ electrons}}{1 \text{ atom}} = 3.013 \times 10^{24}$ electrons

Total = $3(3.013 \times 10^{24}) = 9.039 \times 10^{24}$ particles

---

18. Refer to Sections 3.1 and 3.2.

Recall that the number of neutrons = mass number – atomic number, and that only the # electons (not the number of neutrons) changes going from atom to ion.

a. $10 \text{ Cl - 37 atoms} \times \dfrac{20 \text{ neutrons}}{1 \text{ Cl - 37 atom}} = 200$ neutrons

b. $10 \text{ Cl - 37 ions} \times \dfrac{20 \text{ neutrons}}{1 \text{ Cl - 37 ion}} = 200$ neutrons

c. $1 \text{ mole Cl-37 ions} \times \dfrac{6.022 \times 10^{23} \text{ ions}}{1 \text{ mole}} \times \dfrac{20 \text{ neutrons}}{1 \text{ Cl-37 ion}} = 1.204 \times 10^{25} \text{ neutrons}$

d. $1.000 \text{ g Cl-37 ions} \times \dfrac{1 \text{ mol.}}{36.97 \text{ g}} \times \dfrac{6.022 \times 10^{23} \text{ ions}}{1 \text{ mole}} \times \dfrac{20 \text{ neutrons}}{1 \text{ Cl-37 ion}} = 3.258 \times 10^{23} \text{ neutrons}$

---

20. Refer to Section 3.2 and Examples 3.3 and 3.4.

Multiply the number of each atom in the molecule by the atomic mass of that atom, and then add up the masses to get the molar mass of the molecule.

a. $C_{12}H_{22}O_{11}$: $12(12.011) + 22(1.0079) + 11(15.9994) = 342.30$ g/mol.

b. N_2O: $2(14.0067) + 1(15.9994) = 44.0128$ g/mol.

c. $C_{20}H_{30}O$: $20(12.011) + 30(1.008) + 1(16.00) = 286.44$ g/mol.

---

22. Refer to Sections 3.2 and 3.3.

a. CF_2Cl_2: $12.01 + 2(19.00) + 2(35.45) = 120.91$ g/mol.
$$35.00 \text{ g } CF_2Cl_2 \times \dfrac{1 \text{ mol. } CF_2Cl_2}{120.91 \text{ g } CF_2Cl_2} = 0.2895 \text{ mol. } CF_2Cl_2$$

b. iron(II) sulfate ($FeSO_4$): $55.85 + 32.07 + 4(16.00) = 151.92$ g/mol.
$$100.0 \text{ mg } FeSO_4 \times \dfrac{1 \text{ g}}{1000 \text{ mg}} \times \dfrac{1 \text{ mol. } FeSO_4}{151.92 \text{ g } FeSO_4} = 6.582 \times 10^{-4} \text{ mol. } FeSO_4$$

c. $C_{15}H_{13}ClN_2O$: $15(12.01) + 13(1.008) + 35.45 + 2(14.01) + 16.00 = 272.7$ g/mol.
$$2.000 \text{ g } C_{15}H_{13}ClN_2O \times \dfrac{1 \text{ mol. } C_{15}H_{13}ClN_2O}{272.7 \text{ g } C_{15}H_{13}ClN_2O} = 7.33 \times 10^{-3} \text{ mol. } C_{15}H_{13}ClN_2O$$

---

24. Refer to Section 3.2 and Example 3.4.

Determine or calculate the atomic or molar mass of the species of interest, then use that atomic or molar mass as a conversion factor to convert the given moles to grams.

a. $1 \text{ mole N} = 14.01$ g

$$17.5 \text{ mol.} \times \dfrac{14.01 \text{ g}}{1 \text{ mol.}} = 245 \text{ g}$$

b. 1 mole N_2 = 2(14.01) = 28.02 g

$$17.5 \, \text{mol.} \times \frac{28.02 \, \text{g}}{1 \, \text{mol.}} = 490 \, \text{g}$$

c. 1 mole NH_3 = 14.01 + 3(1.008) = 17.03 g

$$17.5 \, \text{mol.} \times \frac{17.03 \, \text{g}}{1 \, \text{mol.}} = 298 \, \text{g}$$

---

26. Refer to Section 3.2 and Problems 14 and 24 (above).

Number of Grams	Number of Moles	Number of Molecules	Number of N Atoms
172.2	**0.5601**	**3.373 x 10^{23}**	**1.012 x 10^{24}**
210.2	0.9254	5.573 x 10^{23}	1.672 x 10^{24}
4.68 x 10^6	**2.06 x 10^4**	1.24 x 10^{28}	**3.72 x 10^{28}**
9.4	**0.042**	**2.5 x 10^{22}**	7.5 x 10^{22}

Bear in mind that $(NO_2)_3$ means there are 3N's and 6O's in TNT.
$C_7H_5(NO_2)_3$: 7(12.01) + 5(1.008) + 3(14.01) + 6(16.00) = 227.10 g/mol. TNT.

a. $$127.2 \, \text{g} \times \frac{1 \, \text{mol.}}{227.10 \, \text{g}} = 0.5601 \, \text{mol.}$$

$$0.5601 \, \text{mol.} \times \frac{6.022 \times 10^{23} \, \text{molecules}}{1 \, \text{mol.}} = 3.373 \times 10^{23} \, \text{molecules}$$

$$3.373 \times 10^{23} \, \text{molecules} \times \frac{3 \, \text{nitrogen atoms}}{1 \, \text{molecule}} = 1.012 \times 10^{24} \, \text{nitrogen atoms}$$

b. $$0.9254 \, \text{mol.} \times \frac{227.10 \, \text{g}}{1 \, \text{mol.}} = 210.2 \, \text{g}$$

$$0.9254 \, \text{mol.} \times \frac{6.022 \times 10^{23} \, \text{molecules}}{1 \, \text{mol.}} = 5.573 \times 10^{23} \, \text{molecules}$$

$$5.573 \times 10^{23} \, \text{molecules} \times \frac{3 \, \text{nitrogen atoms}}{1 \, \text{molecule}} = 1.672 \times 10^{24} \, \text{nitrogen atoms}$$

c. $$1.24 \times 10^{28} \, \text{molecules} \times \frac{1 \, \text{mole}}{6.022 \times 10^{23} \, \text{molecules}} = 2.06 \times 10^4 \, \text{moles}$$

$$2.06 \times 10^4 \, \text{moles} \times \frac{227.10 \, \text{g}}{1 \, \text{mol.}} = 4.68 \times 10^6 \, \text{g}$$

$$1.24 \times 10^{28} \, \text{molecules} \times \frac{3 \, \text{nitrogen atoms}}{1 \, \text{molecule}} = 3.72 \times 10^{28} \, \text{nitrogen atoms}$$

d. 7.5×10^{22} nitrogen atoms $\times \dfrac{1\,\text{molecule}}{3\,\text{nitrogen atoms}} = 2.5 \times 10^{22}$ molecules

2.5×10^{22} molecules $\times \dfrac{1\,\text{mole}}{6.022 \times 10^{23}\,\text{molecules}} = 0.042$ moles

0.042 moles $\times \dfrac{227.10\,\text{g}}{1\,\text{mol.}} = 9.4\,\text{g}$

28. Refer to Section 3.3 and Example 3.5.

Find the mass of each element in $Al_2(OH)_5Cl$ and the total mass, then calculate the mass percents of each element.

Al: 2(26.98) =   53.96 g
O: 5(16.00) =   80.00 g
H: 5(1.008) =   5.040 g
Cl: 1(35.45) =   35.45 g
        Total mass   174.45 g

mass % Al $= \dfrac{53.96}{174.45} \times 100\% = 30.93\%$

mass % O $= \dfrac{80.00}{174.45} \times 100\% = 45.86\%$

mass % H $= \dfrac{5.040}{174.45} \times 100\% = 2.889\%$

mass % Cl $= \dfrac{35.45}{174.45} \times 100\% = 20.32\%$

Note that the mass percentages should equal 100%.

$30.93 + 45.86 + 2.886 + 20.32 = 100.00\%$

30. Refer to Section 3.3 and Example 3.6.

Find the mass % of Pt. Then use that percentage to calculate the mass of Pt in 250.0 mg. Bear in mind that $(NH_3)_2$ means there are 2N's and 6H's in cisplatin.

Formula mass of $Pt(NH_3)_2Cl_2 = 195.1 + 2(14.01) + 6(1.008) + 2(35.45) = 300.1$ g/mol.

mass % Pt $= \dfrac{195.1}{300.1} \times 100\% = 65.02\%$

This means there are 65.02 g Pt per 100 g $Pt(NH_3)_2Cl_2$, or 65.02 mg Pt per 100 mg $Pt(NH_3)_2Cl_2$.

$$250.0 \text{ mg } Pt(NH_3)_2Cl_2 \times \frac{65.02 \text{ mg Pt}}{100 \text{ mg } Pt(NH_3)_2Cl_2} = 162.6 \text{ mg Pt}$$

---

32. Refer to Section 3.3.

Calculate the mass % of acetaminophen in tylenol, then the mass % of N in acetaminophen. Finally, calculate mass of N in acetaminophen.

a. $\text{mass \% acetaminophen} = \dfrac{0.251 \text{ g } C_8H_9NO_2}{0.611 \text{ g Tylenol}} \times 100\% = 41.1\%$

This means every 100 g Tylenol contains 41.1 g $C_8H_9NO_2$.

b. Formula mass $C_8H_9NO_2 = 8(12.01) + 9(1.008) + 14.01 + 2(16.00) = 151.16$ g

$\text{mass \% of N (in } C_8H_9NO_2) = \dfrac{14.01 \text{ g N}}{151.16 \text{ g } C_8H_9NO_2} \times 100\% = 9.268\%$

This means every 100 g $C_8H_9NO_2$ contains 9.268 g N.

Putting it all together:

$$0.611 \text{ g Tylenol} \times \frac{41.1 \text{ g } C_8H_9NO_2}{100 \text{ g Tylenol}} \times \frac{9.268 \text{ g N}}{100 \text{ g } C_8H_9NO_2} = 0.0233 \text{ g N}$$

---

34. Refer to Section 3.3 and Example 3.8.

Remember that we use grams of CO_2 to determine the mass percent of C, grams H_2O to determine the mass percent of hydrogen and grams of Cl_2 to determine the mass percent of Cl, while oxygen is determined by difference.

$$1.407 \text{ g } CO_2 \times \frac{1 \text{ mol. } CO_2}{44.01 \text{ g}} \times \frac{1 \text{ mol. C}}{1 \text{ mol. } CO_2} \times \frac{12.01 \text{ g C}}{1 \text{ mol. C}} = 0.3840 \text{ g C}$$

$$\text{mass \% C} = \frac{0.3840 \text{ g C}}{1.000 \text{ g sample}} \times 100\% = 38.40 \text{ \% C}$$

$$0.134 \text{ g } H_2O \times \frac{1 \text{ mol. } H_2O}{18.02 \text{ g}} \times \frac{2 \text{ mol. H}}{1 \text{ mol. } H_2O} \times \frac{1.008 \text{ g H}}{1 \text{ mol. H}} = 0.0150 \text{ g H}$$

$$\text{mass \% H} = \frac{0.0150 \text{ g H}}{1.000 \text{ g sample}} \times 100\% = 1.50 \text{ \% H}$$

$$0.523 \text{ g } Cl_2 \times \frac{1 \text{ mol. } Cl_2}{70.90 \text{ g}} \times \frac{2 \text{ mol. Cl}}{1 \text{ mol. } Cl_2} \times \frac{35.45 \text{ g Cl}}{1 \text{ mol. Cl}} = 0.523 \text{ g Cl}$$

$$\text{mass \% Cl} = \frac{0.523 \text{ g Cl}}{1.000 \text{ g sample}} \times 100\% = 52.3 \text{ \% Cl}$$

mass O = 1.000 − 0.3840 − 0.0150 − 0.523 = 0.0780 g

$$\text{mass \% O} = \frac{0.0780\,\text{g O}}{1.000\,\text{g sample}} \times 100\% = 7.80\,\%\,\text{O}$$

---

36. Refer to Section 3.2 and Example 3.7.

Convert the masses of As and Cl to moles, calculate the mole ratio by dividing each of the moles by the smaller of the two values, and then name the resulting compound.

$$1.587\,\text{g As} \times \frac{1\,\text{mol. As}}{74.92\,\text{g As}} = 0.02118\,\text{mol. As}$$

$$3.755\,\text{g Cl} \times \frac{1\,\text{mol. Cl}}{35.45\,\text{g Cl}} = 0.1059\,\text{moles Cl}$$

0.02118 mol. / 0.02118 mol. = 1 mole As
0.1059 mol. / 0.02118 mol. = 5 moles Cl
Thus, 5 moles Cl per 1 mole As: $AsCl_5$
arsenic pentachloride

---

38. Refer to Section 3.3 and Problem 36 (above).

Use the percentages as masses to calculate the moles of each element and then calculate the mole ratios of the elements.

a. $$35.51\,\text{g C} \times \frac{1\,\text{mol. C}}{12.01\,\text{g C}} = 2.957\,\text{mol. C}$$

$$4.77\,\text{g H} \times \frac{1\,\text{mol. H}}{1.008\,\text{g H}} = 4.73\,\text{mol. H}$$

$$8.29\,\text{g N} \times \frac{1\,\text{mol. N}}{14.01\,\text{g N}} = 0.592\,\text{mol. N}$$

$$37.85\,\text{g O} \times \frac{1\,\text{mol. O}}{16.00\,\text{g O}} = 2.366\,\text{mol. O}$$

$$13.60\,\text{g Na} \times \frac{1\,\text{mol. Na}}{22.99\,\text{g Na}} = 0.5916\,\text{mol. Na}$$

2.957 mol. / 0.5916 = 5 mol. C
4.73 mol. / 0.5916 = 8 mol. H
0.592 mol. / 0.5916 = 1 mol. N
2.366 mol. / 0.5916 = 4 mol. O
0.5916 mol. / 0.5916 = 1 mol. Na

Simplest formula: $C_5H_8NO_4Na$

b. $34.91\,g\,O \times \dfrac{1\,mol.\,O}{16.00\,g\,O} = 2.182\,mol.\,O$

$15.32\,g\,Si \times \dfrac{1\,mol.\,Si}{28.09\,g\,Si} = 0.5454\,mol.\,S$

$49.76\,g\,Zr \times \dfrac{1\,mol.\,Zr}{91.22\,g\,Zr} = 0.5455\,mol.\,Zr$

2.182 mol. / 0.5454 mol. = 4 mol. O
0.5454 mol. / 0.5454 mol. = 1 mol. Si
0.5454 mol. / 0.5455 mol. = 1 mol. Zr

Simplest formula: O_4SiZr

c. $74.0\,g\,C \times \dfrac{1\,mol.\,C}{12.01\,g\,C} = 6.16\,mol.\,C$

$8.65\,g\,H \times \dfrac{1\,mol.\,H}{1.008\,g\,H} = 8.58\,mol.\,H$

$17.4\,g\,N \times \dfrac{1\,mol.\,N}{14.01\,g\,N} = 1.24\,mol.\,N$

6.16 mol. / 1.24 mol. = 5 mol. C
8.65 mol. / 1.24 mol. = 7 mol. H
1.24 mol. / 1.24 mol. = 1 mol. N

Simplest formula: C_5H_7N

---

40. Refer to Section 3.3 and Example 3.8.

We use grams of CO_2 to determine the mass of C and grams H_2O to determine the mass of hydrogen, while oxygen is determined by difference. Once the mass of each element is determined, calculate the moles of each and then the mole ratios.

$12.24\,g\,CO_2 \times \dfrac{1\,mol.\,CO_2}{44.01\,g\,CO_2} \times \dfrac{1\,mol.\,C}{1\,mol.\,CO_2} \times \dfrac{12.01\,g\,C}{1\,mol.\,C} = 3.340\,g\,C$

$2.505\,g\,H_2O \times \dfrac{1\,mol.\,H_2O}{18.02\,g} \times \dfrac{2\,mol.\,H}{1\,mol.\,H_2O} \times \dfrac{1.008\,g\,H}{1\,mol.\,H} = 0.2802\,g\,H$

mass % O = 5.287 g – 3.340 g – 0.2802 g = 1.667 g

$3.340\,g\,C \times \dfrac{1\,mol.\,C}{12.01\,g\,C} = 0.2781\,mol.\,C$

$0.2802\,g\,H \times \dfrac{1\,mol.\,H}{1.008\,g\,H} = 0.2780\,mol.\,H$

$1.667\,g\,O \times \dfrac{1\,mol.\,O}{16.00\,g\,O} = 0.1042\,mol.\,O$

0.2781 mol. / 0.1042 mol. = 2.669 mol. C
0.2780 mol. / 0.1042 mol. = 2.668 mol. H
0.1042 mol. / 0.1042 mol. = 1.00 mol. O

0.67 is the decimal equivalent of 2/3; thus, to get whole numbers, multiply by a factor of 3.

2.669 mol. C x 3 = 8 mol. C
2.668 mol. H x 3 = 8 mol. H
1.00 mol. O x 3 = 3 mol. O

Simplest formula: $C_8H_8O_3$

---

42. Refer to Section 3.3, Example 3.8 and Problem 40 (above).

We use grams of CO_2 to determine the mass of C, grams H_2O to determine the mass of hydrogen, and mass percent of N to determine mass of N, while oxygen is determined by difference. Once the mass of each element is determined, calculate the moles of each and then the mole ratios.

$$28.73\,g\,CO_2 \times \frac{1\,mol.\,CO_2}{44.01\,g\,CO_2} \times \frac{1\,mol.\,C}{1\,mol.\,CO_2} \times \frac{12.01\,g\,C}{1\,mol.\,C} = 7.840\,g\,C$$

$$5.386\,g\,H_2O \times \frac{1\,mol.\,H_2O}{18.02\,g} \times \frac{2\,mol.\,H}{1\,mol.\,H_2O} \times \frac{1.008\,g\,H}{1\,mol.\,H} = 0.6026\,g\,H$$

$$10.94\,g\,sample \times \frac{6.963\,g\,N}{100\,g\,sample} = 0.7618\,g\,N$$

mass O = 10.94 g – 7.840 g – 0.6026 g – 0.7618 = 1.74 g O

$$7.840\,g\,C \times \frac{1\,mol.\,C}{12.01\,g\,C} = 0.6528\,mol.\,C$$

$$0.6026\,g\,H \times \frac{1\,mol.\,H}{1.008\,g\,H} = 0.5978\,mol.\,H$$

$$0.7618\,g\,N \times \frac{1\,mol.\,N}{14.01\,g\,N} = 0.05438\,mol.\,N$$

$$1.74\,g\,O \times \frac{1\,mol.\,O}{16.00\,g\,O} = 0.109\,mol.\,O$$

0.6528 mol. / 0.05438 mol. = 12 mol. C
0.5978 mol. / 0.05438 mol. = 11 mol. H
0.05438 mol. / 0.05438 mol. = 1 mol. N
0.109 mol. / 0.05438 mol. = 2 mol. O

Simplest formula: $C_{12}H_{11}NO_2$

We use grams of CO_2 to determine the mass of C and grams H_2O to determine the mass of hydrogen, while nitrogen is determined by difference. Once the mass of each element is determined, calculate the moles of each and then the mole ratios to determine the simplest formula. Then calculate the molar mass of that formula and divide the actual molar mass by the simplest molar mass to determine the multiplier for the simplest formula.

$$14.36 \text{ g CO}_2 \times \frac{1 \text{ mol. CO}_2}{44.01 \text{ g CO}_2} \times \frac{1 \text{ mol. C}}{1 \text{ mol. CO}_2} \times \frac{12.01 \text{ g C}}{1 \text{ mol. C}} = 3.919 \text{ g C}$$

$$7.832 \text{ g H}_2\text{O} \times \frac{1 \text{ mol. H}_2\text{O}}{18.02 \text{ g}} \times \frac{2 \text{ mol. H}}{1 \text{ mol. H}_2\text{O}} \times \frac{1.008 \text{ g H}}{1 \text{ mol. H}} = 0.8762 \text{ g H}$$

mass of N = 6.315 g – 3.919 g - 0.8762g = 1.520 g N

$$3.919 \text{ g C} \times \frac{1 \text{ mol. C}}{12.01 \text{ g C}} = 0.3263 \text{ mol. C}$$

$$0.8762 \text{ g H} \times \frac{1 \text{ mol. H}}{1.008 \text{ g H}} = 0.8692 \text{ mol. H}$$

$$1.520 \text{ g N} \times \frac{1 \text{ mol. N}}{14.01 \text{ g N}} = 0.1085 \text{ mol. N}$$

0.3263 mol. / 0.1085 mol. = 3 mol. C
0.8692 mol. / 0.1085 mol. = 8 mol. H
0.1085 mol. / 0.1085 mol. = 1 mol. N

Simplest formula: C_3H_8N

Formula mass (of the simplest formula) = 3(12.01) + 8(1.008) + 14.01 = 58.10 g/mol.

The actual molar mass 116.2 g/mol.

$\frac{116.2}{58.10} = 2$. Thus the molecular formula = $C_{3x2}H_{8x2}N_{1x2} = C_6H_{16}N_2$

Calculate the mass % of anhydrous salt in the hydrated sample using molar masses. Then use that value to determine the mass of anhydrous salt in the sample.

molar mass of $MgSO_4$ and H_2O :   246.51 g
molar mass of $MgSO_4$:   120.37 g
molar mass of H_2O :   126.14 g

$$\text{mass \% of water} = \frac{126.14 \text{ g H}_2\text{O}}{246.51 \text{ g MgSO}_4 \cdot 7\text{H}_2\text{O}} \times 100\% = 51.17\%$$

$$\text{mass \% of anhydrous salt in hydrate} = \frac{120.37 \text{ g MgSO}_4}{246.51 \text{ g MgSO}_4 \cdot 7\text{H}_2\text{O}} \times 100\% = 48.83\%$$

$$\text{mass of anhydrous salt} = 7.834 \text{ g MgSO}_4 \cdot 7\text{H}_2\text{O} \times \frac{48.83 \text{ g MgSO}_4}{100 \text{ g MgSO}_4 \cdot 7\text{H}_2\text{O}} = 3.825 \text{ g MgSO}_4$$

---

48. Refer to Section 3.4 and Example 3.10.

Compare the reactant and product sides of the equation, starting with elements that appear only once on each side. Balance the elements by adjusting the coefficients (*never the subscripts!*). When you have a choice, it is usually advantageous to start with the element that is present in greatest number.

a. $TiO_2(s) + Cl_2(g) + C(s) \rightarrow TiCl_4(l) + CO(g)$

    $TiO_2(s) + 2Cl_2(g) + C(s) \rightarrow TiCl_4(l) + CO(g)$ balance Cl's

    $TiO_2(s) + 2Cl_2(g) + C(s) \rightarrow TiCl_4(l) + 2CO(g)$ balance O's

    $TiO_2(s) + 2Cl_2(g) + 2C(s) \rightarrow TiCl_4(l) + 2CO(g)$ balance C's

b. $Br_2(l) + I_2(s) \rightarrow IBr_3(g)$

    $Br_2(l) + I_2(s) \rightarrow 2IBr_3(g)$ balance I's

    $3Br_2(l) + I_2(s) \rightarrow 2IBr_3(g)$ balance Br's

c. $C_2H_8N_2(s) + N_2O_4(g) \rightarrow N_2(g) + CO_2(g) + H_2O(g)$

    $C_2H_8N_2(s) + N_2O_4(g) \rightarrow N_2(g) + 2CO_2(g) + H_2O(g)$ balance C's

    $C_2H_8N_2(s) + N_2O_4(g) \rightarrow N_2(g) + 2CO_2(g) + 4H_2O(g)$ balance H's

    $C_2H_8N_2(s) + 2N_2O_4(g) \rightarrow N_2(g) + 2CO_2(g) + 4H_2O(g)$ balance O's

    $C_2H_8N_2(s) + 2N_2O_4(g) \rightarrow 3N_2(g) + 2CO_2(g) + 4H_2O(g)$ balance N's

---

50. Refer to Sections 2.5, 2.6 and 3.4 and Figure 2.10.

Compare the reactant and product sides of the equation, starting with elements that appear only once on each side. Balance the elements by adjusting the coefficients (*never the subscripts!*). When you have a choice, it is usually advantageous to start with the element that is most abundant. Note that elemental sulfur is S_8.

a. $K(s) + S_8(s) \rightarrow K_2S(s)$

    $K(s) + S_8(s) \rightarrow 8K_2S(s)$ balance S's

    $16K(s) + S_8(s) \rightarrow 8K_2S(s)$ balance K's

b. $Mg(s) + S_8(s) \rightarrow MgS(s)$

    $Mg(s) + S_8(s) \rightarrow 8MgS(s)$ balance S's

    $8Mg(s) + S_8(s) \rightarrow 8MgS(s)$ balance Mg's

c. $Al(s) + S_8(s) \rightarrow Al_2S_3(s)$
    $Al(s) + 3S_8(s) \rightarrow 8Al_2S_3(s)$ balance S's
    $16Al(s) + 3S_8(s) \rightarrow 8Al_2S_3(s)$ balance Al's

d. $Ca(s) + S_8(s) \rightarrow CaS(s)$
    $Ca(s) + S_8(s) \rightarrow 8CaS(s)$ balance S's
    $8Ca(s) + S_8(s) \rightarrow 8CaS(s)$ balance Ca's

e. $Fe(s) + S_8(s) \rightarrow FeS(s)$
    $Fe(s) + S_8(s) \rightarrow 8FeS(s)$ balance S's
    $8Fe(s) + S_8(s) \rightarrow 8FeS(s)$ balance Fe's

---

52. *Refer to Sections 2.5, 2.6 and 3.4, Example 3.10, Figure 2.10 and Problem 50 (above).*

Compare the reactant and products side of the equation, starting with elements that appear only once on each side. Balance the elements by adjusting the coefficients (*never the subscripts!*). When you have a choice, it is usually advantageous to start with the element that is present in greatest number.

a. $F_2(g) + H_2O(l) \rightarrow OF_2(g) + HF(g)$
    $F_2(g) + H_2O(l) \rightarrow OF_2(g) + 2HF(g)$ balance H's
    $2F_2(g) + H_2O(l) \rightarrow OF_2(g) + 2HF(g)$ balance F's

b. $O_2(g) + NH_3(g) \rightarrow NO_2(g) + H_2O(l)$
    $O_2(g) + 2NH_3(g) \rightarrow NO_2(g) + 3H_2O(l)$ balance H's
    $O_2(g) + 2NH_3(g) \rightarrow 2NO_2(g) + 3H_2O(l)$ balance N's
    $(7/2)O_2(g) + 2NH_3(g) \rightarrow 2NO_2(g) + 3H_2O(l)$ balance O's
        Since fractions are generally not allowed, multiply all coefficients by 2.
    $7O_2(g) + 4NH_3(g) \rightarrow 4NO_2(g) + 6H_2O(l)$

c. $Au_2S_3(s) + H_2(g) \rightarrow Au(s) + H_2S(g)$
    $Au_2S_3(s) + H_2(g) \rightarrow Au(s) + 3H_2S(g)$ balance the S's
    $Au_2S_3(s) + H_2(g) \rightarrow 2Au(s) + 3H_2S(g)$ balance the Au's
    $Au_2S_3(s) + 3H_2(g) \rightarrow 2Au(s) + 3H_2S(g)$ balance the H's

d. $NaHCO_3(s) \rightarrow Na_2CO_3(s) + H_2O(l) + CO_2(s)$
    $2NaHCO_3(s) \rightarrow Na_2CO_3(s) + H_2O(l) + CO_2(s)$ balance the Na's

e. $SO_2(g) + HF(l) \rightarrow SF_4(g) + H_2O(l)$
    $SO_2(g) + 4HF(l) \rightarrow SF_4(g) + H_2O(l)$ balance F's
    $SO_2(g) + 4HF(l) \rightarrow SF_4(g) + 2H_2O(l)$ balance H's

37

54. Refer to Section 3.4 and Example 3.11.

Use the coefficients from the balanced equation to establish the mole ratios of the species in question.

a. $12.7 \text{ mol. O}_2 \times \dfrac{4 \text{ mol. PH}_3}{8 \text{ mol. O}_2} = 6.35 \text{ mol. PH}_3$

b. $5.43 \text{ mol. PH}_3 \times \dfrac{1 \text{ mol. P}_4\text{O}_{10}}{4 \text{ mol.PH}_3} = 1.36 \text{ mol. P}_4\text{O}_{10}$

c. $1.003 \text{ mol. H}_2\text{O} \times \dfrac{8 \text{ mol.O}_2}{6 \text{ mol.H}_2\text{O}} = 1.337 \text{ mol.O}_2$

d. $1.76 \text{ mol. O}_2 \times \dfrac{1 \text{ mol. P}_4\text{O}_{10}}{8 \text{ mol. O}_2} = 0.220 \text{ mol. P}_4\text{O}_{10}$

56. Refer to Section 3.4, Example 3.11 and Problem 54 (above).

Use the coefficients from the balanced equation to establish the mole ratios of the species in question. Don't forget your mass to mole and/or mole to gram conversions.

a. $12.43 \text{ mol. PH}_3 \times \dfrac{1 \text{ mol. P}_4\text{O}_{10}}{4 \text{ mol.PH}_3} \times \dfrac{283.88 \text{ g P}_4\text{O}_{10}}{1 \text{ mol. P}_4\text{O}_{10}} = 882.2 \text{ g P}_4\text{O}_{10}$

b. $0.739 \text{ mol. H}_2\text{O} \times \dfrac{4 \text{ mol.PH}_3}{6 \text{ mol.H}_2\text{O}} \times \dfrac{33.994 \text{ g PH}_3}{1 \text{ mol.PH}_3} = 16.7 \text{ g PH}_3$

c. $1.000 \text{ g H}_2\text{O} \times \dfrac{1 \text{ mol. H}_2\text{O}}{18.02 \text{ g H}_2\text{O}} \times \dfrac{8 \text{ mol. O}_2}{6 \text{ mol.H}_2\text{O}} \times \dfrac{32.00 \text{ g O}_2}{1 \text{ mol.O}_2} = 2.368 \text{ g O}_2$

d. $20.50 \text{ g PH}_3 \times \dfrac{1 \text{ mol. PH}_3}{33.99 \text{ g PH}_3} \times \dfrac{8 \text{ mol. O}_2}{4 \text{ mol. PH}_3} \times \dfrac{32.00 \text{ g O}_2}{1 \text{ mol.O}_2} = 38.60 \text{ g O}_2$

58. Refer to Sections 2.5, 2.6 and 3.4 and Examples 3.10 amd 3.11.

Balance the equation to determine the mole ratios, then calculate the requested masses. Use the coefficients from the balanced equation to establish the mole ratios of the species in question. Don't forget your mass to mole and/or mole to gram conversions.

a. $SiO_2(s) + C(s) \rightarrow Si(s) + CO(g)$
   $SiO_2(s) + C(s) \rightarrow Si(s) + 2CO(g)$ balance O's
   $SiO_2(s) + 2C(s) \rightarrow Si(s) + 2CO(g)$ balance C's

b. $12.72 \text{ g Si} \times \dfrac{1 \text{ mol. Si}}{28.09 \text{ g Si}} \times \dfrac{1 \text{ mol. SiO}_2}{1 \text{ mol.Si}} = 0.4528 \text{ mol. SiO}_2$

c. $44.99 \, g \, Si \times \dfrac{1 \, mol. \, Si}{28.09 \, g \, Si} \times \dfrac{2 \, mol. \, CO}{1 \, mol. Si} \times \dfrac{28.01 \, g \, CO}{1 \, mol. \, CO} = 89.72 \, g \, CO$

---

60. Refer to Section 3.4.

Calculate the mass of C_2H_5OH (using the mass percent) in the wine cooler. Use that value to calculate mass of $C_6H_{12}O_6$ and, from that, the volume of CO_2.

a. $10.00 \, kg \, wine \, cooler \times \dfrac{1000 \, g}{1 \, kg} \times \dfrac{4.5 \, g \, C_2H_5OH}{100 \, g \, wine \, cooler} = 4.5 \times 10^2 \, g \, C_2H_5OH$

$4.5 \times 10^2 \, g \, C_2H_5OH \times \dfrac{1 \, mol. \, C_2H_5OH}{46.07 \, g \, C_2H_5OH} \times \dfrac{1 \, mol. \, C_6H_{12}O_6}{2 \, mol. \, C_2H_5OH} \times \dfrac{180.16 \, g \, C_6H_{12}O_6}{1 \, mol. \, C_6H_{12}O_6}$

$= 8.8 \times 10^2 \, g \, C_6H_{12}O_6$

b. $8.8 \times 10^2 \, g \, C_6H_{12}O_6 \times \dfrac{1 \, mol. \, C_6H_{12}O_6}{180.16 \, g \, C_6H_{12}O_6} \times \dfrac{2 \, mol. \, CO_2}{1 \, mol. \, C_6H_{12}O_6} = 9.8 \, mol. \, CO_2$

$9.8 \, mol. \, CO_2 \times \dfrac{44.01 \, g \, CO_2}{1 \, mol. \, CO_2} \times \dfrac{1 \, L \, CO_2}{1.80 \, g \, CO_2} = 2.4 \times 10^2 \, L \, CO_2$

---

62. Refer to Section 3.4.

Use the mass percent of sulfur to calculate the mass of sulfur in the oil. Then you can convert from mass of S to moles S, to moles SO_2, to mass SO_2, to volume SO_2.

$1.00 \times 10^4 \, kg \, oil \times \dfrac{1000 \, g}{1 \, kg} \times \dfrac{1.2 \, g \, S}{100 \, g \, oil} = 1.2 \times 10^5 \, g \, S$

$1.2 \times 10^5 \, g \, S \times \dfrac{1 \, mol. \, S}{32.07 \, g \, S} \times \dfrac{1 \, mol. \, SO_2}{1 \, mol. \, S} \times \dfrac{64.07 \, g \, SO_2}{1 \, mol. \, SO_2} \times \dfrac{1 \, L \, SO_2}{2.60 \, g \, SO_2} = 9.2 \times 10^4 \, L$

---

64. Refer to Section 3.4 and Example 3.12.

Use the balanced equation to calculate the moles of Al_2S_3 based on the amount of Al, and also based on the amount of S. The one giving the lower value will be the limiting reactant, and the yield calculated will also be the theoretical yield.

a. $2Al(s) + 3S(s) \rightarrow Al_2S_3(s)$

39

b. $1.18 \, \text{mol. Al} \times \dfrac{1 \, \text{mol. Al}_2\text{S}_3}{2 \, \text{mol. Al}} = 0.590 \, \text{mol. Al}_2\text{S}_3$

$2.25 \, \text{mol. S} \times \dfrac{1 \, \text{mol. Al}_2\text{S}_3}{3 \, \text{mol. S}} = 0.750 \, \text{mol. Al}_2\text{S}_3$

Al produces the least amount of Al_2S_3, and is therefore the limiting reagent.

c. Theoretical yield, based on the limiting reagent, is calculated above as 0.590 mol. Al_2S_3.

d. Calculate the amount of S that would be consumed making the theoretical amount of Al_2S_3 and subtract that from the initial amount.

$0.590 \, \text{mol. Al}_2\text{S}_3 \times \dfrac{3 \, \text{mol. S}}{1 \, \text{mol. Al}_2\text{S}_3} = 1.77 \, \text{mol. S (used in the reaction)}$

2.25 mol. (initial) – 1.77 mol. (reacted) = 0.48 mol. S (unreacted).

---

66. Refer to Section 3.4 and Example 3.13.

Use the equation for percent yield to solve for theoretical yield. Then (from the theoretical yield) calculate moles of Si_3N_4; using the mole ratio (from the balanced equation) calculate moles of N_2, and then mass of N_2.

a. $Si(s) + N_2(g) \rightarrow Si_3N_4(s)$
$3Si(s) + 2N_2(g) \rightarrow Si_3N_4(s)$

b. $\% \, \text{yield} = \dfrac{\text{experimental yield}}{\text{theoretical yield}} \times 100\% \Rightarrow 72\% = \dfrac{2.00 \times 10^2 \, \text{kg}}{\text{theoretical yield}} \times 100\%$

theor. yield $= (2.00 \times 10^2 \, \text{kg})(100\%) / 72\% = 2.8 \times 10^2 \, \text{kg}$

$2.8 \times 10^2 \, \text{kg Si}_3\text{N}_4 \times \dfrac{1000 \, \text{g}}{1 \, \text{kg}} \times \dfrac{1 \, \text{mol. Si}_3\text{N}_4}{140.31 \, \text{g Si}_3\text{N}_4} \times \dfrac{2 \, \text{mol. N}_2}{1 \, \text{mol. Si}_3\text{N}_4} \times \dfrac{28.02 \, \text{g N}_2}{1 \, \text{mol. N}_2}$

$= 1.1 \times 10^5 \, \text{g N}_2 = 1.1 \times 10^2 \, \text{kg N}_2$

---

68. Refer to Section 3.4, Example 3.13 and Problem 64 (above).

---

a. $175 \, \text{g C}_2\text{H}_2 \times \dfrac{1 \, \text{mol. C}_2\text{H}_2}{26.04 \, \text{g C}_2\text{H}_2} \times \dfrac{4 \, \text{mol. CO}_2}{2 \, \text{mol. C}_2\text{H}_2} \times \dfrac{44.01 \, \text{g CO}_2}{1 \, \text{mol. CO}_2} = 592 \, \text{g CO}_2$

$175 \, \text{g O}_2 \times \dfrac{1 \, \text{mol. O}_2}{32.00 \, \text{g O}_2} \times \dfrac{4 \, \text{mol. CO}_2}{5 \, \text{mol. O}_2} \times \dfrac{44.01 \, \text{g CO}_2}{1 \, \text{mol. CO}_2} = 193 \, \text{g CO}_2$

The O_2 gives the lesser amount of CO_2 and must therefore be the limiting reagent. Thus the theoretical yield of CO_2 is 193 g.

b. $193 \text{ g } CO_2 \times \dfrac{1 \text{ L } CO_2}{1.80 \text{ g } CO_2} = 107 \text{ L } CO_2$

$\% \text{ yield} = \dfrac{\text{experimental yield}}{\text{theoretical yield}} \times 100\% = \dfrac{68.5 \text{ L}}{107 \text{ L}} \times 100\% = 63.9\%$

c. $193 \text{ g } CO_2 \times \dfrac{1 \text{ mol. } CO_2}{44.01 \text{ g } CO_2} \times \dfrac{2 \text{ mol. } C_2H_2}{4 \text{ mol. } CO_2} \times \dfrac{26.04 \text{ g } C_2H_2}{1 \text{ mol. } C_2H_2} = 57.1 \text{ g } C_2H_2$

175 g (initial) – 57.1 g (used) = 118 g C_2H_2 (unused)

---

70. Refer to Section 3.4 and Problem 66 (above).

$\% \text{ yield} = \dfrac{\text{experimental yield}}{\text{theoretical yield}} \times 100\% \implies 75.0\% = \dfrac{0.250 \text{ L}}{\text{theoretical yield}} \times 100\%$

theor. yield = (0.250)(100%) / 75.0% = 0.333 L H_3PO_4.

$0.333 \text{ L } H_3PO_4 \times \dfrac{1000 \text{ mL}}{1 \text{ L}} \times \dfrac{1 \text{ cm}^3}{1 \text{ mL}} \times \dfrac{1.651 \text{ g } H_3PO_4}{1 \text{ cm}^3 \, H_3PO_4} = 550 \text{ g } H_3PO_4$

$550 \text{ g } H_3PO_4 \times \dfrac{1 \text{ mol. } H_3PO_4}{81.99 \text{ g } H_3PO_4} \times \dfrac{1 \text{ mol. } PI_3}{1 \text{ mol. } H_3PO_4} \times \dfrac{411.67 \text{ g } PI_3}{1 \text{ mol. } PI_3} = 2.76 \times 10^3 \text{ g } PI_3$

Weigh out 2.76×10^3 g PI_3.

$550 \text{ g } H_3PO_4 \times \dfrac{1 \text{ mol. } H_3PO_4}{81.99 \text{ g } H_3PO_4} \times \dfrac{3 \text{ mol. } H_2O}{1 \text{ mol. } H_3PO_4} \times \dfrac{18.02 \text{ g } H_2O}{1 \text{ mol. } H_2O} = 363 \text{ g } H_2O$

$363 \text{ g } H_2O \times \dfrac{1 \text{ cm}^3 \, H_2O}{1.00 \text{ g } H_2O} \times \dfrac{1 \text{ mL}}{1 \text{ cm}^3} = 363 \text{ mL } H_2O$

need 45% excess, thus: 363 mL + (0.45)(363 mL) = 526 mL H_2O

---

72. Refer to Section 3.4.

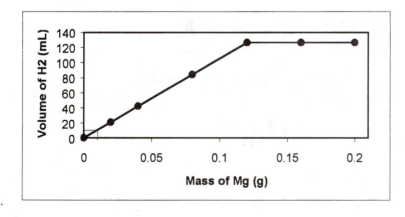

a.

b. Since the amount of Mg determines the amount of product, **Mg** is the limiting reagent in experiment 1.

c. Since the amount of Mg determines the amount of product, **Mg** is the limiting reagent in experiment 3.

d. After adding 0.120 g Mg, the amount of H_2 does not change when additional Mg is added. Therefore, the limiting reagent is the **acid** in experiment 6.

e. Since **experiment 4** represents the point at which the slope changes to zero, it is the point at which stoichiometric amounts of acid and Mg are present.

f. These values are determined from the graph.

0.300 g Mg corresponds to 122 mL H_2 gas (all masses of Mg above 0.120 g correspond with **122 mL H_2**).

0.010 g Mg corresponds to ~**11 mL** of H_2 gas (note lines on the graph in part (a) above.)

Alternatively, it can be calculated using simple ratios $\dfrac{0.020\,g}{21\,mL} = \dfrac{0.020\,g}{x}$; and $x = 11$.

---

74. Refer to Sections 1.2, 3.2 and 3.3 and Table 1.3.

$$6.00 \text{ oz. salami} \times \frac{1\,g}{0.03527\,oz.} \times \frac{0.090 \text{ g NaC}_7\text{H}_5\text{O}_2}{100 \text{ g salami}} = 15 \text{ g NaC}_7\text{H}_5\text{O}_2$$

$$15 \text{ g NaC}_7\text{H}_5\text{O}_2 \times \frac{1 \text{ mol. NaC}_7\text{H}_5\text{O}_2}{144.10 \text{ g NaC}_7\text{H}_5\text{O}_2} \times \frac{6.022 \times 10^{23} \text{ molecules}}{1 \text{ mol.}} = 6.4 \times 10^{20} \text{ molecules}$$

$$6.4 \times 10^{20} \text{ molecules} \times \frac{1 \text{ Na atom}}{1 \text{ NaC}_7\text{H}_5\text{O}_2 \text{ molecule}} = 6.4 \times 10^{20} \text{ Na atoms}$$

---

76. Refer to Section 3.4 and Problems 60 and 62 (above).

$$725 \text{ mL wine} \times \frac{11.0 \text{ mL C}_2\text{H}_5\text{OH}}{100 \text{ mL wine}} \times \frac{1 \text{ cm}^3}{1 \text{ mL}} \times \frac{0.789 \text{ g C}_2\text{H}_5\text{OH}}{1 \text{ cm}^3 \text{ C}_2\text{H}_5\text{OH}} = 62.9 \text{ g C}_2\text{H}_5\text{OH}$$

$$62.9 \text{ g C}_2\text{H}_5\text{OH} \times \frac{1 \text{ mol. C}_2\text{H}_5\text{OH}}{46.07 \text{ g C}_2\text{H}_5\text{OH}} \times \frac{1 \text{ mol. C}_6\text{H}_{12}\text{O}_6}{2 \text{ mol. C}_2\text{H}_5\text{OH}} \times \frac{180.16 \text{ g C}_6\text{H}_{12}\text{O}_6}{1 \text{ mol. C}_6\text{H}_{12}\text{O}_6} = 123 \text{ g C}_6\text{H}_{12}\text{O}_6$$

78. Refer to Section 3.4.

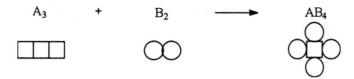

From the product, it is evident that we need 4 ◯'s for each ▢. Thus we need 6 ◯◯'s for one ▢▢▢.

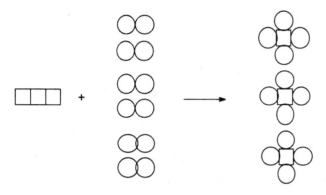

Thus: $A_3 + 6B_2 \rightarrow 3AB_4$

---

80. Refer to Section 3.4.

Subtract unreacted products from the reactants to get the balanced equation, then simplify the coefficients to their least common denominator.

a. $5▢ + 10◯ \rightarrow 3▢O_3 + 2▢ + ◯$
Thus the balanced reaction is:
$$3▢ + 9◯ \rightarrow 3▢O_3$$
which simplifies to:
$$▢ + 3◯ \rightarrow ▢O_3$$
thus: $X + 3Y \rightarrow XY_3$

b.   Started with 5 moles of X and 10 moles Y.

c.   The reaction yielded 3 moles product (XY_3) with 2 moles X and 1 mole Y left unreacted.

---

82. Refer to Section 3.2.

a.   **False.** theoretical yield $= 4.0 \, \text{mol. CCl}_4 \times \dfrac{1 \, \text{mol. CCl}_2 F_2}{1 \, \text{mol. CCl}_4} = 4.0 \, \text{mol. CCl}_2 F_2$

b. **False**. theoretical yield $= 4.0\,\text{mol. CCl}_4 \times \dfrac{2\,\text{mol. HCl}}{1\,\text{mol. CCl}_4} \times \dfrac{36.46\,\text{g HCl}}{1\,\text{mol. HCl}} = 2.9 \times 10^2\,\text{g HCl}$

c. **True**. $\%\text{ yield} = \dfrac{\text{experimental yield}}{\text{theoretical yield}} \times 100\% = \dfrac{3.0\,\text{mol. CCl}_2\text{F}_2}{4.0\,\text{mol. CCl}_2\text{F}_2} = 75\%$

d. **False**. The theoretical yield is based on the limiting reagent, which must be CCl_4 since the text states that HF is used in excess.

e. **True**. We do not know how much HF was used (only that an excess was used), so we cannot calculate how much was left unreacted.

f. **False**. Law of conservation of mass states that the mass of products will equal the mass of reactants, but this does not hold true for moles; consider the reaction given in problem 76.

g. **False**. Two moles of HF are consumed for every one mole of CCl_4 used.

h. **True**. As the limiting reactant, CCl_4, is completely consumed.

---

84. *Refer to Section 3.1 and Table 19.3 (page 558).*

Calculate the mass ratio of Si to Li_3N using standard masses. Then use that ratio to calculate the mass of Li_3N assuming a molar mass of 10 for Si.

$$\text{ratio} = \dfrac{34.83\,\text{g Li}_3\text{N}}{27.96924\,\text{g Si} - 28} = 1.245$$

$$1.245 = \dfrac{\text{mass of Li}_3\text{N}}{10.00\,\text{g Si} - 28} \implies \text{mass of Li}_3\text{N} = 12.45\,\text{g}$$

---

85. *Refer to Section 3.3.*

Calculate the mass of chlorophyll in one mole: Use the chlorophyll:Mg mole ratio as a conversion factor to calculate the moles of Mg in one mole of chlorophyll. Convert moles of Mg to grams and then calculate mass of chlorophyll using the mass % of Mg in chlorophyll.

$$1\,\text{mol. chlorophyll} \times \dfrac{1\,\text{mol. Mg}}{1\,\text{mol. chlorophyll}} \times \dfrac{24.30\,\text{g Mg}}{1\,\text{mol. Mg}} \times \dfrac{100\,\text{g chlorophyll}}{2.72\,\text{g Mg}} = 893\,\text{g chlorophyll}$$

Thus the molar mass of chlorophyll $= 893$ g/mol.

86. Refer to Sections 1.2, 1.3 and 3.2.

volume of a cube $= s^3$

$$s = 0.409 \text{ nm} \times \frac{1 \text{ m}}{1 \times 10^9 \text{ nm}} \times \frac{100 \text{ cm}}{1 \text{ m}} = 4.09 \times 10^{-8} \text{ cm}$$

$V = (4.09 \times 10^{-8} \text{ cm})^3 = 6.84 \times 10^{-23} \text{ cm}^3$

$$\frac{4 \text{ atoms}}{6.84 \times 10^{-23} \text{ cm}^3} \times \frac{1 \text{ cm3}}{10.5 \text{ g}} \times \frac{107.87 \text{ g Ag}}{1 \text{ mole Ag}} = 6.01 \times 10^{23} \text{ atoms/mole}$$

(Avogadro was right.)

87. Refer to Sections 3.3 and 3.4.

First write balanced chemical equations. Then calculate the amount of CaO formed and the amount of Ca needed to form the CaO. The remaining Ca will have formed the Ca_3N_2.

Reactions: $2Ca + O_2 \rightarrow 2CaO$
$3Ca + N_2 \rightarrow Ca_3N_2$
$CaO + H_2O \rightarrow Ca(OH)_2$

$$4.832 \text{ g Ca(OH)}_2 \times \frac{1 \text{ mol. Ca(OH)}_2}{74.10 \text{ g Ca(OH)}_2} \times \frac{1 \text{ mol. CaO}}{1 \text{ mol. Ca(OH)}_2} \times \frac{56.08 \text{ g CaO}}{1 \text{ mol. CaO}}$$

$$= 3.657 \text{ g CaO}$$

$$3.657 \text{ g CaO} \times \frac{1 \text{ mol. CaO}}{56.08 \text{ g CaO}} \times \frac{2 \text{ mol. Ca}}{2 \text{ mol. CaO}} \times \frac{40.08 \text{ g Ca}}{1 \text{ mol. Ca}} = 2.614 \text{ g Ca}$$

5.025 g Ca (original) - 2.614 g Ca (to form CaO) = 2.411 g Ca (to form Ca_3N_2)

$$2.411 \text{ g Ca} \times \frac{1 \text{ mol. Ca}}{40.08 \text{ g Ca}} \times \frac{1 \text{ mole Ca}_3\text{N}_2}{3 \text{ mol. Ca}} \times \frac{148.26 \text{ g Ca}_3\text{N}_2}{1 \text{ mol. Ca}_3\text{N}_2} = 2.973 \text{ g Ca}_3\text{N}_2$$

88. Refer to Sections 3.3 and 3.4.

Calculate the mass lost on converting KBr to KCl. This corresponds to the difference in masses between Br and Cl. Then calculate the mass the of KBr and the percentage.

79.90 g/mol. Br - 35.45 g/mol. Cl = 44.45 g/mol (of Br replaced with Cl)

3.595 g (original mixture) - 3.129 g (final KCl) = 0.466 g

$$0.466 \text{ g} \times \frac{1 \text{ mol.}}{44.45 \text{ g}} = 0.0105 \text{ mol. KBr converted to KCl.}$$

$$0.0105 \text{ mol. KBr} \times \frac{119.0 \text{ g KBr}}{1 \text{ mol. KBr}} = 1.25 \text{ g KBr}$$

$$\text{mass } \% = \frac{1.25 \text{ g}}{3.595 \text{ g}} \times 100\% = 34.8\%$$

89. Refer to Sections 3.3 and 3.4.

a. Both oxides contain 2.573 g V. The first also contains (4.589 g VO_X - 2.573 g V)
2.016 g O, while the second contains (3.782 g VO_Y - 2.573 g V) 1.209 g O.

$$2.573 \text{ g V} \times \frac{1 \text{ mol. V}}{50.94 \text{ g V}} = 0.05051 \text{ mol. V}$$

$$2.016 \text{ g O} \times \frac{1 \text{ mol. O}}{16.00 \text{ g O}} = 0.1260 \text{ mol. O} \quad \text{(1st oxide)}$$

$$1.209 \text{ g O} \times \frac{1 \text{ mol. O}}{16.00 \text{ g O}} = 0.07556 \text{ mol. O} \text{ (2nd oxide)}$$

0.05051 mol. / 0.05051 mol. = 1.000 mol. V
0.07556 mol. / 0.05051 mol. = 1.496 mol. O
0.1260 mol. / 0.05051 mol. = 2.495 mol. O

1st oxide: V_2O_3
2nd oxide: V_2O_5

b. Calculate the mass of H_2O formed using the mass of O lost. The second heating converted all the oxygen to water.

$$2.016 \text{ g O} \times \frac{1 \text{ mol. O}}{16.00 \text{ g O}} \times \frac{1 \text{ mol. H}_2\text{O}}{1 \text{ mol. O}} \times \frac{18.02 \text{ g H}_2\text{O}}{1 \text{ mol. H}_2\text{O}} = 2.271 \text{ g H}_2\text{O}$$

90. Refer to Sections 3.3 and 3.4.

Calculate the amount of CO_2 expected from burning cocaine and sucrose using variables for the masses of those compounds. Remember that cocaine has 17 C atoms and will thus produce 17 CO_2 atoms; similarly, sucrose will produce 12 CO_2's Then set up a algebraic equation and solve for the mass of cocaine in the sample. Then calculate the mass percent.

Molar mass of $C_{17}H_{21}O_4N$ = 303.39 g/mol.

Molar mass of $C_{12}H_{22}O_{11}$ = 342.34 g/mol.

$$\text{mass of } CO_2 = 1.00 \text{ mL} \times \frac{1 \text{ L}}{1000 \text{ mL}} \times \frac{1.80 \text{ g}}{1 \text{ L}} = 1.80 \times 10^{-3} \text{ g}$$

mass of sugar + cocaine $= 1.00 \text{ mg} \times \dfrac{1 \text{ g}}{1000 \text{ mg}} = 1.00 \times 10^{-3} \text{ g}$

Let X = mass of cocaine, then

$(1.0 \times 10^{-3} - X)$ = mass of sucrose

$$X \text{ g cocaine} \times \dfrac{1 \text{ mol. cocaine}}{303.39 \text{ g cocaine}} \times \dfrac{17 \text{ mol. CO}_2}{1 \text{ mol. cocaine}} \times \dfrac{44.01 \text{ g CO}_2}{1 \text{ mol. CO}_2} = 2.466 X \text{ g CO}_2$$

$$(1.00 \times 10^{-3} - X) \text{ g sucrose} \times \dfrac{1 \text{ mol. sucrose}}{342.34 \text{ g sucrose}} \times \dfrac{12 \text{ mol. CO}_2}{1 \text{ mol. sucrose}} \times \dfrac{44.01 \text{ g CO}_2}{1 \text{ mol. CO}_2}$$

$$= (1.54 \times 10^{-3} - 1.54 X) \text{ g CO}_2$$

total CO_2 = CO_2 from cocaine + CO_2 from sucrose

$1.80 \times 10^{-3} = 2.466 X + 1.54 \times 10^{-3} - 1.54 X$

$2.60 \times 10^{-4} = 0.93 X$

$X = 2.8 \times 10^{-4}$ g cocaine

mass % cocaine $= \dfrac{2.8 \times 10^{-4} \text{ g}}{1.00 \times 10^{-3} \text{ g}} \times 100\% = 28\%$

Chapter 4: Reactions in Aqueous Solution

2. Refer to Section 4.1, Example 4.1 and Figure 4.1.

Use molarity to convert from liters of solution to moles of solute and then convert to mass.

a. $0.400 \, L \times \dfrac{0.155 \, mol. \, Sr(OH)_2}{1 \, L} \times \dfrac{121.64 \, g \, Sr(OH)_2}{1 \, mol. \, Sr(OH)_2} = 7.54 \, g \, Sr(OH)_2$

Dissolve 7.54 g $Sr(OH)_2$ in sufficient water to make 0.400 L of solution.

b. $1.75 \, L \times \dfrac{0.333 \, mol. \, (NH_4)_2CO_3}{1 \, L} \times \dfrac{96.09 \, g \, (NH_4)_2CO_3}{1 \, mol. \, (NH_4)_2CO_3} = 56.0 \, g \, (NH_4)_2CO_3$

Dissolve 56.0 g $(NH_4)_2CO_3$ in sufficient water to make 1.75 L of solution.

4. Refer to Section 4.1 and Example 4.1.

a. Convert from mL to L, then use molarity to convert from L to moles.

$45.6 \, mL \times \dfrac{1 \, L}{1000 \, mL} \times \dfrac{0.450 \, mol. \, K_2CO_3}{1 \, L} = 0.0205 \, mol. \, K_2CO_3$

b. Use molarity to convert from moles to L, then convert L to mL.

$0.800 \, mol. \, K_2CO_3 \times \dfrac{1 \, L}{0.450 \, mol. \, K_2CO_3} \times \dfrac{1000 \, mL}{1 \, L} = 1.78 \times 10^3 \, mL$

c. Calculate moles K_2CO_3 present in the 0.450 M solution and the moles needed to make 1 M solution. The difference is the moles needed to bring the solution up to 1 M.

$2.00 \, L \times \dfrac{0.450 \, mol. \, K_2CO_3}{1 \, L} = 0.900 \, mol. \, K_2CO_3$

$2.00 \, L \times \dfrac{1.000 \, mol. \, K_2CO_3}{1 \, L} = 2.00 \, mol. \, K_2CO_3$

moles needed = 2.00 mol. – 0.900 mol. = 1.10 mol.

$1.10 \, mol. \, K_2CO_3 \times \dfrac{138.21 \, g \, K_2CO_3}{1 \, mol. \, K_2CO_3} = 152 \, g \, K_2CO_3$

d. Calculate moles in 50.0 mL of solution, then recalculate molarity with the new volume.

$$500 \text{ mL} \times \frac{1 \text{ L}}{1000 \text{ mL}} \times \frac{0.450 \text{ mol. K}_2\text{CO}_3}{1 \text{ L}} = 0.0225 \text{ mol. K}_2\text{CO}_3$$

$$M = \frac{0.0225 \text{ mol. K}_2\text{CO}_3}{0.125 \text{ L}} = 0.180 \, M \text{ K}_2\text{CO}_3$$

Alternatively, you could use the equation: $(M_{init})(V_{init}) = (M_{final})(V_{final})$.

$(0.450 \, M)(50.0 \text{ mL}) = (M_{final})(125 \text{ mL})$

$(M_{final}) = 0.180 \, M.$

---

6. **Refer to Sections 2.6 and 4.1.**

---

a. $MgCl_2 \rightarrow Mg^{+2} + 2Cl^-$

$$0.4132 \text{ mol. MgCl}_2 \times \frac{1 \text{ mol. Mg}^{+2}}{1 \text{ mol. MgCl}_2} = 0.4132 \text{ mol. Mg}^{+2}$$

$$0.4132 \text{ mol. MgCl}_2 \times \frac{2 \text{ mol. Cl}^-}{1 \text{ mol. MgCl}_2} = 0.8264 \text{ mol. Cl}^-$$

Total moles of ions = 0.4132 + 0.8264 = 1.2396 mol.

b. $Mn_3N_2 \rightarrow 3Mn^{+2} + 2N^{-3}$

$$0.4132 \text{ mol. Mn}_3\text{N}_2 \times \frac{3 \text{ mol. Mn}^{+2}}{1 \text{ mol. Mn}_3\text{N}_2} = 1.240 \text{ mol. Mn}^{+2}$$

$$0.4132 \text{ mol. Mn}_3\text{N}_2 \times \frac{2 \text{ mol. N}^{-3}}{1 \text{ mol. Mn}_3\text{N}_2} = 0.8264 \text{ mol. N}^{-3}$$

Total moles of ions = 1.240 + 0.8264 = 2.066 mol.

c. $Co(NO_3)_3 \rightarrow Co^{+3} + 3NO_3^-$

$$0.4132 \text{ mol. Co(NO}_3)_3 \times \frac{1 \text{ mol. Co}^{+3}}{1 \text{ mol. Co(NO}_3)_3} = 0.4132 \text{ mol. Co}^{+3}$$

$$0.4132 \text{ mol. Co(NO}_3)_3 \times \frac{3 \text{ mol. NO}_3^-}{1 \text{ mol. Co(NO}_3)_3} = 1.240 \text{ mol. NO}_3^-$$

Total moles of ions = 0.4132 + 1.240 = 1.653 mol.

d. $Fe(MnO_4)_2 \rightarrow Fe^{+2} + 2MnO_4^-$

$$0.4132 \text{ mol. } Fe(MnO_4)_2 \times \frac{1 \text{ mol. } Fe^{+2}}{1 \text{ mol. } Fe(MnO_4)_2} = 0.4132 \text{ mol. } Fe^{+2}$$

$$0.4132 \text{ mol. } Fe(MnO_4)_2 \times \frac{2 \text{ mol. } MnO_4^-}{1 \text{ mol. } Fe(MnO_4)_2} = 0.8264 \text{ mol. } MnO_4^-$$

Total moles of ions = 0.4132 + 0.8264 = 1.2396 mol.

---

8. Refer to Sections 2.6 and 4.2 and Figure 4.4.

a. Na_2SO_4: **Soluble** – Most sulfates are soluble, and this is *not* one of the exceptions.

b. $Fe(NO_3)_3$: **Soluble** – All nitrates are soluble.

c. $AgCl$: **Insoluble** – Most chlorides are soluble, but this *is* one of the exceptions.

d. $Cr(OH)_3$: **Insoluble** – Most transition metal hydroxides are insoluble, and this is *not* one of the exceptions.

---

10. Refer to Sections 2.6 and 4.2, Figure 4.4 and Example 4.4.

$CoCl_3 \rightarrow Co^{3+} + 3Cl^-$

a. To precipitate $CoPO_4$, add a soluble phosphate such as sodium phosphate (Na_3PO_4).

b. To precipitate $Co_2(CO_3)_3$, add a soluble carbonate such as sodium carbonate (Na_2CO_3).

c. To precipitate $Co(OH)_3$, add a soluble hydroxide such as sodium hydroxide (NaOH).

---

12. Refer to Sections 2.6, 4.1 and 4.2 and Figures 4.4 and 4.5.

Recall that soluble salts ionize when dissolved. Write the reactions for the ionizations. Look at the resulting ions. If there are pairs that would result in insoluble salts, these salts would form and precipitate from solution.

a. $FeCl_3 \rightarrow Fe^{+3}(aq) + 3Cl^-(aq)$
$NaOH \rightarrow Na^+(aq) + OH^-(aq)$
$Fe^{+3}(aq) + 3Cl^-(aq) + 3Na^+(aq) + 3OH^-(aq) \rightarrow Fe(OH)_3(s) + 3Na^+(aq) + 3Cl^-(aq)$
$Fe^{+3}(aq) + 3OH^-(aq) \rightarrow Fe(OH)_3(s)$ (*net ionic equation*)
Thus, $Fe(OH)_3$ would precipitate from solution (NaCl is soluble and remains in solution).

b. $AgOH \rightarrow Ag^+(aq) + OH^-(aq)$
$CuCl_2 \rightarrow Cu^{+2}(aq) + 2Cl^-(aq)$
$2Ag^+(aq) + 2OH^-(aq) + Cu^{+2}(aq) + 2Cl^-(aq) \rightarrow Cu(OH)_2(s) + 2AgCl(s)$
Thus, $Cu(OH)_2$ and $AgCl$ would both precipitate from solution.

---

14. Refer to Sections 2.6, 4.1 and 4.2, Figures 4.4 and 4.5 and Example 4.3.

Recall that soluble salts ionize when dissolved. Write the reactions for the ionizations. Look at the resulting ions. If there are pairs that would result in insoluble salts, these salts would form and precipitate from solution.

a. $KNO_3 \rightarrow K^+(aq) + NO_3^-(aq)$
$MgSO_4 \rightarrow Mg^{+2}(aq) + SO_4^{-2}(aq)$
Nothing would precipitate from solution(K_2SO_4 and $Mg(NO_3)_2$ are both soluble and remain in solution).

b. $AgNO_3 \rightarrow Ag^+(aq) + NO_3^-(aq)$
$K_2CO_3 \rightarrow 2K^+(aq) + CO_3^{-2}(aq)$
$2Ag^+(aq) + 2NO_3^-(aq) + 2K^+(aq) + CO_3^{-2}(aq) \rightarrow Ag_2CO_3(s) + 2K^+(aq) + 2NO_3^-(aq)$
$2Ag^+(aq) + CO_3^{-2}(aq) \rightarrow Ag_2CO_3(s)$ *(net ionic equation)*
Ag_2CO_3 would precipitate from solution(KNO_3 is soluble and remains in solution).

c. $(NH_4)_2CO_3 \rightarrow 2NH_4^+(aq) + CO_3^{-2}(aq)$
$CoCl_3 \rightarrow Co^{+3}(aq) + 3Cl^-(aq)$
$6NH_4^+(aq) + 3CO_3^{-2}(aq) + 2Co^{+3}(aq) + 6Cl^-(aq)$
     $\rightarrow Co_2(CO_3)_3(s) + 6NH_4^+(aq) + 2NO_3^-(aq)$
$2Co^{+3}(aq) + 3CO_3^{-2}(aq) \rightarrow Co_2(CO_3)_3(s)$ *(net ionic equation)*
$Co_2(CO_3)_3$ would precipitate from solution (NH_4NO_3 is soluble and remains in solution).

d. $Na_3PO_4 \rightarrow 3Na^+(aq) + PO_4^{-3}(aq)$
$Ba(OH)_2 \rightarrow Ba^{+2}(aq) + 2OH^-(aq)$
$6Na^+(aq) + 2PO_4^{-3}(aq) + 3Ba^{+2}(aq) + 6OH^-(aq)$
     $\rightarrow Ba_3(PO_4)_2(s) + 6Na^+(aq) + 6OH^-(aq)$
$3Ba^{+2}(aq) + 2PO_4^{-3}(aq) \rightarrow Ba_3(PO_4)_2(s)$ *(net ionic equation)*
$Ba_3(PO_4)_2$ would precipitate from solution ($NaOH$ is soluble and remains in solution).

e. $Ba(NO_3)_2 \rightarrow Ba^{+2}(aq) + 2NO_3^-(aq)$
$KOH \rightarrow K^+(aq) + OH^-(aq)$
Nothing would precipitate from solution ($Ba(OH)_2$ and KNO_3 are both soluble and remain in solution).

Recall that soluble salts ionize when dissolved. Write the reactions for the ionizations. Look at the resulting ions. If there are pairs that would result in insoluble salts, these salts would form and precipitate from solution.

a. $Na_3PO_4 \rightarrow 3Na^+(aq) + PO_4^{-3}(aq)$
$BaCl_2 \rightarrow Ba^{+2}(aq) + 2Cl^-(aq)$
$6Na^+(aq) + 2PO_4^{-3}(aq) + 3Ba^{+2}(aq) + 6Cl^-(aq) \rightarrow Ba_3(PO_4)_2(s) + 6Na^+(aq) + 6Cl^-(aq)$
$3Ba^{+2}(aq) + 2PO_4^{-3}(aq) \rightarrow Ba_3(PO_4)_2(s)$ (*net ionic equation*)
$Ba_3(PO_4)_2$ would precipitate from solution (NaCl is soluble and remains in solution).

b. $ZnSO_4 \rightarrow Zn^{+2}(aq) + SO_4^{-2}(aq)$
$KOH \rightarrow K^+(aq) + OH^-(aq)$
$Zn^{+2}(aq) + SO_4^{-2}(aq) + 2K^+(aq) + 2OH^-(aq) \rightarrow Zn(OH)_2(s) + 2K^+(aq) + SO_4^{-2}(aq)$
$Zn^{+2}(aq) + 2OH^-(aq) \rightarrow Zn(OH)_2(s)$ (*net ionic equation*)
$Zn(OH)_2$ would precipitate from solution (K_2SO_4 is soluble and remains in solution).

c. $(NH_4)_2SO_4 \rightarrow 2NH_4^+(aq) + SO_4^{-2}(aq)$
$NaCl \rightarrow Na^+(aq) + Cl^-(aq)$
Nothing would precipitate from solution (NH_4Cl and Na_2SO_4 are both soluble and remain in solution).

d. $Na_3PO_4 \rightarrow 3Na^+(aq) + PO_4^{-3}(aq)$
$Co(NO_3)_3 \rightarrow Co^{+3}(aq) + 3NO_3^-(aq)$
$3Na^+(aq) + PO_4^{-3}(aq) + Co^{+3}(aq) + 3NO_3^-(aq) \rightarrow CoPO_4(s) + 3Na^+(aq) + 3NO_3^-(aq)$
$Co^{+3}(aq) + PO_4^{-3}(aq) \rightarrow CoPO_4(s)$ (*net ionic equation*)
$CoPO_4$ would precipitate from solution ($NaNO_3$ is soluble and remains in solution).

Recall that soluble salts ionize when dissolved. Write the reactions for the ionizations and the net ionic equation. Use the volume and molarity of the given solution to calculate moles, then use the mole ratios of the various ions to calculate the moles of $Ba(NO_3)_2$. Finally, use the moles of $Ba(NO_3)_2$ and the molarity to calculate the volume of $Ba(NO_3)_2$ solution.

a. $Co_2(SO_4)_3 \rightarrow 2Co^{+3}(aq) + 3SO_4^{-2}(aq)$
$Ba(NO_3)_2 \rightarrow Ba^{+2}(aq) + 2NO_3^-(aq)$
$Ba^{+2}(aq) + SO_4^{-2}(aq) \rightarrow BaSO_4(s)$ (*net ionic equation*)

$$20.0\,\text{mL} \times \frac{1\,\text{L}}{1000\,\text{mL}} \times \frac{0.0937\,\text{mol. Co}_2(\text{SO}_4)_3}{1\,\text{L}} \times \frac{3\,\text{mol. SO}_4^{-2}}{1\,\text{mol. Co}_2(\text{SO}_4)_3} = 5.62 \times 10^{-3}\,\text{mol. SO}_4^{-2}$$

$$5.62 \times 10^{-3}\,\text{mol. SO}_4^{-2} \times \frac{1\,\text{mol. Ba}^{+2}}{1\,\text{mol. SO}_4^{-2}} \times \frac{1\,\text{mol. Ba(NO}_3)_2}{1\,\text{mol. Ba}^{+2}} = 5.62 \times 10^{-3}\,\text{mol. Ba(NO}_3)_2$$

$$M = \frac{\text{moles}}{\text{volume (L)}} \implies 0.773\,\text{mol./L} = \frac{5.62 \times 10^{-3}\,\text{mol. Ba(NO}_3)_2}{\text{volume}}$$

volume = 7.27×10^{-3} L = 7.27 mL

b. $K_2CO_3 \rightarrow 2K^+(aq) + CO_3^{-2}(aq)$
   $Ba(NO_3)_2 \rightarrow Ba^{+2}(aq) + 2NO_3^-(aq)$
   $Ba^{+2}(aq) + CO_3^{-2}(aq) \rightarrow BaCO_3(s)$ (*net ionic equation*)

$$43.3\,\text{mL} \times \frac{1\,\text{L}}{1000\,\text{mL}} \times \frac{0.396\,\text{mol. K}_2\text{CO}_3}{1\,\text{L}} \times \frac{1\,\text{mol. CO}_3^{-2}}{1\,\text{mol. K}_2\text{CO}_3} = 0.0171\,\text{mol. CO}_3^{-2}$$

$$0.0171\,\text{mol. CO}_3^{-2} \times \frac{1\,\text{mol. Ba}^{+2}}{1\,\text{mol. CO}_3^{-2}} \times \frac{1\,\text{mol. Ba(NO}_3)_2}{1\,\text{mol. Ba}^{+2}} = 0.0171\,\text{mol. Ba(NO}_3)_2$$

$$M = \frac{\text{moles}}{\text{volume (L)}} \implies 0.773\,\text{mol./L} = \frac{0.0171\,\text{mol. Ba(NO}_3)_3}{\text{volume}}$$

volume = 0.0222 L = 22.2 mL

c. $(NH_4)_3PO_4 \rightarrow 3NH_4^+(aq) + PO_4^{-3}(aq)$
   $Ba(NO_3)_2 \rightarrow Ba^{+2}(aq) + 2NO_3^-(aq)$
   $3Ba^{+2}(aq) + 2PO_4^{-3}(aq) \rightarrow Ba_3(PO_4)_2(s)$ (*net ionic equation*)

$$0.0295\,\text{L} \times \frac{0.631\,\text{mol. (NH}_4)_3\text{PO}_4}{1\,\text{L}} \times \frac{1\,\text{mol. PO}_4^{-3}}{1\,\text{mol. (NH}_4)_3\text{PO}_4} = 0.0186\,\text{mol. PO}_4^{-3}$$

$$0.0186\,\text{mol. PO}_4^{-3} \times \frac{3\,\text{mol. Ba}^{+2}}{2\,\text{mol. PO}_4^{-3}} \times \frac{1\,\text{mol. Ba(NO}_3)_3}{1\,\text{mol. Ba}^{+2}} = 0.0279\,\text{mol. Ba(NO}_3)_3$$

$$M = \frac{\text{moles}}{\text{volume (L)}} \implies 0.773\,\text{mol/L} = \frac{0.0279\,\text{mol. Ba(NO}_3)_3}{\text{volume}}$$

volume = 0.0361 L = 36.1 mL

20. Refer to Section 4.2, Figure 4.4 and Examples 4.5 and 4.6.

a. $CrCl_3 \rightarrow Cr^{+3}(aq) + 3Cl^-(aq)$
    $AgNO_3 \rightarrow Ag^+(aq) + NO_3^-(aq)$
    $Ag^+(aq) + Cl^-(aq) \rightarrow AgCl(s)$ (*net ionic equation*)

$$0.05000\,L \times \frac{0.0250\,mol.\,AgNO_3}{1\,L} \times \frac{1\,mol.Ag^+}{1\,mol.\,AgNO_3} = 1.25 \times 10^{-3}\,mol.\,Ag^+$$

$$1.25 \times 10^{-3}\,mol.\,Ag^+ \times \frac{1\,mol.\,Cl^-}{1\,mol.\,Ag^+} \times \frac{1\,mol.\,CrCl_3}{3\,mol.\,Cl^-} = 1.25 \times 10^{-3}\,mol.\,CrCl_3$$

$$M = \frac{moles}{volume\,(L)} \Rightarrow 0.0400\,M = \frac{4.17 \times 10^{-4}\,mol.\,CrCl_3}{volume}$$

volume = 0.0104 L = 10.4 mL

b. $$1.25 \times 10^{-3}\,mol.\,Ag^+ \times \frac{1\,mol.\,AgCl}{1\,mol.\,Ag^+} \times \frac{143.35\,g\,AgCl}{1\,mol.\,AgCl} = 0.179\,g\,AgCl$$

---

22. Refer to Sections 2.6 and 4.3 and Tables 2.3 and 4.1.

Acids not in Table 4.1 should be considered to be weak acids. Also, acids containing only C, H and O are organic acids and should also be considered weak.

a. Sulfurous acid (H_2SO_3) is a weak acid, the reacting species is H_2SO_3.

b. Chlorous acid ($HClO_2$) is a weak acid, the reacting species is $HClO_2$.

c. Perchloric acid ($HClO_4$) is a strong acid, the reacting species is H^+.

d. Sulfuric acid (H_2SO_4) is a strong acid, the reacting species is H^+.

e. Formic acid ($HCHO_2$) is a weak acid, the reacting species is $HCHO_2$.

---

24. Refer to Sections 2.6 and 4.3 and Tables 2.3 and 4.1.

Bases not in Table 4.1 should be considered to be weak bases. Also, bases containing only C, H and N are organic bases and should also be considered weak.

a. Potassium hydroxide (KOH) is a strong base, the reacting species is OH^-.

b. Aniline ($C_6H_5NH_2$) is a weak base, the reacting species is $C_6H_5NH_2$.

c. Dimethyl amine ($(CH_3)_2NH$) is a weak base, the reacting species is $(CH_3)_2NH$.

d. Barium hydroxide ($Ba(OH)_2$) is a strong base, the reacting species is OH^-.

26. Refer to Sections 2.6 and 4.3, Tables 2.3 and 4.1 and Problems 22 and 24 above.

    a. $HClO_4$ is a strong acid

    b. CsOH is a strong base

    c. H_2CO_3 is a weak acid

    d. $C_2H_5NH_2$ is a weak organic base

28. Refer to Section 2.6 and 4.3, Tables 2.3 and 4.1 and Example 4.7.

    Identify the acid and the base as weak or strong, then identify the reacting species. Then write the reaction for the two reacting species.

    a. Acetic acid ($HC_2H_3O_2$) is a weak acid, the reacting species is $HC_2H_3O_2$.
       Strontium hydroxide ($Sr(OH)_2$) is a strong base, the reacting species is OH^-.
       $HC_2H_3O_2(aq) + OH^-(aq) \rightarrow H_2O(l) + C_2H_3O_2^-(aq)$

    b. Sulfuric acid (H_2SO_4) is a strong acid, the reacting species is H^+.
       Diethyl amine ($(C_2H_5)_2NH$) is a weak base, the reacting species is $(C_2H_5)_2NH$.
       $H^+(aq) + (C_2H_5)_2NH(aq) \rightarrow (C_2H_5)_2NH_2^+(aq)$

    c. Hydrogen cyanide (HCN) is a weak acid, the reacting species is HCN.
       Sodium hydroxide (NaOH) is a strong base, the reacting species is OH^-.
       $HCN(aq) + OH^-(aq) \rightarrow H_2O(l) + CN^-(aq)$

30. Refer to Section 2.6 and 4.3 and Tables 2.3 and 4.1.

    For the net ionic equation $H^+(aq) + OH^-(aq) \rightarrow H_2O(l)$ to be correct, the reactants must be a strong acid and a strong base.

    a. The equation is **not** correct.
       Nitric acid (HNO_3) is a strong acid, the reacting species is H^+.
       Ethyl amine ($C_2H_5NH_2$) is a weak base, the reacting species is $C_2H_5NH_2$.
       $H^+(aq) + C_2H_5NH_2(aq) \rightarrow C_2H_5NH_3^+(aq)$

    b. The equation **is** correct.
       Perchloric acid ($HClO_4$) is a strong acid, the reacting species is H^+.
       Cesium hydroxide (CsOH) is a strong base, the reacting species is OH^-.
       $H^+(aq) + OH^-(aq) \rightarrow H_2O(l)$

c. The equation is **not** correct.
Acetic acid ($HC_2H_3O_2$) is a weak acid, the reacting species is $HC_2H_3O_2$.
Lithium hydroxide (LiOH) is a strong base, the reacting species is OH^-.
$$HC_2H_3O_2(aq) + OH^-(aq) \rightarrow H_2O(l) + C_2H_3O_2^-(aq)$$

d. The equation **is** correct.
Sulfuric acid (H_2SO_4) is a strong acid, the reacting species is H^+.
Calcium hydroxide ($Ca(OH)_2$) is a strong base, the reacting species is OH^-.
$$H^+(aq) + OH^-(aq) \rightarrow H_2O(l)$$

e. The equation **is** correct.
Nitric acid (HNO_3) is a strong acid, the reacting species is H^+.
Barium hydroxide ($Ba(OH)_2$) is a strong base, the reacting species is OH^-.
$$H^+(aq) + OH^-(aq) \rightarrow H_2O(l)$$

---

32. Refer to Section 4.3, Table 4.1 and Examples 4.7 and 4.8.

Recall that strong acids and bases ionize in solution. Write the reactions for the ionizations and the overall equation for the reacting species. Use the mass of $Ba(OH)_2$ to calculate moles, then use the mole ratios of the various ions to calculate the moles of HNO_3. Finally, use the moles of HNO_3 and the volume of solution to calculate the molarity.

$$HNO_3(aq) \rightarrow H^+(aq) + NO_3^-(aq)$$
$$Ba(OH)_2(aq) \rightarrow Ba^{+2}(aq) + 2OH^-(aq)$$
$$H^+(aq) + OH^-(aq) \rightarrow H_2O(l)$$

$$0.216\,g\ Ba(OH)_2 \times \frac{1\,mol.\ Ba(OH)_2}{171.3\,g\ Ba(OH)_2} \times \frac{2\,mol.\ OH^-}{1\,mol.\ Ba(OH)_2} = 2.52 \times 10^{-3}\ mol.\ OH^-$$

$$2.52 \times 10^{-3}\ mol.\ OH^- \times \frac{1\,mol.\ H^+}{1\,mol.\ OH^-} \times \frac{1\,mol.\ HNO_3}{1\,mol.\ H^+} = 2.52 \times 10^{-3}\ mol.\ HNO_3$$

$$M = \frac{moles}{volume\ (L)} = \frac{2.52 \times 10^{-3}\ mol.\ HNO_3}{0.02000\ L} = 0.126\,M\ HNO_3$$

---

34. Refer to Section 4.3 and Example 4.8.

Calculate the moles of the given base, use the mole ratio of HCl to base to calculate the moles of HCl and then use molarity to calculate the volume of HCl.

a. $HCl + NH_3 \rightarrow NH_4^+ + Cl^-$

$$0.02500\,\text{L} \times \frac{0.288\,\text{mol. NH}_3}{1\,\text{L}} \times \frac{1\,\text{mol. HCl}}{1\,\text{mol. NH}_3} \times \frac{1\,\text{L}}{0.885\,\text{mol. HCl}} = 8.14 \times 10^{-3}\,\text{L HCl}$$

$$= 8.14\,\text{mL HCl}$$

b. $NaOH + HCl \rightarrow NaCl + H_2O$

$$10.00\,\text{g NaOH} \times \frac{1\,\text{mol. NaOH}}{40.00\,\text{g NaOH}} \times \frac{1\,\text{mol. HCl}}{1\,\text{mol. NaOH}} = 0.2500\,\text{mol. HCl}$$

$$M = \frac{\text{moles}}{\text{volume (L)}} \implies 0.885\,M = \frac{0.2500\,\text{mol. HCl}}{\text{volume}}$$

volume = 0.2825 L = 282.5 mL

c. Use density to convert volume of solution to mass and then use the mass percent to calculate the mass of methyl amine in solution. Then proceed as above.

$$CH_3NH_2 + HCl \rightarrow CH_3NH_3^+ + Cl^-$$

$$25.00\,\text{mL} \times \frac{1\,\text{cm}^3}{1\,\text{mL}} \times \frac{0.928\,\text{g solution}}{1\,\text{cm}^3} \times \frac{10\,\text{g CH}_3\text{NH}_2}{100\,\text{g solution}} = 2.32\,\text{g CH}_3\text{NH}_2$$

$$2.32\,\text{g CH}_3\text{NH}_2 \times \frac{1\,\text{mol. CH}_3\text{NH}_2}{31.06\,\text{gCH}_3\text{NH}_2} \times \frac{1\,\text{mol. HCl}}{1\,\text{mol. CH}_3\text{NH}_2} = 0.0747\,\text{mol. HCl}$$

$$M = \frac{\text{moles}}{\text{volume (L)}} \implies 0.885\,M = \frac{0.0747\,\text{mol. HCl}}{\text{volume}}$$

volume = 0.0844 L = 84.4 mL

---

36. *Refer to Section 4.3 and Problem 34 (above).*

Calculate the moles of HCl, use the mole ratio of HCl to $NaHCO_3$ to calculate the moles of $NaHCO_3$ and then use molar mass to calculate the mass of $NaHCO_3$.

$$HCl + NaHCO_3 \rightarrow H_2CO_3 + NaCl$$

$$0.075\,\text{L} \times \frac{3.00\,\text{mol. HCl}}{1\,\text{L}} \times \frac{1\,\text{mol. NaHCO}_3}{1\,\text{mol. HCl}} \times \frac{84.01\,\text{g}}{1\,\text{mol. NaHCO}_3} = 19\,\text{g NaHCO}_3$$

38. *Refer to Section 4.3, Example 4.9 and Problem 36 (above).*

Calculate mass of $NaHCO_3$ as in problem 36 above, then calculate the mass percent.

$HCl + NaHCO_3 \rightarrow H_2CO_3 + NaCl$

$$0.0155\,L \times \frac{0.275\,mol.\ HCl}{1\,L} \times \frac{1\,mol.\ NaHCO_3}{1\,mol.\ HCl} \times \frac{84.01g}{1\,mol.\ NaHCO_3} = 0.359\,g\ NaHCO_3$$

$$\%\,NaHCO_3 = \frac{mass\ of\ NaHCO_3}{mass\ of\ sample} \times 100\% = \frac{0.359\,g\ NaHCO_3}{0.500\,g\ sample} \times 100\% = 71.6\%$$

40. *Refer to Section 4.3.*

Determine the mole ratio of lactic acid to NaOH by calculating the moles of each.

$$0.100\,g\ C_3H_6O_3 \times \frac{1\,mol.\ C_3H_6O_3}{90.08\,g\ C_3H_6O_3} = 1.11 \times 10^{-3}\ mol.\ C_3H_6O_3$$

$$0.01295\,L \times \frac{0.0857\,mol.\ NaOH}{1\,L} = 1.11 \times 10^{-3}\ mol.\ NaOH$$

mole ratio = 1:1 (one mole of NaOH is required to neutralize one mole of $C_3H_6O_3$)

42. *Refer to Sections 2.6 and 4.4 and Example 4.10.*

To calculate the oxidation number, recall that oxygen has an oxidation number of -2 (O^{-2}) and that the sum of the oxidation numbers must equal the overall charge.

a. N_2O_3 [2(N) + 3(O) = charge]
                 [2(N) + 3(-2) = 0]
                 N = +3

b. SO_2 [1(S) + 2(O) = charge]
                 [S + 2(-2) = 0]
                 S = +4

c. $Cr_2O_7^{-2}$ [2(Cr) + 7(O) = charge]
                 [2(Cr) + 7(-2) = -2]
                 Cr = +6

d. ClO^- [1(Cl) + 1(O) = charge]
                 [Cl + 1(-2) = -1]
                 Cl = +1

Recall that the sum of the oxidation numbers must equal the overall charge.

a. P_2O_5 [2(P) + 5(O) = charge]
             [2(P) + 5(-2) = 0]
             P = +5

b. NH_3 [1(N) + 3(H) = charge]
             [N + 3(+1) = 0]
             N = -3

c. CO_3^{-2} [1(C) + 3(O) = charge]
             [C + 3(-2) = -2]
             C = +4

d. $S_2O_3^{-2}$ [2(S) + 3(O) = charge]
             [2(S) + 3(-2) = -2]
             S = +2

e. N_2H_4 [2(N) + 4(H) = charge]
             [2(N) + 4(+1) = 0]
             N = -2

46. Refer to Section 4.4 and Problem 44 (above).

Determine the oxidation number of the species of interest in each of the half reactions. If the oxidation number decreases, there is a reduction, if it increases, there is an oxidation.

a. TiO_2: Ti^{+4}
    Ti^{+3}
    oxidation number *de*creased from +4 to +3, so this is a **reduction**.

b. Zn^{+2}
    Zn^0
    oxidation number *de*creased from +2 to 0, so this is a **reduction**.

c. NH_4^+: N^{-3}
    N_2: N^0
    oxidation number *in*creased from -3 to 0, so this is an **oxidation**.

d. CH$_3$OH: [1(C) + 4(H) + 1(O) = charge]
               [1(C) + 4(+1) + (-2) = 0]
               C^{-2}

   CH$_2$O: [1(C) + 2(H) + 1(O) = charge]
               [1(C) + 2(+1) + (-2) = 0]
               C^0

oxidation number *in*creased from -2 to 0, so this is an **oxidation**.

---

48. Refer to Section 4. , Example 4.11 and Problem 46 (above)4.

Begin by determining the oxidation states of the element being reduced or oxidized. Balance the electrons for that element and then balance the element. Finish by adding H$^+$ (if acidic)or OH$^-$ (if basic) to balance the charges, and H$_2$O to balance H and O.

a. TiO$_2$(s) → Ti^{+3}(aq)   [basic]
    Ti^{+4} + e$^-$ → Ti^{+3} electrons balanced
    TiO$_2$(s) + e$^-$ → Ti^{+3}(aq)   Ti balanced
    TiO$_2$(s) + e$^-$ → Ti^{+3}(aq) + 4OH$^-$(aq)   charges balanced
    TiO$_2$(s) + e$^-$ + 2H$_2$O(l) → Ti^{+3}(aq) + 4OH$^-$(aq)   O and H balanced

b. Zn^{+2}(aq) → Zn(s)   [basic]
    Zn^{+2} + 2e$^-$ → Zn^0 electrons balanced
    Zn^{+2}(aq) + 2e$^-$ → Zn(s)   completely balanced

c. NH$_4^+$(aq) → N$_2$(g)   [acidic]
    N^{-3} → N^0 + 3e$^-$ electrons balanced
    2NH$_4^+$(aq) → N$_2$(g) + 6e$^-$ N balanced
    2NH$_4^+$(aq) → N$_2$(g) + 6e$^-$ + 8H$^+$(aq)   charges, O and H balanced

d. CH$_3$OH(aq) → CH$_2$O(aq)   [acidic]
    C^{-2} → C^0 + 2e$^-$ electrons balanced
    CH$_3$OH(aq) → CH$_2$O(aq) + 2e$^-$ C balanced
    CH$_3$OH(aq) → CH$_2$O(aq) + 2e$^-$ + 2H$^+$(aq)   charges, O and H balanced

---

50. Refer to Section 4.4, Example 4.11 and Problem 48 (above).

a. Mn^{+2}(aq) → MnO$_4^-$(aq)   [acidic]   **oxidation**
    Mn^{+2} → Mn^{+7} + 5e$^-$ electrons balanced
    Mn^{+2}(aq) → MnO$_4^-$(aq) + 5e$^-$ Mn balanced
    Mn^{+2}(aq) → MnO$_4^-$(aq) + 5e$^-$ + 8H$^+$(aq)   charges balanced
    4H$_2$O(l) + Mn^{+2}(aq) → MnO$_4^-$(aq) + 5e$^-$ + 8H$^+$(aq)   H and O balanced

b. $CrO_4^{-2}(aq) \rightarrow Cr^{+3}(aq)$ [basic]   **reduction**

     $Cr^{+6} + 3e^- \rightarrow Cr^{+3}$ electrons balanced

     $CrO_4^{-2}(aq) + 3e^- \rightarrow Cr^{+3}(aq)$ Cr balanced

     $CrO_4^{-2}(aq) + 3e^- \rightarrow Cr^{+3}(aq) + 8OH^-(aq)$ charges balanced

     $4H_2O(l) + CrO_4^{-2}(aq) + 3e^- \rightarrow Cr^{+3}(aq) + 8OH^-(aq)$ H and O balanced

c. $PbO_2(s) \rightarrow Pb^{+2}(aq)$ [basic]   **reduction**

     $Pb^{+4} + 2e^- \rightarrow Pb^{+2}$ electrons balanced

     $PbO_2(s) + 2e^- \rightarrow Pb^{+2}(aq)$ Pb balanced

     $PbO_2(s) + 2e^- \rightarrow Pb^{+2}(aq) + 4OH^-(aq)$ charges balanced

     $2H_2O(l) + PbO_2(s) + 2e^- \rightarrow Pb^{+2}(aq) + 4OH^-(aq)$ H and O balanced

d. $ClO_2^-(aq) \rightarrow ClO^-(aq)$ [acidic]   **reduction**

     $Cl^{+3} + 2e^- \rightarrow Cl^+$ electrons balanced

     $ClO_2^-(aq) + 2e^- \rightarrow ClO^-(aq)$ Mn balanced

     $ClO_2^-(aq) + 2e^- + 2H^+(aq) \rightarrow ClO^-(aq)$ charges balanced

     $ClO_2^-(aq) + 2e^- + 2H^+(aq) \rightarrow ClO^-(aq) + H_2O(l)$ H and O balanced

---

52. Refer to Section 4.4 and Example 4.12.

a. $H_2O_2(aq) \rightarrow H_2O(l)$

     $O^- \rightarrow O^{-2}$

     Oxidation number has decreased, so O is reduced and H_2O_2 is the oxidizing agent.

     $Ni^{+2}(aq) \rightarrow Ni^{+3}(aq)$

     Oxidation number has increased, so Ni is oxidized and Ni^{+2} is the reducing agent.

b. $Cr_2O_7^{-2}(aq) \rightarrow Cr^{+3}(l)$

     $Cr^{+6} \rightarrow Cr^{+3}$

     Oxidation number has decreased, so Cr is reduced and $Cr_2O_7^{-2}$ is the oxidizing agent.

     $Sn^{+2}(aq) \rightarrow Sn^{+4}(aq)$

     Oxidation number has increased, so Sn is oxidized and Sn^{+2} is the reducing agent.

---

54. Refer to Sections 4.4, Example 4.12 and Problem 52 (above).

     Balance the two half reactions. If the number of electrons in the balanced half reactions are not equal, multiply the equations through with the appropriate coefficient. Combine the two half reactions and cancel those species which appear on both sides of the reaction.

a. $H_2O_2(aq) \rightarrow H_2O(l)$ [acidic]

     $O^- + e^- \rightarrow O^{-2}$ electrons balanced

     $H_2O_2(aq) + 2e^- \rightarrow 2H_2O(l)$ O balanced

     $H_2O_2(aq) + 2e^- + 2H^+ \rightarrow 2H_2O(l)$ charges, H and O balanced

$Ni^{+2}(aq) \rightarrow Ni^{+3}(aq)$

$Ni^{+2}(aq) \rightarrow Ni^{+3}(aq) + e^-$ everything is balanced

$2Ni^{+2}(aq) \rightarrow 2Ni^{+3}(aq) + 2e^-$ balance electrons in half-reactions

$2Ni^{+2}(aq) + H_2O_2(aq) + 2e^- + 2H^+(aq) \rightarrow 2H_2O(l) + 2Ni^{+3}(aq) + 2e^-$

$2Ni^{+2}(aq) + H_2O_2(aq) + 2H^+(aq) \rightarrow 2H_2O(l) + 2Ni^{+3}(aq)$

b. $Cr_2O_7^{-2}(aq) \rightarrow Cr^{+3}(aq)$

$Cr^{+6} + 3e^- \rightarrow Cr^{+3}$ electrons balanced

$Cr_2O_7^{-2}(aq) + 6e^- \rightarrow 2Cr^{+3}(aq)$ Cr balanced

$Cr_2O_7^{-2}(aq) + 6e^- + 14H^+(aq) \rightarrow 2Cr^{+3}(aq)$ charges balanced

$Cr_2O_7^{-2}(aq) + 6e^- + 14H^+(aq) \rightarrow 2Cr^{+3}(aq) + 7H_2O(l)$ H and O balanced

$Sn^{+2}(aq) \rightarrow Sn^{+4}(aq)$

$Sn^{+2} \rightarrow Sn^{+4} + 2e^-$ everything is balanced

$3Sn^{+2}(aq) \rightarrow 3Sn^{+4}(aq) + 6e^-$ balance electrons in half-reactions

$3Sn^{+2}(aq) + Cr_2O_7^{-2}(aq) + 6e^- + 14H^+(aq) \rightarrow 2Cr^{+3}(aq) + 7H_2O(l) + 3Sn^{+4}(aq) + 6e^-$

$3Sn^{+2}(aq) + Cr_2O_7^{-2}(aq) + 14H^+(aq) \rightarrow 2Cr^{+3}(aq) + 7H_2O(l) + 3Sn^{+4}(aq)$

---

56. Refer to Sections 4.4, Example 4.12 and Problem 54 (above).

---

a. $Ni^{+2}(aq) \rightarrow Ni^{+3}(aq)$

$Ni^{+2}(aq) \rightarrow Ni^{+3}(aq) + e^-$ everything is balanced

$IO_4^-(aq) \rightarrow I^-(aq)$

$I^{+7} + 8e^- \rightarrow I^-$ electrons balanced

$IO_4^-(aq) + 8e^- \rightarrow I^-(aq)$ I balanced

$IO_4^-(aq) + 8e^- + 8H^+(aq) \rightarrow I^-(aq)$ charges balanced

$IO_4^-(aq) + 8e^- + 8H^+(aq) \rightarrow I^-(aq) + 4H_2O(l)$ H and O balanced

$8Ni^{+2}(aq) \rightarrow 8Ni^{+3}(aq) + 8e^-$ balance electrons in half-reactions

$8Ni^{+2}(aq) + IO_4^-(aq) + 8e^- + 8H^+(aq) \rightarrow I^-(aq) + 4H_2O(l) + 8Ni^{+3}(aq) + 8e^-$

$8Ni^{+2}(aq) + IO_4^-(aq) + 8H^+(aq) \rightarrow I^-(aq) + 4H_2O(l) + 8Ni^{+3}(aq)$

b. $O_2(g) \rightarrow H_2O(l)$

$O^0(aq) + 2e^- \rightarrow O^{-2}(aq)$ electrons balanced

$O_2(g) + 4e^- \rightarrow 2H_2O(l)$ O balanced

$O_2(g) + 4e^- + 4H^+(aq) \rightarrow 2H_2O(l)$ charges, H and O balanced

$Br^-(aq) \rightarrow Br_2(l)$

$Br^- \rightarrow Br^0 + e^-$ electrons balanced

$2Br^-(aq) \rightarrow Br_2(l) + 2e^-$ everything is balanced

$4Br^-(aq) \rightarrow 2Br_2(l) + 4e^-$ balance electrons in half-reactions

$4Br^-(aq) + O_2(g) + 4e^- + 4H^+(aq) \rightarrow 2H_2O(l) + 2Br_2(l) + 4e^-$
$4Br^-(aq) + O_2(g) + 4H^+(aq) \rightarrow 2H_2O(l) + 2Br_2(l)$

c. $Ca(s) \rightarrow Ca^{+2}(aq)$
$Ca^0 \rightarrow Ca^{+2} + 2e^-$ everything is balanced

$Cr_2O_7^{-2}(aq) \rightarrow Cr^{+3}(aq)$
$Cr^{+6} + 3e^- \rightarrow Cr^{+3}$ electrons balanced
$Cr_2O_7^{-2}(aq) + 6e^- \rightarrow 2Cr^{+3}(aq)$ Cr balanced
$Cr_2O_7^{-2}(aq) + 6e^- + 14H^+(aq) \rightarrow 2Cr^{+3}(aq)$ charges balanced
$Cr_2O_7^{-2}(aq) + 6e^- + 14H^+(aq) \rightarrow 2Cr^{+3}(aq) + 7H_2O(l)$ H and O balanced

$3Ca(s) \rightarrow 3Ca^{+2}(aq) + 6e^-$ balance electrons in half-reactions

$3Ca(s) + Cr_2O_7^{-2}(aq) + 6e^- + 14H^+(aq) \rightarrow 2Cr^{+3}(aq) + 7H_2O(l) + 3Ca^{+2}(aq) + 6e^-$
$3Ca(s) + Cr_2O_7^{-2}(aq) + 14H^+(aq) \rightarrow 2Cr^{+3}(aq) + 7H_2O(l) + 3Ca^{+2}(aq)$

d. $IO_3^-(aq) \rightarrow I^-(aq)$
$I^{+5} + 6e^- \rightarrow I^-$ electrons balanced
$IO_3^-(aq) + 6e^- \rightarrow I^-(aq)$ I balanced
$IO_3^-(aq) + 6e^- + 6H^+(aq) \rightarrow I^-(aq)$ charges balanced
$IO_3^-(aq) + 6e^- + 6H^+(aq) \rightarrow I^-(aq) + 3H_2O(l)$ H and O balanced

$Mn^{+2}(aq) \rightarrow MnO_2(s)$
$Mn^{+2} \rightarrow Mn^{+4} + 2e^-$ electrons balanced
$Mn^{+2}(aq) \rightarrow MnO_2(s) + 2e^-$ Mn balanced
$Mn^{+2}(aq) \rightarrow MnO_2(s) + 2e^- + 4H^+(aq)$ charges balanced
$Mn^{+2}(aq) + 2H_2O(l) \rightarrow MnO_2(s) + 2e^- + 4H^+(aq)$ H and O balanced

$3Mn^{+2}(aq) + 6H_2O(l) \rightarrow 3MnO_2(s) + 6e^- + 12H^+(aq)$ balance electrons in half-reactions

$IO_3^-(aq) + 6e^- + 6H^+(aq) + 3Mn^{+2}(aq) + 6H_2O(l)$
$\qquad \rightarrow 3MnO_2(s) + 6e^- + 12H^+(aq) + I^-(aq) + 3H_2O(l)$
$IO_3^-(aq) + 3Mn^{+2}(aq) + 3H_2O(l) \rightarrow 3MnO_2(s) + 6H^+(aq) + I^-(aq)$

---

58. Refer to Sections 4.4, Example 4.12 and Problem 48 (above).

---

a. $Ni(OH)_2(s) \rightarrow Ni(s)$
$Ni^{+2} + 2e^- \rightarrow Ni^0$ electrons balanced
$Ni(OH)_2(s) + 2e^- \rightarrow Ni(s)$ Ni balanced
$Ni(OH)_2(s) + 2e^- \rightarrow Ni(s) + 2OH^-$ charges, H and O balanced

$N_2H_4(aq) \rightarrow N_2(g)$

$N^{-2} \rightarrow N^0 + 2e^-$ electrons balanced

$N_2H_4(aq) \rightarrow N_2(g) + 4e^-$ N balanced

$N_2H_4(aq) + 4OH^-(aq) \rightarrow N_2(g) + 4\,e^- + 4H_2O(l)$ charges and H balanced

$2Ni(OH)_2(s) + 4e^- \rightarrow 2Ni(s) + 4OH^-(aq)$ balance electrons in half-reactions

$N_2H_4(aq) + 4OH^-(aq) + 2Ni(OH)_2(s) + 4e^- \rightarrow 2Ni(s) + 4OH^-(aq) + N_2(g) + 4e^- + 4H_2O(l)$

$N_2H_4(aq) + 2Ni(OH)_2(s) \rightarrow 2Ni(s) + N_2(g) + 4H_2O(l)$

b. $Fe(OH)_3(s) \rightarrow Fe(OH)_2(s)$

    $Fe^{+3} + e^- \rightarrow Fe^{+2}$ electrons balanced

    $Fe(OH)_3(s) + e^- \rightarrow Fe(OH)_2(s)$ Fe balanced

    $Fe(OH)_3(s) + e^- \rightarrow Fe(OH)_2(s) + OH^-(aq)$ charges, H and O balanced

    $Cr^{+3}(aq) \rightarrow CrO_4^{-2}(aq)$

    $Cr^{+3}(aq) \rightarrow Cr^{+6} + 3e^-$ electrons balanced

    $Cr^{+3}(aq) \rightarrow CrO_4^{-2}(aq) + 3\,e^-$ Cr balanced

    $Cr^{+3}(aq) + 8OH^-(aq) \rightarrow CrO_4^{-2}(aq) + 3e^-$ charges balanced

    $Cr^{+3}(aq) + 8OH^-(aq) \rightarrow CrO_4^{-2}(aq) + 3e^- + 4H_2O(l)$ H and O balanced

    $3Fe(OH)_3(s) + 3e^- \rightarrow 3Fe(OH)_2(s) + 3OH^-(aq)$ balance electrons in half-reactions

    $3Fe(OH)_3(s) + 3e^- + Cr^{+3}(aq) + 8OH^-(aq)$
              $\rightarrow CrO_4^{-2}(aq) + 3e^- + 4H_2O(l) + 3Fe(OH)_2(s) + 3OH^-(aq)$

    $3Fe(OH)_3(s) + Cr^{+3}(aq) + 5OH^- \rightarrow CrO_4^{-2}(aq) + 4H_2O(l) + 3Fe(OH)_2(s)$

c. $MnO_4^-(aq) \rightarrow MnO_2(s)$

    $Mn^{+7} + 3e^- \rightarrow Mn^{+4}$ electrons balanced

    $MnO_4^-(aq) + 3e^- \rightarrow MnO_2(s)$ Mn balanced

    $MnO_4^-(aq) + 3e^- + 2H_2O(l) \rightarrow MnO_2(s) + 4OH^-(aq)$ charges, H and O balanced

    $BrO_3^-(aq) \rightarrow BrO_4^-(aq)$

    $Br^{+5} \rightarrow Br^{+7} + 2e^-$ electrons balanced

    $BrO_3^-(aq) \rightarrow BrO_4^-(aq) + 2e^-$ Br balanced

    $BrO_3^-(aq) + 2OH^-(aq) \rightarrow BrO_4^-(aq) + 2e^-$ charges balanced

    $BrO_3^-(aq) + 2OH^-(aq) \rightarrow BrO_4^-(aq) + 2e^- + H_2O(l)$ H and O balanced

                             balance electrons in half-reactions

    $2MnO_4^-(aq) + 6e^- + 4H_2O(l) \rightarrow 2MnO_2(s) + 8OH^-(aq)$

    $3BrO_3^-(aq) + 6OH^-(aq) \rightarrow 3BrO_4^-(aq) + 6e^- + 3H_2O(l)$

    $2MnO_4^-(aq) + 6e^- + 4H_2O(l) + 3BrO_3^-(aq) + 6OH^-(aq)$
              $\rightarrow 3BrO_4^-(aq) + 6e^- + 3H_2O(l) + 2MnO_2(s) + 8OH^-(aq)$

    $2MnO_4^-(aq) + 3BrO_3^-(aq) + H_2O(l) \rightarrow 3BrO_4^-(aq) + 2MnO_2(s) + 2OH^-(aq)$

d. $H_2O_2(aq) \rightarrow O_2(g)$

$\quad\quad O^- \rightarrow O^0 + e^-$ electrons balanced

$\quad\quad H_2O_2(aq) \rightarrow O_2(g) + 2e^-$ O balanced

$\quad\quad H_2O_2(aq) + 2OH^-(aq) \rightarrow O_2(g) + 2e^-$ charges balanced

$\quad\quad H_2O_2(aq) + 2OH^-(aq) \rightarrow O_2(g) + 2e^- + 2H_2O(l)$ H and O balanced

$\quad\quad IO_4^-(aq) \rightarrow IO_2^-(aq)$

$\quad\quad I^{+7} + 4e^- \rightarrow I^{+3}$ electrons balanced

$\quad\quad IO_4^-(aq) + 4e^- \rightarrow IO_2^-(aq)$ I balanced

$\quad\quad IO_4^-(aq) + 4e^- \rightarrow IO_2^-(aq) + 4OH^-(aq)$ charges balanced

$\quad\quad IO_4^-(aq) + 4e^- + 2H_2O(l) \rightarrow IO_2^-(aq) + 4OH^-(aq)$ H and O balanced

$\quad\quad 2H_2O_2(aq) + 4OH^-(aq) \rightarrow 2O_2(g) + 4e^- + 4H_2O(l)$ balance electrons in half-reactions

$\quad\quad 2H_2O_2(aq) + 4OH^-(aq) + IO_4^-(aq) + 4e^- + 2H_2O(l)$
$\quad\quad\quad\quad \rightarrow IO_2^-(aq) + 4OH^-(aq) + 2O_2(g) + 4e^- + 4H_2O(l)$
$\quad\quad 2H_2O_2(aq) + IO_4^-(aq) \rightarrow IO_2^-(aq) + 2O_2(g) + 2H_2O(l)$

---

60. Refer to Sections 2.6, 4.2 and 4.4, Example 4.12.

Rewrite the reaction, deleting those species that do not change oxidation state, then balance the two half reactions. If the number of electrons in the balanced half reactions are not equal, multiply the equations through with the appropriate coefficient. Combine the two half reactions and cancel those species which appear on both sides of the reaction.

a. $Zn(s) + HCl(aq) \rightarrow ZnCl_2(aq) + H_2(g)$

$\quad\quad Zn(s) + H^+(aq) \rightarrow Zn^{+2}(aq) + H_2(g)$ (unbalanced ionic equation)

$\quad\quad Zn(s) \rightarrow ZnCl_2(aq)$

$\quad\quad Zn^0 \rightarrow Zn^{+2}$ electrons balanced

$\quad\quad Zn(s) \rightarrow Zn^{+2}(aq) + 2e^-$ everything is balanced

$\quad\quad HCl(aq) \rightarrow H_2(g)$

$\quad\quad H^+ + e^- \rightarrow H^0$ electrons balanced

$\quad\quad 2H^+(aq) + 2e^- \rightarrow H_2(g)$ everything is balanced

$\quad\quad Zn(s) + 2H^+(aq) + 2e^- \rightarrow Zn^{+2}(aq) + 2e^- + H_2(g)$

$\quad\quad Zn(s) + 2H^+(aq) \rightarrow Zn^{+2}(aq) + H_2(g)$

b. $CuS(s) + HNO_3(aq) \rightarrow Cu(NO_3)_2(aq) + S_8(s) + NO(g)$

$\quad\quad S^{-2}(aq) + NO_3^-(aq) \rightarrow S_8(s) + NO(g)$ (unbalanced ionic equation)

$\quad\quad S^{-2}(aq) \rightarrow S_8(s)$

$\quad\quad S^{-2} \rightarrow S^0 + 2e^-$ electrons balanced

$\quad\quad 8S^{-2}(aq) \rightarrow S_8(s) + 16e^-$ everything is balanced

$NO_3^-(aq) \rightarrow NO(g)$

$N^{+5} + 3e^- \rightarrow N^{+2}$ electrons balanced

$NO_3^-(aq) + 3e^- \rightarrow NO(g)$ N balanced

$NO_3^-(aq) + 3e^- + 4H^+(aq) \rightarrow NO(g)$ charges balanced

$NO_3^-(aq) + 3e^- + 4H^+(aq) \rightarrow NO(g) + 2H_2O(l)$ H and O balanced

$24S^{-2}(aq) \rightarrow 3S_8(s) + 48e^-$ balance electrons in half-reactions

$16NO_3^-(aq) + 48e^- + 64H^+(aq) \rightarrow 16NO(g) + 32H_2O(l)$

$16NO_3^-(aq) + 48e^- + 64H^+(aq) + 24S^{-2}(aq) \rightarrow 3S_8(s) + 48e^- + 16NO(g) + 32H_2O(l)$

$16NO_3^-(aq) + 64H^+(aq) + 24S^{-2}(aq) \rightarrow 3S_8(s) + 16NO(g) + 32H_2O(l)$

c. $Sb^{+3}(aq) + IO_4^-(aq) \rightarrow Sb^{+5}(aq) + I^-(aq)$ (unbalanced ionic equation)

$Sb^{+3}(aq) \rightarrow Sb^{+5}(aq)$

$Sb^{+3} \rightarrow Sb^{+5} + 2e^-$ electrons balanced

$Sb^{+3}(aq) \rightarrow Sb^{+5}(aq) + 2e^-$ everything is balanced

$IO_4^-(aq) \rightarrow I^-(aq)$

$I^{+7} + 8e^- \rightarrow I^-$ electrons balanced

$IO_4^-(aq) + 8e^- \rightarrow I^-(aq)$ I balanced

$IO_4^-(aq) + 8e^- + 8H^+(aq) \rightarrow I^-(aq)$ charges balanced

$IO_4^-(aq) + 8e^- + 8H^+(aq) \rightarrow I^-(aq) + 4H_2O(l)$ H and O balanced

$4Sb^{+3}(aq) \rightarrow 4Sb^{+5}(aq) + 8e^-$ balance electrons in half-reactions

$IO_4^-(aq) + 8e^- + 8H^+(aq) + 4Sb^{+3}(aq) \rightarrow 4Sb^{+5}(aq) + 8e^- + I^-(aq) + 4H_2O(l)$

$IO_4^-(aq) + 8H^+(aq) + 4Sb^{+3}(aq) \rightarrow 4Sb^{+5}(aq) + I^-(aq) + 4H_2O(l)$

---

62. Refer to Sections 4.4, Example 4.13 and Problem 60 (above).

---

a. $KMnO_4(aq) + H_2C_2O_4(aq) \rightarrow MnO_2(s) + CO_2(g)$

$MnO_4^-(aq) \rightarrow MnO_2(s)$

$Mn^{+7} + 3e^- \rightarrow Mn^{+4}$ electrons balanced

$MnO_4^-(aq) + 3e^- \rightarrow MnO_2(s)$ Mn balanced

$MnO_4^-(aq) + 3e^- + 4H^+(aq) \rightarrow MnO_2(s)$ charges balanced

$MnO_4^-(aq) + 3e^- + 4H^+(aq) \rightarrow MnO_2(s) + 2H_2O(l)$ H and O balanced

$H_2C_2O_4(aq) \rightarrow CO_2(g)$

$C^{+3} \rightarrow C^{+4} + e^-$ electrons balanced

$H_2C_2O_4(aq) \rightarrow 2CO_2(g) + 2e^-$ C balanced

$H_2C_2O_4(aq) \rightarrow 2CO_2(g) + 2e^- + 2H^+$ charges, H and O balanced

$3H_2C_2O_4(aq) \rightarrow 6CO_2(g) + 6e^- + 6H^+$ balance electrons in half-reactions

$2MnO_4^-(aq) + 6e^- + 8H^+(aq) \rightarrow 2MnO_2(s) + 4H_2O(l)$

$2MnO_4^-(aq) + 6e^- + 8H^+(aq) + 3H_2C_2O_4(aq)$
$$\rightarrow 6CO_2(g) + 6e^- + 6H^+ + 2MnO_2(s) + 4H_2O(l)$$
$2MnO_4^-(aq) + 2H^+(aq) + 3H_2C_2O_4(aq) \rightarrow 6CO_2(g) + 2MnO_2(s) + 4H_2O(l)$

b. $0.0200\,L \times \dfrac{0.300\,mol.\,KMnO_4}{1\,L} \times \dfrac{1\,mol.\,MnO_4^-}{1\,mol.\,KMnO_4} \times \dfrac{3\,mol.\,H_2C_2O_4}{2\,mol.\,MnO_4^-}$

$$= 9.00 \times 10^{-3}\,mol.\,H_2C_2O_4$$

$$M = \frac{moles}{volume\,(L)} = \frac{9.00 \times 10^{-3}\,mol.\,H_2C_2O_4}{0.0137\,L} = 0.657\,M\,H_2C_2O_4$$

c. $9.00 \times 10^{-3}\,mol.\,H_2C_2O_4 \times \dfrac{2\,mol.\,MnO_2}{3\,mol.\,H_2C_2O_4} \times \dfrac{86.94\,g\,MnO_2}{1\,mol.\,MnO_2} = 0.522\,g\,MnO_2$

---

64. Refer to Section 4.4 Example 4.13 and Problem 62 (above).

$Ag(s) \rightarrow Ag^+(aq)$

$Ag^0 \rightarrow Ag^+ + e^-$ electrons balanced
$Ag(s) \rightarrow Ag^+(aq) + e^-$ everything balanced

$NO_3^-(aq) \rightarrow NO_2(g)$
$N^{+5} + e^- \rightarrow N^{+4}$ electrons balanced
$NO_3^-(aq) + e^- \rightarrow NO_2(g)$ N balanced
$NO_3^-(aq) + e^- + 2H^+(aq) \rightarrow NO_2(g)$ charges balanced
$NO_3^-(aq) + e^- + 2H^+(aq) \rightarrow NO_2(g) + H_2O(l)$ H and O balanced

$Ag(s) + NO_3^-(aq) + e^- + 2H^+(aq) \rightarrow NO_2(g) + H_2O(l) + Ag^+(aq) + e^-$
$Ag(s) + NO_3^-(aq) + 2H^+(aq) \rightarrow NO_2(g) + H_2O(l) + Ag^+(aq)$

$0.04250\,L \times \dfrac{12.0\,mol.\,HNO_3}{1\,L} \times \dfrac{1\,mol.\,H^+}{1\,mol.\,HNO_3} \times \dfrac{1\,mol.\,Ag}{2\,mol.\,H^+} \times \dfrac{107.9\,g\,Ag}{1\,mol.\,Ag} = 27.5\,g\,Ag$

---

66. Refer to Section 4.4 Example 4.13 and Problem 62 (above).

$H_2O_2(aq) \rightarrow O_2(g)$
$O^- \rightarrow O^0 + e^-$ electrons balanced
$H_2O_2(aq) \rightarrow O_2(g) + 2e^-$ O balanced
$H_2O_2(aq) \rightarrow O_2(g) + 2e^- + 2H^+(aq)$ charges, H and O balanced

$$Cr_2O_7^{-2}(aq) \rightarrow Cr^{+3}(aq)$$
$$Cr^{+6} + 3e^- \rightarrow Cr^{+3} \qquad \text{electrons balanced}$$
$$Cr_2O_7^{-2}(aq) + 6e^- \rightarrow 2Cr^{+3}(aq) \qquad \text{Cr balanced}$$
$$Cr_2O_7^{-2}(aq) + 6e^- + 14H^+(aq) \rightarrow 2Cr^{+3}(aq) \qquad \text{charges balanced}$$
$$Cr_2O_7^{-2}(aq) + 6e^- + 14H^+(aq) \rightarrow 2Cr^{+3}(aq) + 7H_2O(l) \qquad \text{H and O balanced}$$

$$3H_2O_2(aq) \rightarrow 3O_2(g) + 6e^- + 6H^+(aq) \qquad \text{balance electrons in half-reactions}$$

$$Cr_2O_7^{-2}(aq) + 6e^- + 14H^+(aq) + 3H_2O_2(aq)$$
$$\rightarrow 3O_2(g) + 6e^- + 6H^+(aq) + 2Cr^{+3}(aq) + 7H_2O(l)$$
$$Cr_2O_7^{-2}(aq) + 8H^+(aq) + 3H_2O_2(aq) \rightarrow 3O_2(g) + 2Cr^{+3}(aq) + 7H_2O(l)$$

$$0.0546 \, L \times \frac{0.715 \, mol. \, K_2Cr_2O_7}{1 \, L} \times \frac{1 \, mol. \, Cr_2O_7^{-2}}{1 \, mol. \, K_2Cr_2O_7} \times \frac{3 \, mol. \, H_2O_2}{1 \, mol. \, Cr_2O_7^{-2}} \times \frac{34.02 \, g \, H_2O_2}{1 \, mol. \, H_2O_2}$$

$$= 3.98 \, g \, H_2O_2$$

$$mass \, \% = \frac{3.98 \, g \, H_2O_2}{30.0 \, g \, bleach} \times 100\% = 13.3\%$$

---

68. Refer to Section 4.4 and Example 4.13.

$$50.0 \, mL \, bleach \times \frac{1 \, cm^3}{1 \, mL} \times \frac{1.02 \, g}{1 \, cm^3} = 51.0 \, g \, bleach$$

$$4.95 \, g \, AgCl \times \frac{1 \, mol. \, AgCl}{143.4 \, g} \times \frac{1 \, mol. \, Cl^-}{1 \, mol. \, AgCl} \times \frac{1 \, mol. \, ClO^-}{1 \, mol. \, Cl^-} \times \frac{1 \, mol. \, NaClO}{1 \, mol. \, ClO^-} \times \frac{74.44 \, g \, NaClO}{1 \, mol. \, NaClO}$$

$$= 2.57 \, g \, NaClO$$

$$mass \, \% = \frac{2.57 \, g \, NaClO}{51.0 \, g \, bleach} \times 100\% = 5.04\%$$

---

70. Refer to Section 4.4, Example 4.13 and Problem 68 (above).

$$MnO_4^-(aq) \rightarrow Mn^{+2}(aq)$$
$$Mn^{+7} + 5e^- \rightarrow Mn^{+2} \qquad \text{electrons balanced}$$
$$MnO_4^-(aq) + 5e^- \rightarrow Mn^{+2}(aq) \qquad \text{Mn balanced}$$
$$MnO_4^-(aq) + 5e^- + 8H^+(aq) \rightarrow Mn^{+2}(aq) \qquad \text{charges balanced}$$
$$MnO_4^-(aq) + 5e^- + 8H^+(aq) \rightarrow Mn^{+2}(aq) + 4H_2O(l) \qquad \text{H and O balanced}$$

$$Fe^{+2}(aq) \rightarrow Fe^{+3}(aq)$$
$$Fe^{+2}(aq) \rightarrow Fe^{+3}(aq) + e^- \qquad \text{electrons balanced}$$

$$5Fe^{+2}(aq) \rightarrow 5Fe^{+3}(aq) + 5e^-$$ balance electrons in half-reactions

$$MnO_4^-(aq) + 5e^- + 8H^+(aq) + 5Fe^{+2}(aq) \rightarrow 5Fe^{+3}(aq) + 5e^- + Mn^{+2}(aq) + 4H_2O(l)$$
$$MnO_4^-(aq) + 8H^+(aq) + 5Fe^{+2}(aq) \rightarrow 5Fe^{+3}(aq) + Mn^{+2}(aq) + 4H_2O(l)$$

$$32.3 \, mL \times \frac{1 \, L}{1000 \, mL} \times \frac{0.002100 \, mol. \, KMnO_4}{1 \, L} \times \frac{1 \, mol. \, MnO_4^-}{1 \, mol. \, KMnO_4} = 6.78 \times 10^{-5} \, mol. \, MnO_4^-$$

$$6.78 \times 10^{-5} \, mol. \, MnO_4^- \times \frac{5 \, mol. \, Fe^{+2}}{1 \, mol. \, MnO_4^-} \times \frac{55.85 \, g \, Fe^{+2}}{1 \, mol. \, Fe^{+2}} = 0.0189 \, g \, Fe^{+2}$$

$$mass \, \% = \frac{0.0189 \, g \, Fe^{+2}}{5.00 \, g \, hemoglobin} \times 100\% = 0.378\%$$

---

72. Refer to Sections 4.1 and 4.3.

Calculate the initial and unreacted moles of H^+, then the moles of reacted H^+. From that, calculate the moles and then grams of NH_3. Convert volume of air to mass of air. Finally, calculate the mass percent of NH_3 in the 100.0 mL sample of air.

$$100.0 \, mL \times \frac{1 \, L}{1000 \, mL} \times \frac{0.02500 \, mol. \, H^+}{1 \, L} = 2.500 \times 10^{-3} \, mol. \, H^+ \quad (initial \, moles)$$

$$57.00 \, mL \times \frac{1 \, L}{1000 \, mL} \times \frac{0.03500 \, mol. \, NaOH}{1 \, L} \times \frac{1 \, mol. \, H^+}{1 \, mol. \, NaOH} = 1.995 \times 10^{-3} \, mol. \, H^+$$

mol. reacted H^+ = (mol. initial) – (mol. reacted) = $2.500 \times 10^{-3} - 1.995 \times 10^{-3}$

$$= 0.505 \times 10^{-3} = 5.05 \times 10^{-4} \, mol. \, H^+$$

$$5.05 \times 10^{-4} \, mol. \, H^+ \times \frac{1 \, mol. \, NH_3}{1 \, mol. \, H^+} \times \frac{17.03 \, g \, NH_3}{1 \, mol. \, NH_3} = 8.60 \times 10^{-3} \, g \, NH_3$$

$$100.0 \, mL \, air \times \frac{1 \, L}{1000 \, mL} \times \frac{1.19 \, g \, air}{1 \, L} = 0.119 \, g \, air$$

$$mass \, \% = \frac{mass \, NH_3}{total \, mass \, (of \, air)} \times 100\% = \frac{8.60 \times 10^{-3} \, g \, NH_3}{0.119 \, g \, air} \times 100\% = 7.23\%$$

7.23% is greater than $5.00 \times 10^{-3}\%$, therefore it is **not** in compliance.

74. *Refer to Section 4.2.*

Recall that NaCl is soluble and AgCl is insoluble

a. $Na^+ + Cl^-$:

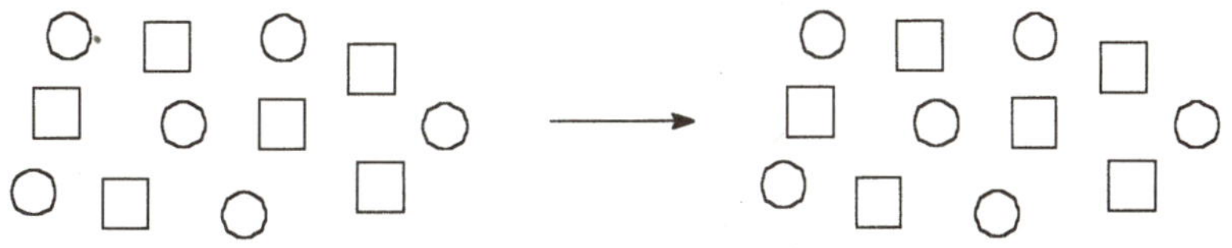

b. $Ag^+ + Cl^-$:

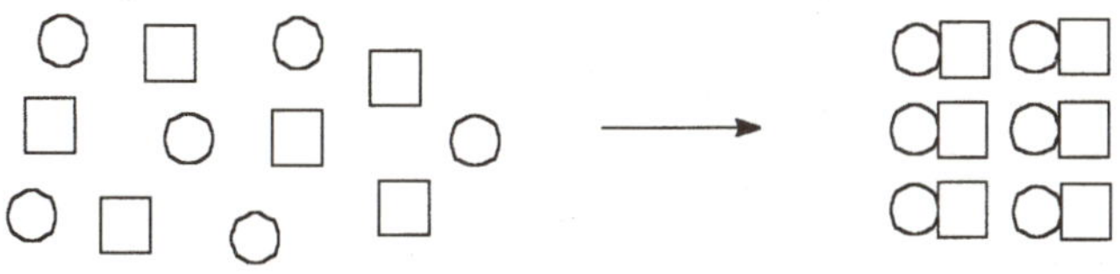

76. *Refer to Section 4.3.*

a. HCl

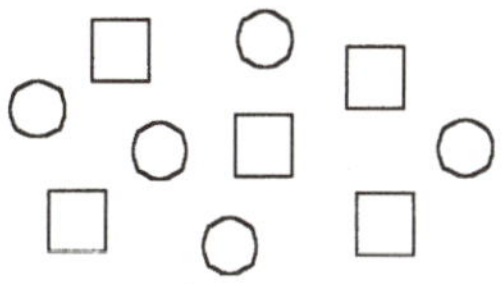

HCl is a strong acid and is consequently completely dissociated in solution.

b. HF

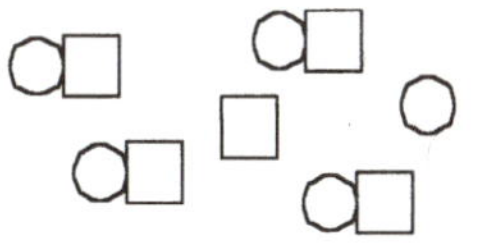

HF is a weak acid, therefore most (but not all) of the acid remains **un**dissociated.

c. KOH

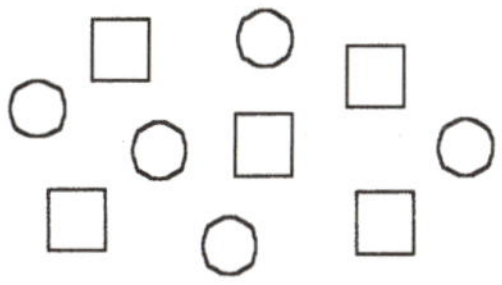

KOH is a strong base and is consequently completely dissociated in solution.

d. HNO_2

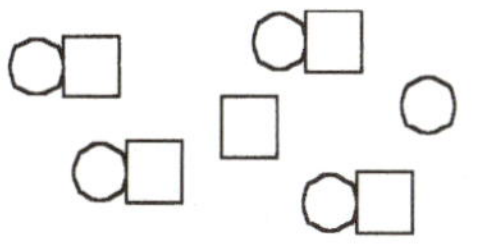

HNO_2 is a weak acid, therefore most (but not all) of the acid remains **un**dissociated. Note that NO_2 is represented with a single circle for the anion.

78. *Refer to Section 4.2 and Figure 4.4.*

No precipitate with Cl^- indicates the metal ion is not Ag^+.
A precipitate with CO_3^{-2} indicates the metal ion is not Na^+.
A precipitate with SO_4^{-2} indicates the metal ion is not Ni^{+2}.
Thus, by elimination, the ion must be $\mathbf{Ba^{+2}}$.

80. *Refer to Section 4.4 and Problem 79.*

Relative strength as oxidizing agent (from problem 79): $Y^+ > Z^+ > W^+ > X^+$. Thus all three reactions will occur.

a. $Y^+ + W \rightarrow Y + W^+$

b. $Z^+ + X \rightarrow Z + X^+$

c. $W^+ + X \rightarrow W + X^+$

81. *Refer to Section 4.4 and Example 4.13*

First write a balanced redox equation for the oxidation of oxalate to CO_2. Then use the amount of $KMnO_4^-$ to calculate the amount of CaC_2O_4 present in the sample.

$CaC_2O_4(s) + 2H^+(aq) \rightarrow H_2C_2O_4^{2-}(aq) + Ca^{2+}(aq)$
$H_2C_2O_4(aq) + MnO_4^-(aq) \rightarrow CO_2(g) + Mn^{2+}(aq)$

$C^{3+} \rightarrow C^{4+} + e^-$ Oxidation half-reaction
$H_2C_2O_4(aq) \rightarrow 2CO_2(g) + 2e^-$
$H_2C_2O_4(aq) \rightarrow 2CO_2(g) + 2e^- + 2H^+(aq)$ eq. 1

$Mn^{7+} + 5e^- \rightarrow Mn^{2+}$ Reduction half-reaction
$MnO_4^-(aq) + 5e^- \rightarrow Mn^{2+}(aq)$
$8H^+(aq) + MnO_4^-(aq) + 5e^- \rightarrow Mn^{2+}(aq)$
$8H^+(aq) + MnO_4^-(aq) + 5e^- \rightarrow Mn^{2+}(aq) + 4H_2O(l)$ eq. 2

To balance the electrons, multiply eq. 1 by 5 and eq. 2 by 2 and add the results.

$16H^+(aq) + 2MnO_4^-(aq) + 10e^- + 5H_2C_2O_4(aq)$
$\rightarrow 10CO_2(g) + 10e^- + 10H^+(aq) + 2Mn^{2+}(aq) + 8H_2O(l)$

$6H^+(aq) + 2MnO_4^-(aq) + 5H_2C_2O_4(aq) \rightarrow 10CO_2(g) + 2Mn^{2+}(aq) + 8H_2O(l)$

$26.2 \text{ mL} \times \dfrac{1 \text{ L}}{1000 \text{ mL}} \times \dfrac{0.0946 \text{ mol. KMnO}_4}{1 \text{ L}} \times \dfrac{1 \text{ mol. MnO}_4^-}{1 \text{ mol. KMnO}_4} \times \dfrac{5 \text{ mol. C}_2\text{O}_4^{2-}}{2 \text{ mol. MnO}_4^-}$

$\times \dfrac{1 \text{ mol. CaC}_2\text{O}_4}{1 \text{ mol. C}_2\text{O}_4^{2-}} \times \dfrac{128.10 \text{ g CaC}_2\text{O}_4}{1 \text{ mol. CaC}_2\text{O}_4} = 0.794 \text{ g CaC}_2\text{O}_4$

$0.794 \text{ g} \times \dfrac{1 \text{ mol. CaC}_2\text{O}_4}{128.10 \text{ g}} \times \dfrac{1 \text{ mol. Ca}^{2+}}{1 \text{ mol. CaC}_2\text{O}_4} \times \dfrac{40.08 \text{ g Ca}^{2+}}{1 \text{ mol. Ca}^{2+}} = 0.248 \text{ g Ca}^{2+}$

$= 248 \text{ mg Ca}^{2+}$ This is within the normal range.

Calculate the amount of $Mg(OH)_2$ and $NaHCO_3$ present and the moles of acid each could neutralize. Then calculate the volume of acid this corresponds to.

330 mg = 0.330 g
mass $Mg(OH)_2$ = 0.410 x 0.330 g = 0.135 g
mass $NaHCO_3$ = 0.362 x 0.330 g = 0.119 g

$$0.135 \text{ g Mg(OH)}_2 \times \frac{1 \text{ mol. Mg(OH)}_2}{58.3 \text{ g Mg(OH)}_2} \times \frac{2 \text{ mol. H}^+}{1 \text{ mol. Mg(OH)}_2} = 0.00463 \text{ mol. H}^+$$

$$0.119 \text{ g NaHCO}_3 \times \frac{1 \text{ mol. NaHCO}_3}{84.0 \text{ g NaHCO}_3} \times \frac{1 \text{ mol. H}^+}{1 \text{ mol. NaHCO}_3} = 0.00142 \text{ mol. H}^+$$

Total moles H^+ neutralized = 0.00463 + 0.00142 = 0.00605 mol. H^+

$$0.00605 \text{ mol. H}^+ \times \frac{1 \text{ L acid}}{0.020 \text{ mol. H}^+} = 0.30 \text{ L acid}$$

The difference in mass between the original copper strip and the silver coated strip corresponds to the mass of copper lost and the mass of silver gained. Bear in mind that two moles of silver are gained for each mole of copper lost. Set up an algebraic equation.

Let x = mass of Cu lost,
then $2.00 - x$ = mass of Cu remaining.

$$\text{mass of Ag} = x \text{ g Cu} \times \frac{1 \text{ mol. Cu}}{63.55 \text{ g Cu}} \times \frac{2 \text{ mol. Ag}}{1 \text{ mol. Cu}} \times \frac{107.9 \text{ g Ag}}{1 \text{ mol. Ag}} = 3.40 x$$

4.18 g = $(2.00 - x) + 3.40x$
4.18 g = $2.00 + 2.40x$
$x = 0.908$ g

Mass of Cu remaining = 2.00 - 0.908 = 1.09 g.
Mass of Ag remaining = 3.40 x 0.908 = 3.09 g.

The first permanganate titration tells us the concentration of ferrous ions, while the second titration tells us the total iron concentration. The difference is the ferric ion concentration.

$MnO_4^-(aq) + Fe^{2+}(aq) \rightarrow Fe^{3+}(aq) + Mn^{2+}(aq)$

$Fe^{2+} \rightarrow Fe^{3+} + e^-$ Oxidation half-reaction

$MnO_4^- + 5e^- \rightarrow Mn^{2+}$ Reduction half-reaction

$8H^+(aq) + MnO_4^-(aq) + 5e^- \rightarrow Mn^{2+}(aq) + 4H_2O(l)$ (see Problem 70)

To balance the electrons, multiply the oxidation half-reaction by 5 and add the result to the balanced permanganate half-reaction.

$8H^+(aq) + MnO_4^-(aq) + 5e^- + 5Fe^{2+}(aq) \rightarrow 5Fe^{3+}(aq) + 5e^- + Mn^{2+}(aq) + 4H_2O(l)$

$8H^+(aq) + MnO_4^-(aq) + 5Fe^{2+}(aq) \rightarrow 5Fe^{3+}(aq) + Mn^{2+}(aq) + 4H_2O(l)$

$$35.0\,\text{mL} \times \frac{1\,\text{L}}{1000\,\text{mL}} \times \frac{0.0280\,\text{mol. KMnO}_4}{1\,\text{L}} \times \frac{1\,\text{mol MnO}_4^-}{1\,\text{mol. KMnO}_4} \times \frac{5\,\text{mol. Fe}^{2+}}{1\,\text{mol. MnO}_4^-} = 0.00490\,\text{mol. Fe}^{2+}$$

$$48.0\,\text{mL} \times \frac{1\,\text{L}}{1000\,\text{mL}} \times \frac{0.0280\,\text{mol. KMnO}_4}{1\,\text{L}} \times \frac{1\,\text{mol MnO}_4^-}{1\,\text{mol. KMnO}_4} \times \frac{5\,\text{mol. Fe}^{3+}}{1\,\text{mol. MnO}_4^-} = 0.00672\,\text{mol. Fe}^{3+}$$

mol. Fe^{3+} = 0.00672 - 0.00490 = 0.00182 mol. Fe^{3+}

$$[Fe^{2+}] = \frac{0.00490\,\text{mol. Fe}^{2+}}{0.05000\,\text{L}} = 0.0980\,M$$

$$[Fe^{3+}] = \frac{0.00182\,\text{mol. Fe}^{3+}}{0.05000\,\text{L}} = 0.0364\,M$$

Chapter 5: Gases

2. *Refer to Sections 1.2 and 5.1, Example 5.1 and Table 1.3.*

$$10 \text{ gal} \times \frac{4 \text{ qt}}{1 \text{ gal}} \times \frac{1 \text{ L}}{1.057 \text{ qt}} = 37.84 \text{ L}$$

$$1.234 \text{ mol. CH}_4 \times \frac{16.04 \text{ g}}{1 \text{ mol. CH}_4} = 19.94 \text{ g CH}_4$$

$$t_{\circ C} = \frac{74 - 32}{1.8} = 23^\circ C$$

$$23^\circ C + 273 = 296 \text{ K}$$

4. *Refer to Section 5.1*

mm Hg	atmospheres	kilopascals	bar
1215	**1.600**	**161.9**	**1.619**
543	0.714	**72.3**	**0.723**
1.07 x 10³	**1.41**	143	**1.43**
678	**0.892**	**90.4**	0.904

$$1215 \text{ mm Hg} \times \frac{1 \text{ atm}}{760.0 \text{ mm Hg}} = 1.600 \text{ atm}$$

$$1.600 \text{ atm} \times \frac{1.013 \text{ bar}}{1 \text{ atm}} = 1.619 \text{ bar}$$

$$1.619 \text{ bar} \times \frac{1 \times 10^5 \text{ Pa}}{1 \text{ atm}} \times \frac{1 \text{ kPa}}{1000 \text{ Pa}} = 161.9 \text{ kPa}$$

$$0.714 \text{ atm} \times \frac{760 \text{ mm Hg}}{1 \text{ atm}} = 543 \text{ mm Hg}$$

$$0.714 \text{ atm} \times \frac{1.013 \times 10^5 \text{ Pa}}{1 \text{ atm}} \times \frac{1 \text{ kPa}}{1000 \text{ Pa}} = 72.3 \text{ kPa}$$

$$0.714 \text{ atm} \times \frac{1.013 \text{ bar}}{1 \text{ atm}} = 0.723 \text{ bar}$$

$$143 \text{ kPa} \times \frac{1000 \text{ Pa}}{1 \text{ kPa}} \times \frac{1 \text{ bar}}{1.01 \times 10^5 \text{ Pa}} = 1.43 \text{ bar}$$

$$1.43 \text{ bar} \times \frac{1 \text{ atm}}{1.013 \text{ bar}} = 1.41 \text{ atm}$$

$$1.41 \text{ atm} \times \frac{760 \text{ mm Hg}}{1 \text{ atm}} = 1.07 \times 10^3 \text{ mm Hg}$$

$$0.904 \text{ bar} \times \frac{1 \text{ atm}}{1.013 \text{ bar}} = 0.892 \text{ atm}$$

$$0.904 \text{ bar} \times \frac{1 \times 10^5 \text{ Pa}}{1 \text{ atm}} \times \frac{1 \text{ kPa}}{1000 \text{ Pa}} = 90.4 \text{ kPa}$$

$$0.892 \text{ atm} \times \frac{760 \text{ mm Hg}}{1 \text{ atm}} = 678 \text{ mm Hg}$$

---

6.   *Refer to Section 5.3 and Example 5.2.*

a. $T_1 = 27°C = 27 + 273 = 300 \text{ K}$
     $T_2 = -10°C = -10 + 273 = 263 \text{ K}$

$$\frac{P_1}{P_2} = \frac{T_1}{T_2} \implies \frac{1.00 \text{ atm}}{P_2} = \frac{300 \text{ K}}{263 \text{ K}} \implies P_2 = 0.877 \text{ atm}$$

b. $T_1 = 27°C = 27 + 273 = 300 \text{ K}$
     $T_2 = 40°C = 40 + 273 = 313 \text{ K}$

$$\frac{P_1}{P_2} = \frac{T_1}{T_2} \implies \frac{1.00 \text{ atm}}{P_2} = \frac{300 \text{ K}}{313 \text{ K}} \implies P_2 = 1.04 \text{ atm}$$

---

8.   *Refer to Section 5.3 and Example 5.2.*

a. $T_1 = 22°C = 22 + 273 = 295 \text{ K}$
     $T_2 = 2(T_1 \text{ (in °C)}) = 2(22°C) = 44°C$
     $T_2 = 44°C = 44 + 273 = 317 \text{ K}$

$$\frac{V_1}{V_2} = \frac{T_1}{T_2} \implies \frac{V_1}{V_2} = \frac{295 \text{ K}}{313 \text{ K}} = 0.931$$

b. $T_1 = 22°C = 22 + 273 = 295 \text{ K}$
     $T_2 = 2(T_1 \text{ (in K)}) = 2(295) = 590 \text{ K}$

$$\frac{V_1}{V_2} = \frac{T_1}{T_2} \implies \frac{V_1}{V_2} = \frac{295 \text{ K}}{590 \text{ K}} = 0.500$$

First convert the temperatures to Kelvins and calculate the actual pressure in the tire. Then calculate the new pressure using the ratios of pressures to temperatures.

$$T_1 = \frac{71°F - 32}{1.8} = 22°C = 22 + 273 = 295 \text{ K}$$

$$T_2 = \frac{115°F - 32}{1.8} = 46°C = 46 + 273 = 319 \text{ K}$$

$$P_{(actual)} = P_{(gauge)} + 14.7 = 28.0 + 14.7 = 42.7 \text{ psi}$$

$$\frac{P_1}{P_2} = \frac{T_1}{T_2} \implies \frac{42.7 \text{ psi}}{P_2} = \frac{295 \text{ K}}{319 \text{ K}} \implies P_2 = 46.2 \text{ psi} \qquad \text{(final actual pressure)}$$

$$P_{(gauge)} = P_{(actual)} - 14.7 = 46.2 - 14.7 = 31.5 \text{ psi} \qquad \text{(final gauge pressure)}$$

Since R is a constant, we can solve this problem using the ideal gas law, by solving for R, setting the equations equal to one another and solving for V_2. (Remember to convert temperature to K and pressure to atm.)

$$T_1 = -6°C = -6 + 273 = 267 \text{ K}$$
$$T_2 = 37°C = 37 + 273 = 310 \text{ K}$$

$$P_1 = 751 \text{ mm Hg} \times \frac{1 \text{ atm}}{760.0 \text{ mm Hg}} = 0.988 \text{ atm}$$

$$\frac{PV}{nT} = R, \text{ thus } \frac{P_1 V_1}{n_1 T_1} = R = \frac{P_2 V_2}{n_2 T_2}. \quad \text{Solving for } V_2, \text{ we get: } V_2 = \frac{P_1 V_1 n_2 T_2}{n_1 T_1 P_2}.$$

$$\text{Since } n_1 = n_2, \; V_2 = \frac{P_1 V_1 T_2}{P_2 T_1} = \frac{(0.988 \text{ atm})(0.50 \text{ L})(310 \text{ K})}{(1.00 \text{ atm})(267 \text{ K})} = 0.57 \text{ L}$$

Since R is a constant, we can solve this problem using the ideal gas law, by solving for R, setting the equations equal to one another and solving for V_2. (Remember to convert temperature to K.)

$$T_1 = 31°C + 273 = 304 \text{ K}$$

$$T_2 = -25°C + 273 = 248 \text{ K}$$

$$\frac{PV}{nT} = R, \text{ thus } \frac{P_1 V_1}{n_1 T_1} = R = \frac{P_2 V_2}{n_2 T_2}. \text{ Solving for } V_2, \text{ we get: } V_2 = \frac{P_1 V_1 n_2 T_2}{n_1 T_1 P_2}.$$

$$\text{Since } n_1 = n_2, \; V_2 = \frac{P_1 V_1 T_2}{P_2 T_1} = \frac{(0.998 \text{ atm})(1.28 \times 10^3 \text{ L})(248 \text{ K})}{(0.753 \text{ atm})(304 \text{ K})} = 1.38 \times 10^3 \text{ L}$$

---

16. Refer to Section 5.3 and Example 5.3.

Calculate the moles of O_2 from the mass, then use the ideal gas law to calculate pressure.

$$T = 37°C + 273 = 310 \text{ K}$$

$$n = \frac{0.25 \text{ mg } O_2}{1 \text{ mL}} \times \frac{1000 \text{ mL}}{1 \text{ L}} \times \frac{1 \text{ g}}{1000 \text{ mg}} \times \frac{1 \text{ mol. } O_2}{32.00 \text{ g } O_2} = 7.8 \times 10^{-3} \text{ mol. } O_2 \text{ per liter of blood}$$

$$P = \frac{nRT}{V} = \frac{(7.8 \times 10^{-3} \text{ mol.})(0.0821 \text{ L} \cdot \text{atm/mol.} \cdot \text{K})(310 \text{ K})}{(1.00 \text{ L})} = 0.20 \text{ atm}$$

---

18. Refer to Section 5.3 and Example 5.3.

First calculate moles of N_2 and moles of He. Then solve the ideal gas law for R and set the equations (one for N_2 and one for He) equal to one another and solve for P_{He}. (Remember to convert temperature to K.)

Mass of He = Mass of N_2 = 764.9 g – 732.2 g = 32.7 g

$$n_{N_2} = 32.7 \text{ g} \times \frac{1 \text{ mol. } N_2}{28.02 \text{ g } N_2} = 1.17 \text{ mol. } N_2$$

$$n_{He} = 32.7 \text{ g} \times \frac{1 \text{ mol. He}}{4.003 \text{ g He}} = 8.17 \text{ mol. He}$$

$$\frac{PV}{nT} = R, \text{ thus } \frac{P_{N_2} V_{N_2}}{n_{N_2} T_{N_2}} = R = \frac{P_{He} V_{He}}{n_{He} T_{He}}. \text{ Solving for } P_{He}, \text{ we get: } P_{He} = \frac{P_{N_2} V_{N_2} n_{He} T_{He}}{n_{N_2} T_{N_2} V_{He}}.$$

Since temperature and volume are constant, these terms cancel out.

$$P_{He} = \frac{P_{N_2} n_{He}}{n_{N_2}} = \frac{(1.25 \text{ atm})(8.17 \text{ mol. He})}{(1.17 \text{ mol. } N_2)} = 8.73 \text{ atm}$$

20. Refer to Section 5.3 and Example 5.3.

Start by converting all units to those used in the ideal gas law. Calculated moles using the ideal gas equation if P, V and T are given, otherwise use the molar mass and the given mass. Then apply the ideal gas law to fill in any other missing data.

Pressure	Volume	Temperature	Moles	Grams
1.77 atm	4.98 L	43.1°C	**0.339**	**31.2**
673 mm Hg	488 mL	**6.73 K**	0.783	72.1
0.899 bar	**7.45 L**	912°C	**0.0679**	6.25
3.28 atm	1.15 L	39°F	0.166	**15.3**

$T = 43.1°C + 273.2 = 316.3$ K

$$n = \frac{PV}{RT} = \frac{(1.77\,\text{atm})(4.98\,\text{L})}{(0.0821\,\text{L} \cdot \text{atm/mol.} \cdot \text{K})(316.3\,\text{K})} = 0.339\,\text{mol. N}_2\text{O}_4$$

$$0.339\,\text{mol. N}_2\text{O}_4 \times \frac{92.02\,\text{g N}_2\text{O}_4}{1\,\text{mol. N}_2\text{O}_4} = 31.2\,\text{g N}_2\text{O}_4$$

$$0.783\,\text{mol. N}_2\text{O}_4 \times \frac{92.02\,\text{g N}_2\text{O}_4}{1\,\text{mol. N}_2\text{O}_4} = 72.1\,\text{g N}_2\text{O}_4$$

$$673\,\text{mm Hg} \times \frac{1\,\text{atm}}{760\,\text{mm Hg}} = 0.886\,\text{atm}$$

$$V = 488\,\text{mL} \times \frac{1\,\text{L}}{1000\,\text{mL}} = 0.488\,\text{L}$$

$$T = \frac{PV}{nR} = \frac{(0.886\,\text{atm})(0.488\,\text{L})}{(0.0821\,\text{L} \cdot \text{atm/mol.} \cdot \text{K})(0.783\,\text{mol.})} = 6.73\,\text{K}$$

$$6.25\,\text{g N}_2\text{O}_4 \times \frac{1\,\text{mol. N}_2\text{O}_4}{92.02\,\text{g N}_2\text{O}_4} = 0.0679\,\text{mol. N}_2\text{O}_4$$

$$0.899\,\text{bar} \times \frac{1\,\text{atm}}{1.013\,\text{bar}} = 0.887\,\text{atm}$$

$T = 912°C + 273 = 1185$ K

$$V = \frac{nRT}{P} = \frac{(0.0679\,\text{mol.})(0.0821\,\text{L} \cdot \text{atm/mol.} \cdot \text{K})(1185\,\text{K})}{(0.887\,\text{atm})} = 7.45\,\text{L}$$

$$0.166 \text{ mol. } N_2O_4 \times \frac{92.02 \text{ g } N_2O_4}{1 \text{ mol. } N_2O_4} = 15.3 \text{ g } N_2O_4$$

$$T = \frac{39°F - 32}{1.8} + 273 = 277 \text{ K}$$

$$P = \frac{nRT}{V} = \frac{(0.166 \text{ mol.})(0.0821 \text{ L} \cdot \text{atm/mol.} \cdot \text{K})(277 \text{ K})}{(1.15 \text{ L})} = 3.28 \text{ atm}$$

---

22. Refer to Sections 5.1-5.3 and Example 5.5.

Calculate the moles of gas in one liter at the given conditions. Then calculate the mass and finally, the density. Remember to convert T to Kelvins and P and atms.

$$T = 97°C + 273 = 370 \text{ K}$$

$$755 \text{ mm Hg} \times \frac{1 \text{ atm}}{760 \text{ mm Hg}} = 0.993 \text{ atm}$$

$$n = \frac{PV}{RT} = \frac{(0.993 \text{ atm})(1.00 \text{ L})}{(0.0821 \text{ L} \cdot \text{atm/mol.} \cdot \text{K})(370 \text{ K})} = 0.0327 \text{ mol. gas}$$

a. $\quad 0.0327 \text{ mol. HCl} \times \dfrac{36.46 \text{ g HCl}}{1 \text{ mol. HCl}} = 1.19 \text{ g HCl}$

$\quad d = \dfrac{\text{g}}{\text{L}} \times \dfrac{1.19 \text{ g HCl}}{1.00 \text{ L}} = 1.19 \text{ g/L}$

b. $\quad 0.0327 \text{ mol. SO}_2 \times \dfrac{64.07 \text{ g SO}_2}{1 \text{ mol. SO}_2} = 2.10 \text{ g SO}_2$

$\quad d = \dfrac{\text{g}}{\text{L}} \times \dfrac{2.10 \text{ g SO}_2}{1 \text{ L}} = 2.10 \text{ g/L}$

c. $\quad 0.0327 \text{ mol. C}_4\text{H}_{10} \times \dfrac{58.12 \text{ g C}_4\text{H}_{10}}{1 \text{ mol. C}_4\text{H}_{10}} = 1.90 \text{ g C}_4\text{H}_{10}$

$\quad d = \dfrac{\text{g}}{\text{L}} \times \dfrac{1.90 \text{ g C}_4\text{H}_{10}}{1 \text{ L}} = 1.90 \text{ g/L}$

24. Refer to Section 5.3 and Example 5.5.

For the given conditions (convert T to Kelvins), calculate the moles and mass of one liter of CO_2 on Earth and on Venus, then calculate the densities.

$T_E = 25°C + 273 = 298$ K
$T_V = 460°C + 273 = 733$ K

$$n_E = \frac{PV}{RT} = \frac{(1.00\,atm)(1.00\,L)}{(0.0821\,L \cdot atm/mol. \cdot K)(298\,K)} = 0.0409\,mol.\,CO_2$$

$$0.0409\,mol.\,CO_2 \times \frac{44.01\,g\,CO_2}{1\,mol.\,CO_2} = 1.80\,g\,CO_2$$

$$d_E = \frac{g}{L} \times \frac{1.80\,g\,CO_2}{1.00\,L} = 1.80\,g/L$$

$$n_V = \frac{PV}{RT} = \frac{(75\,atm)(1.00\,L)}{(0.0821\,L \cdot atm/mol. \cdot K)(733\,K)} = 1.2\,mol.\,CO_2$$

$$1.2\,mol.\,CO_2 \times \frac{44.01\,g\,CO_2}{1\,mol.\,CO_2} = 53\,g\,CO_2$$

$$d_V = \frac{g}{L} \times \frac{53\,g\,CO_2}{1\,L} = 53\,g/L$$

The density of CO_2 on Venus is almost 30 times that on Earth.

26. Refer to Sections 3.3 and 5.3 and Examples 3.7, 3.9, 5.4 and 5.5.

Consider your units. Molar mass has units of g/mol., while density has units of g/L. Therefore, you can calculate the moles of cyclopropane in 1 L, and then divide the mass of 1 L by the number of moles. Finally, use the percent composition and molar mass to determine the simplest and molecular formulae.

a. $T = 25°C + 273 = 298$ K

$$755\,mm\,Hg \times \frac{1\,atm}{760\,mm\,Hg} = 0.993\,atm$$

$$n = \frac{PV}{RT} = \frac{(0.993\,atm)(1.00\,L)}{(0.0821\,L \cdot atm/mol. \cdot K)(298\,K)} = 0.0406\,mol.$$

$$molar\,mass = \frac{1.71\,g}{0.0406\,mol.} = 42.1\,g/mol.$$

b. $85.7 \, g \, C \times \dfrac{1 \, mol. \, C}{12.01 \, g} = 7.14 \, mol. \, C$

$14.3 \, g \, H \times \dfrac{1 \, mol. \, H}{1.008 \, g} = 14.2 \, mol. \, H$

C: 7.14 mol. / 7.14 mol. = 1
H: 14.2 mol. / 7.14 mol. = 2

Thus, the empirical formula is: CH_2 (empirical mass = 12.0 + 2(1.0) = 14.0)
Molar mass = 3(empirical mass), thus
Molecular formula = 3(empirical formula) = $C_{1x3}H_{2x3}$) = C_3H_6

---

28. Refer to Section 5.3.

The molar mass of exhaled air is simply the sum of the molar masses of the individual componenets multiplied by their abundances. Calculate the moles of air in 1 L, then the mass of 1 L to get density.

a. (28.02 g/mol. N_2)(0.745) + (32.00 g/mol. O_2)(0.157) + (44.01 g/mol. CO_2)(0.036)
     + (18.02 g/mol. H_2O)(0.062) = 28.6 g/mol.

b. $T = 37°C + 273 = 310 \, K$

$757 \, mm \, Hg \times \dfrac{1 \, atm}{760 \, mm \, Hg} = 0.996 \, atm$

$n = \dfrac{PV}{RT} = \dfrac{(0.996 \, atm)(1.00 \, L)}{(0.0821 \, L \cdot atm/mol. \cdot K)(310 \, K)} = 0.0391 \, mol.$

$0.0391 \, mol. \times \dfrac{28.6 \, g \, air_{Exhaled}}{1 \, mol. \, air_{Exhaled}} = 1.12 \, g \, air_{Exhaled}$

$d_{air_{Exhaled}} = \dfrac{g}{L} \times \dfrac{1.12 \, g \, air_{Exhaled}}{1.00 \, L} = 1.12 \, g/L$

$0.0391 \, mol. \times \dfrac{29.0 \, g \, air_{ordinary}}{1 \, mol. \, air_{ordinary}} = 1.13 \, g \, air_{ordinary}$

$d_{air_{ordinary}} = \dfrac{g}{L} \times \dfrac{1.13 \, g \, air_{ordinary}}{1.00 \, L} = 1.13 \, g/L$

Ordinary air is slightly more dense than exhaled air.

Use the ideal gas equation to calculate the number of moles in the sample, then the molar mass of the compound. Then calculate the atomic mass of X and you have the identity.

$$T = 35°C + 273 = 308 \text{ K}$$

$$297 \text{ mL} \times \frac{1 \text{ L}}{1000 \text{ mL}} = 0.297 \text{ L}$$

$$769 \text{ mm Hg} \times \frac{1 \text{ atm}}{760 \text{ mm Hg}} = 1.01 \text{ atm}$$

$$n = \frac{PV}{RT} = \frac{(1.01 \text{ atm})(0.297 \text{ L})}{(0.0821 \text{ L} \cdot \text{atm/mol.} \cdot \text{K})(308 \text{ K})} = 0.0119 \text{ mol.}$$

$$\text{molar mass} = \frac{1.58 \text{ g}}{0.0119 \text{ mol.}} = 133 \text{ g/mol.}$$

133 g/mol. = 2(C) + 3(H) + 3(X)
133 g/mol. = 2(12.01 g/mol.) + 3(1.01 g/mol.) + 3(X)
$3X$ = 106 g/mol.
X = 35.3 g/mol.
X = Cl

Since R is a constant, we can solve this problem using the ideal gas law, by solving for R, setting the equations equal to one another and solving for V_{Cl_2}.

$$\frac{PV}{nT} = R, \text{ thus } \frac{P_{SO_2} V_{SO_2}}{n_{SO_2} T_{SO_2}} = R = \frac{P_{Cl_2} V_{Cl_2}}{n_{Cl_2} T_{Cl_2}}. \quad \text{Solving for } V_{Cl_2}, \text{ we get:}$$

$$V_{Cl_2} = \frac{P_{SO_2} V_{SO_2} n_{Cl_2} T_{Cl_2}}{n_{SO_2} T_{SO_2} P_{Cl_2}}.$$

Since $T_{SO_2} = T_{Cl_2}$ and $P_{SO_2} = P_{Cl_2}, V_{Cl_2} = \frac{V_{SO_2} n_{Cl_2}}{n_{SO_2}} = \frac{(21.7 \text{ L})(2.00 \text{ mol. Cl}_2)}{(1.00 \text{ mol. SO}_2)} = 43.4 \text{ L Cl}_2$

34. Refer to Sections 4.1 and 5.4 and Example 5.6.

Calculate moles of N_2O_5 with the ideal gas equation, and moles H^+ using the mole ratio. Then consider the mole ratio of HNO_3 and H^+ and calculate molarity.

a. $T = 25°C + 273 = 298$ K

$$n = \frac{PV}{RT} = \frac{(1.00\,\text{atm})(1.50\,\text{L})}{(0.0821\,\text{L}\cdot\text{atm/mol}\cdot\text{K})(298\,\text{K})} = 0.0613\,\text{mol. } N_2O_5$$

$$0.0613\,\text{mol. } N_2O_5 \times \frac{2\,\text{mol. } H^+}{1\,\text{mol. } N_2O_5} = 0.123\,\text{mol. } H^+$$

b. $HNO_3 \rightarrow H^+ + NO_3^-$ (Thus, moles H^+ = moles HNO_3)

$$M = \frac{0.123\,\text{mol. } HNO_3}{0.437\,\text{L}} = 0.281\,M$$

36. Refer to Sections 3.4 and 5.4 and Example 5.7.

After balancing the equation, calculate the moles of NH_4NO_3 and the total moles of gas (using the mole ratio of NH_4NO_3 to all gases). Calculate the pressure for the given conditions.

a. $NH_4NO_3(s) \rightarrow N_2(g) + O_2(g) + H_2O(g)$ balance H's
   $NH_4NO_3(s) \rightarrow N_2(g) + O_2(g) + 2H_2O(g)$

There are too many O's on the product side. Multiply NH_4NO_3 by two and rebalance.

$2NH_4NO_3(s) \rightarrow 2N_2(g) + O_2(g) + 4H_2O(g)$

b. $1.00\,\text{kg } NH_4NO_3 \times \dfrac{1000\,\text{g}}{1\,\text{kg}} \times \dfrac{1\,\text{mol. } NH_4NO_3}{80.05\,\text{g } NH_4NO_3} \times \dfrac{7\,\text{mol. gas}}{2\,\text{mol. } NH_4NO_3} = 43.7\,\text{mol. gas}$

$T = 787°C + 273 = 1060$ K

$$P = \frac{nRT}{V} = \frac{(43.7\,\text{mol.})(0.0821\,\text{L}\cdot\text{atm/mol}\cdot\text{K})(1060\,\text{K})}{50.0\,\text{L}} = 76.1\,\text{atm}$$

Calculate the moles of each gas, the total moles, then the total pressure. Then calculate the partial pressures of the individual gases.

$$9.00 \text{ g HCl} \times \frac{1 \text{ mol. HCl}}{36.46 \text{ g HCl}} = 0.247 \text{ mol. HCl}$$

$$2.00 \text{ g H}_2 \times \frac{1 \text{ mol. H}_2}{2.016 \text{ g H}_2} = 0.992 \text{ mol. H}_2$$

$$165.0 \text{ g Ne} \times \frac{1 \text{ mol. Ne}}{20.18 \text{ g Ne}} = 8.176 \text{ mol. Ne}$$

Total moles = 0.247 + 0.992 + 8.176 = 9.415 mol.

$$P = \frac{nRT}{V} = \frac{(9.415 \text{ mol.})(0.0821 \text{ L} \cdot \text{atm/mol.} \cdot \text{K})(295 \text{ K})}{75.0 \text{ L}} = \textbf{3.04 atm (total pressure)}$$

$$X_{HCl} = \frac{\text{mol. HCl}}{\text{total mol.}} = \frac{0.247 \text{ mol. HCl}}{9.415 \text{ mol.}} = 0.0262$$

$$X_{H_2} = \frac{\text{mol. H}_2}{\text{total mol.}} = \frac{0.992 \text{ mol. H}_2}{9.415 \text{ mol.}} = 0.0977$$

$$X_{Ne} = \frac{\text{mol. Ne}}{\text{total mol.}} = \frac{8.176 \text{ mol. Ne}}{9.415 \text{ mol.}} = 0.868$$

$P_{HCl} = (X_{HCl})(P_{total}) = (0.0262)(3.04 \text{ atm}) = 0.0796 \text{ atm}$

$P_{N_2} = (X_{N_2})(P_{total}) = (0.0977)(3.04 \text{ atm}) = 0.297 \text{ atm}$

$P_{Ne} = (X_{Ne})(P_{total}) = (0.868)(3.04 \text{ atm}) = 2.64 \text{ atm}$

Therefore **HCl has the lowest partial pressure**.

Note that the sum of the partial pressures (3.02 atm) should equal the total pressure. Indeed, 3.02 versus 3.04 is within the error expected from round-off.

40. Refer to Sections 5.3 and 5.5, and Example 5.8.

First calculate the moles of wet gas and dry gas. The difference gives the moles of water. Then calculate the mole fraction of water, and from that the partial pressure of water.

$$n = \frac{PV}{RT} = \frac{(0.986 \text{ atm})(1.00 \text{ L})}{(0.0821 \text{ L} \cdot \text{atm/mol.} \cdot \text{K})(315 \text{ K})} = 0.0381 \text{ mol. wet gas}$$

$$n = \frac{PV}{RT} = \frac{(1.00 \text{ atm})(1.04 \text{ L})}{(0.0821 \text{ L} \cdot \text{atm/mol.} \cdot \text{K})(363 \text{ K})} = 0.0349 \text{ mol. dry gas}$$

$$n_{\text{water}} = n_{\text{wet}} - n_{\text{dry}} = 0.0381 - 0.0349 = 3.2 \times 10^{-3} \text{ mol. water}$$

$$X_{H_2O} = \frac{\text{mol. } H_2O}{\text{total mol.}} = \frac{3.2 \times 10^{-3} \text{ mol. } H_2O}{0.0381 \text{ mol.}} = 0.084$$

$$P_{H_2O} = (X_{H_2O})(P_{\text{total}}) = (0.084)(0.986 \text{ atm}) = 0.083 \text{ atm}$$

$$0.083 \text{ atm} \times \frac{760 \text{ mm Hg}}{1 \text{ atm}} = 63 \text{ mm Hg}$$

---

42. Refer to Section 5.5.

Calculate the moles of each gas (using any value for T other than zero), then the total moles of gas and the total volume of both bulbs. Then calculate the pressure using the combined volumes and moles (and the same temperature).

$$n_{Ar} = \frac{PV}{RT} = \frac{(2.50 \text{ atm})(4.00 \text{ L})}{(0.0821 \text{ L} \cdot \text{atm/mol.} \cdot \text{K})(300 \text{ K})} = 0.406 \text{ mol. wet gas}$$

$$n_{Cl} = \frac{PV}{RT} = \frac{(1.00 \text{ atm})(1.00 \text{ L})}{(0.0821 \text{ L} \cdot \text{atm/mol.} \cdot \text{K})(300 \text{ K})} = 0.0406 \text{ mol. dry gas}$$

$$n_{\text{total}} = 0.406 + 0.0406 = 0.447$$

$$V_{\text{total}} = 4.00 \text{ L} + 1.00 \text{ L} = 5.00 \text{ L}$$

$$P = \frac{nRT}{V} = \frac{(0.447 \text{ mol.})(0.0821 \text{ L} \cdot \text{atm/mol.} \cdot \text{K})(300 \text{ K})}{5.00 \text{ L}} = 2.20 \text{ atm}$$

---

44. Refer to Section 5.5.

Balance the reaction. Calculate the moles of ethylene chloride, and from that the moles of products. Use the ideal gas equation to calculate the pressure with the given conditions. Finally, calculate the moles of each product gas, mole fractions and partial pressures.

a. $C_2H_3Cl(g) + O_2(g) \rightarrow CO_2(g) + H_2O(g) + HCl(g)$
    $C_2H_3Cl(g) + O_2(g) \rightarrow 2CO_2(g) + H_2O(g) + HCl(g)$ balance C's
    $C_2H_3Cl(g) + {}^5/2 O_2(g) \rightarrow 2CO_2(g) + H_2O(g) + HCl(g)$ balance O's
    $2C_2H_3Cl(g) + 5O_2(g) \rightarrow 4CO_2(g) + 2H_2O(g) + 2HCl(g)$ eliminate fraction.

b. $25.00 \text{ g C}_2\text{H}_3\text{Cl} \times \dfrac{1\,\text{mol. C}_2\text{H}_3\text{Cl}}{62.49\,\text{g C}_2\text{H}_3\text{Cl}} \times \dfrac{8\,\text{mol. product}}{2\,\text{mol. C}_2\text{H}_3\text{Cl}} = 1.600\,\text{mol. products}$

$T = 75°\text{C} + 273 = 348 \text{ K}$

$P = \dfrac{nRT}{V} = \dfrac{(1.600\,\text{mol.})(0.0821\,\text{L}\cdot\text{atm/mol.}\cdot\text{K})(348\,\text{K})}{5.00\,\text{L}} = 9.14\,\text{atm}$

c. $25.00 \text{ g C}_2\text{H}_3\text{Cl} \times \dfrac{1\,\text{mol. C}_2\text{H}_3\text{Cl}}{62.49\,\text{g C}_2\text{H}_3\text{Cl}} \times \dfrac{4\,\text{mol. CO}_2}{2\,\text{mol. C}_2\text{H}_3\text{Cl}} = 0.8000\,\text{mol. CO}_2$

$25.00 \text{ g C}_2\text{H}_3\text{Cl} \times \dfrac{1\,\text{mol. C}_2\text{H}_3\text{Cl}}{62.49\,\text{g C}_2\text{H}_3\text{Cl}} \times \dfrac{2\,\text{mol. H}_2\text{O}}{2\,\text{mol. C}_2\text{H}_3\text{Cl}} = 0.4000\,\text{mol. H}_2\text{O}$

$25.00 \text{ g C}_2\text{H}_3\text{Cl} \times \dfrac{1\,\text{mol. C}_2\text{H}_3\text{Cl}}{62.49\,\text{g C}_2\text{H}_3\text{Cl}} \times \dfrac{2\,\text{mol. HCl}}{2\,\text{mol. C}_2\text{H}_3\text{Cl}} = 0.4000\,\text{mol. HCl}$

$X_{\text{CO}_2} = \dfrac{\text{mol. CO}_2}{\text{total mol.}} = \dfrac{0.8000\,\text{mol. CO}_2}{1.600\,\text{mol.}} = 0.5000$

$X_{\text{H}_2\text{O}} = \dfrac{\text{mol. H}_2\text{O}}{\text{total mol.}} = \dfrac{0.4000\,\text{mol. H}_2\text{O}}{1.600\,\text{mol.}} = 0.2500$

$X_{\text{HCl}} = \dfrac{\text{mol. HCl}}{\text{total mol.}} = \dfrac{0.4000\,\text{mol. HCl}}{1.600\,\text{mol.}} = 0.2500$

$P_{\text{CO}_2} = (X_{\text{CO}_2})(P_{\text{total}}) = (0.5000)(9.14\,\text{atm}) = 4.57\,\text{atm}$

$P_{\text{H}_2\text{O}} = (X_{\text{H}_2\text{O}})(P_{\text{total}}) = (0.2500)(9.14\,\text{atm}) = 2.29\,\text{atm}$

$P_{\text{HCl}} = (X_{\text{HCl}})(P_{\text{total}}) = (0.2500)(9.14\,\text{atm}) = 2.29\,\text{atm}$

---

46. Refer to Section 5.6 and Example 5.11.

a. Since the gas moves faster than propane, then, according to Graham's law, it must be **lighter** than propane.

b. $\dfrac{\text{rate of effusion of B}}{\text{rate of effusion of A}} = \left(\dfrac{M_A}{M_B}\right)^{\frac{1}{2}}$

If we let x = rate of effusion of C_3H_8, then $1.55x$ equals the rate of effusion of the unknown gas.

$\dfrac{1.55\,x}{x} = \left(\dfrac{44.09\,\text{g/mol.}}{M_{\text{unk gas}}}\right)^{\frac{1}{2}}$

$$(1.55)^2 = \frac{44.09 \text{ g/mol.}}{M_{\text{unk gas}}}$$

$$M_{\text{unk. gas}} = 18.4 \text{ g/mol.}$$

48. *Refer to Section 5.6 and Example 5.11.*

Calculate the ratio of the effusion rates of the two gases.

$$\frac{\text{rate of effusion of } H_2}{\text{rate of effusion of } N_2} = \left(\frac{M_{N_2}}{M_{H_2}}\right)^{\frac{1}{2}} = \left(\frac{28.02}{2.016}\right)^{\frac{1}{2}} = 3.73$$

Thus the hydrogen balloon will deflate 3.73 times faster than the nitrogen balloon.

50. *Refer to Section 5.6 and Example 5.10.*

a. $\mu_{Br_2} = \left(\frac{3RT}{M}\right)^{\frac{1}{2}} = \left(\frac{(3)(8.31 \times 10^3 \text{ g} \cdot \text{m}^2/\text{s}^2 \cdot \text{mol.} \cdot \text{K})(301 \text{ K})}{159.8 \text{ g/mol.}}\right)^{\frac{1}{2}} = 217 \text{ m/s}$

b. $\mu_{Kr} = \left(\frac{3RT}{M}\right)^{\frac{1}{2}} = \left(\frac{(3)(8.31 \times 10^3 \text{ g} \cdot \text{m}^2/\text{s}^2 \cdot \text{mol.} \cdot \text{K})(245 \text{ K})}{83.80 \text{ g/mol.}}\right)^{\frac{1}{2}} = 270 \text{ m/s}$

52. *Refer to Section 5.7.*

Deviations from ideal behavior tend to be largest at high pressures and low temperatures.

a. If pressure is reduced from 20 atm to 1 atm, CH_4 should behave **more** ideally.

b. If temperature is reduced from 50°C to –50°C, CH_4 should behave **less** ideally.

54. *Refer to Sections 5.3 and 5.6 and Figure 5.10.*

Use the table in Figure 5.10 to estimate the ratio of the real molar volume to the ideal molar volume. Calculate the ideal molar volume. Calculate the density, substituting the ideal molar volume and the ratio for the real molar volume. Note that a different estimate of the V_m/V_m° ratio will result a slightly different density.

a. According to Figure 5.10, at 100 atm, V_m/V_m° is approximately 0.75.

Thus, $V_m = 0.75 V_m^\circ$.

$$V_m^\circ = \frac{RT}{P} = \frac{(0.0821\,\text{L}\cdot\text{atm/mol.}\cdot\text{K})(298)}{100\,\text{atm}} = 0.245\,\text{L/mol.}$$

$$d = \frac{M}{V_m} = \frac{M}{(0.75)(V_m^\circ)} = \frac{15.03\,\text{g/mol.}}{(0.75)(0.245\,\text{L/mol.})} = 82\,\text{g/L}$$

b. $$d = \frac{MP}{RT} = \frac{(15.03\,\text{g/mol.})(100\,\text{atm})}{(0.0821\,\text{L}\cdot\text{atm/mol.}\cdot\text{K})(298)} = 61.4\,\text{g/L}$$

c. The densities will be equal when $V_m^\circ = V_m$, which occurs when $P = 350$ atm.

---

56. Refer to Sections 3.4 and 5.3, Example 5.7 and Table 1.3.

Balance the reaction. From the mileage, calculate the volume of octane used. Then calculate the number of moles of octane this corresponds to, and from that the moles of CO_2 that would be produced. Finally, calculate the volume of CO_2 using the ideal gas equation.

a. $C_8H_{18}(l) + O_2(g) \rightarrow CO_2(g) + H_2O(l)$

$C_8H_{18}(l) + O_2(g) \rightarrow 8CO_2(g) + H_2O(l)$ balance C's

$C_8H_{18}(l) + O_2(g) \rightarrow 8CO_2(g) + 9H_2O(l)$ balance H's

$C_8H_{18}(l) + 12\tfrac{1}{2}O_2(g) \rightarrow 8CO_2(g) + 9H_2O(l)$ balance O's

$2C_8H_{18}(l) + 25O_2(g) \rightarrow 16CO_2(g) + 18H_2O(l)$ remove fractions

b. $$75\,\text{mi.} \times \frac{1\,\text{gal. }C_8H_{18}}{22\,\text{mi.}} \times \frac{4\,\text{qt.}}{1\,\text{gal.}} \times \frac{1\,\text{L}}{1.057\,\text{qt.}} \times \frac{1000\,\text{mL}}{1\,\text{L}} \times \frac{1\,\text{cm}^3}{1\,\text{mL}} = 1.3 \times 10^4\,\text{cm}^3\,C_8H_{18}$$

$$1.3 \times 10^4\,\text{cm}^3 \times \frac{0.692\,\text{g }C_8H_{18}}{1\,\text{cm}^3\,C_8H_{18}} \times \frac{1\,\text{mol. }C_8H_{18}}{114.22\,\text{g }C_8H_{18}} \times \frac{16\,\text{mol. }CO_2}{2\,\text{mol. }C_8H_{18}} = 630\,\text{mol. }CO_2$$

$$V = \frac{nRT}{P} = \frac{(630\,\text{mol.})(0.0821\,\text{L}\cdot\text{atm/mol.}\cdot\text{K})(298\,\text{K})}{1.00\,\text{atm}} = 1.5 \times 10^4\,\text{L }CO_2$$

58. Refer to Sections 5.5, 5.6 and 5.7.

a. The translational energy depends only on the temperature. Since all the molecules are at the same temperature, they must also have the **same** translational energy.

b. Partial pressure is directly proportional to the number of moles, therefore **He**, which is present in the greater quantity, will have the greater partial pressure.

c. Mole fraction is directly proportional to the number of moles, therefore **He**, which is present in the greater quantity, will have the greater mole fraction.

d. Effusion rate is inversely proportional to the molar mass. Thus the smaller molecule, **He**, will have the greater effusion rate.

60. Refer to Sections 5.3 and 5.4.

Use the ideal gas equation to calculate the moles of air, and then the mole percent to calculate moles of O_2. Then use the mole ratio and the molar mass of gasoline to calculate the mass of gasoline needed.

a. $n_{Air} = \dfrac{PV}{RT} = \dfrac{(1.00\,\text{atm})(0.618\,\text{L})}{(0.0821\,\text{L} \cdot \text{atm/mol.} \cdot \text{K})(348\,\text{K})} = 0.0216\,\text{mol. air}$

$0.0216\,\text{mol. air} \times \dfrac{21.0\,\text{mol. } O_2}{100\,\text{mol. air}} = 4.54 \times 10^{-3}\,\text{mol. } O_2$

b. $4.54 \times 10^{-3}\,\text{mol. } O_2 \times \dfrac{1\,\text{mol. gasoline}}{12\,\text{mol. } O_2} \times \dfrac{1.0 \times 10^2\,\text{g}}{1\,\text{mol. gasoline}} = 0.038\,\text{g gasoline}$

62. Refer to Sections 3.3, 5.3 and 5.4 and Example 3.8.

Use the ideal gas equation to calculate the moles of CO_2. Then calculate the mass of C and H and the mass percent based on the initial amount of glycine. Do the same for N_2 and then calculate the percentage of O by difference. With the mass percents in hand, calculate the empirical formula.

$n_{CO_2} = \dfrac{PV}{RT} = \dfrac{(1.00\,\text{atm})(0.1329\,\text{L})}{(0.0821\,\text{L} \cdot \text{atm/mol.} \cdot \text{K})(298\,\text{K})} = 5.43 \times 10^{-3}\,\text{mol. } CO_2$

$5.43 \times 10^{-3}\,\text{mol. } CO_2 \times \dfrac{1\,\text{mol. C}}{1\,\text{mol. } CO_2} \times \dfrac{12.01\,\text{g C}}{1\,\text{mol. C}} = 0.0652\,\text{g C}$

$$\frac{0.0652\,\text{g C}}{0.2036\,\text{g glycine}} \times 100\% = 32.0\%\,\text{C}$$

$$0.122\,\text{g H}_2\text{O} \times \frac{1\,\text{mol. H}_2\text{O}}{18.02\,\text{g H}_2\text{O}} \times \frac{2\,\text{mol. H}}{1\,\text{mol. H}_2\text{O}} \times \frac{1.008\,\text{g H}}{1\,\text{mol. H}} = 0.0136\,\text{g H}$$

$$\frac{0.0136\,\text{g H}}{0.2036\,\text{g glycine}} \times 100\% = 6.70\%\,\text{H}$$

$$n_{\text{N}_2} = \frac{PV}{RT} = \frac{(1.00\,\text{atm})(0.0408\,\text{L})}{(0.0821\,\text{L}\cdot\text{atm/mol.}\cdot\text{K})(298\,\text{K})} = 1.67 \times 10^{-3}\,\text{mol. N}_2$$

$$1.67 \times 10^{-3}\,\text{mol. N}_2 \times \frac{2\,\text{mol. N}}{1\,\text{mol. N}_2} \times \frac{14.01\,\text{g N}}{1\,\text{mol. N}} = 0.0468\,\text{g N}$$

$$\frac{0.0468\,\text{g N}}{0.2500\,\text{g glycine}} \times 100\% = 18.7\%\,\text{N}$$

mass % O = 100% − 32.0% − 6.68% − 18.7% = 42.6%

$$32.0\,\text{g C} \times \frac{1\,\text{mol. C}}{12.01\,\text{g C}} = 2.66\,\text{mol. C}$$

$$6.70\,\text{g H} \times \frac{1\,\text{mol. H}}{1.008\,\text{g H}} = 6.65\,\text{mol. H}$$

$$18.7\,\text{g N} \times \frac{1\,\text{mol. N}}{14.01\,\text{g N}} = 1.33\,\text{mol. N}$$

$$42.6\,\text{g O} \times \frac{1\,\text{mol. O}}{16.00\,\text{g O}} = 2.66\,\text{mol. O}$$

C: 2.66 / 1.33 = 2
H: 6.65 / 1.33 = 5
N: 1.33 / 1.33 = 1
O: 2.66 / 1.33 = 2

Empirical formula = $C_2H_5NO_2$

64. Refer to Section 5.2.

Since this is a rigid, sealed container, neither the volume nor number of moles can change. Thus, as the temperature is lowered, only the pressure can change, which also drops.

a.

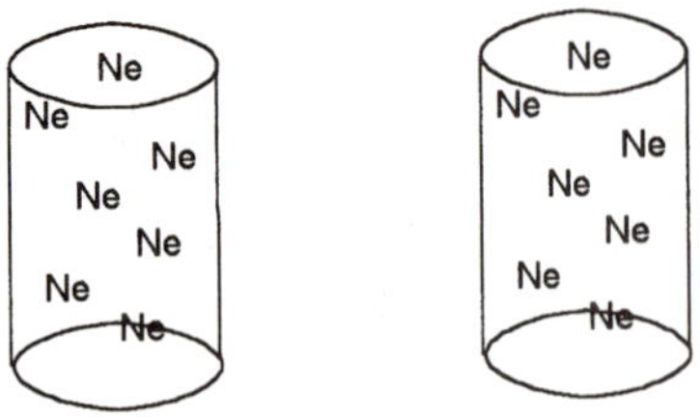

25°C -80°C

b.

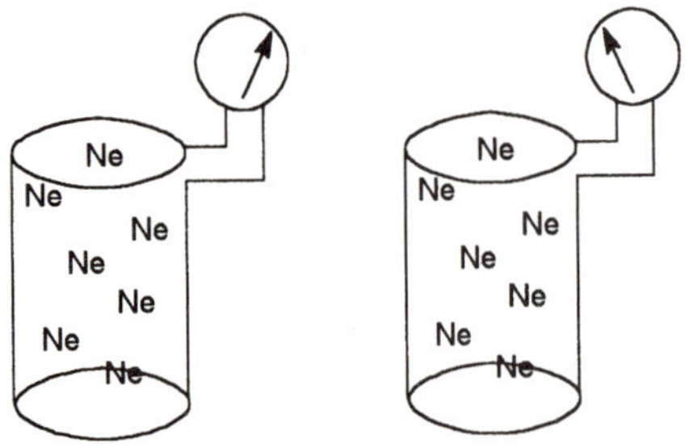

25°C -80°C

66. Refer to Section 5.2.

Since R is a constant, we can solve these problems using the ideal gas law, by solving for R, and setting the equations equal to one another.

$\frac{PV}{nT} = R$, thus $\frac{P_A V_A}{n_A T_A} = R = \frac{P_B V_B}{n_B T_B}$. Since $V_A = V_B$ and $T_A = T_B$, then: $\frac{P_A}{n_A} = \frac{P_B}{n_B}$.

a. If $n_A = n_B$, then: $P_A = P_B$

b. CO_2 has a much larger mass than He, thus if there are equal grams of each, there will be much more He than CO_2, and $n_A < n_B$. Thus $P_A < P_B$.

68. Refer to Sections 5.2 and 5.5.

a. Pressure is directly proportional to the number of moles, so **bulb C** (the one with the fewest molecules) would have the lowest pressure.

b. The relative pressures are directly proportional to the relative moles. Bulb C contains half as many molecules (or moles) as bulb A, so the pressure would be half that of A: **1.00 atm**.

c. As mentioned in part (b) above, the pressure in bulb C is 1.00 atm. From similar reasoning, pressure in bulb B is [(6/8)(2.00)] 1.50 atm.
Total pressure = 2.00 + 1.50 + 1.00 = **4.50 atm**.

d. Bulbs A and B now contain 7 molecules each, so $P_A = P_B = (7/8)(2.00$ atm$) = 1.75$ atm.

$P_A + P_B = 1.75 + 1.75 = \mathbf{3.50\ atm}$

$P_A + P_B + P_C = 1.75 + 1.75 + 1.00 = \mathbf{4.50\ atm}$

Thus the total pressure is unchanged

e. Each bulb now contains 6 molecules, so $P_A = P_B = P_C = (6/8)(2.00$ atm$) = 1.50$ atm.

$P_A + P_B = 1.50 + 1.50 = \mathbf{3.00\ atm}$

$P_A + P_B + P_C = 1.50 + 1.50 + 1.50 = \mathbf{4.50\ atm}$

Thus the total pressure is unchanged

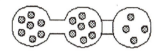

70. *Refer to Sections 5.3 and 5.7 and Appendix 1.*

$$23.76\ \text{mm Hg} \times \frac{1\ \text{atm}}{760\ \text{mm Hg}} = 0.0313\ \text{atm}$$

$$\frac{P_1}{T_1} = \frac{P_2}{T_2} \Rightarrow \frac{0.0313}{298} = \frac{P_2}{313} \Rightarrow P_2 = 0.0329\ \text{atm}$$

$$0.0329\ \text{atm} \times \frac{760\ \text{mm Hg}}{1\ \text{atm}} = 25.0\ \text{mm Hg}$$

$$\frac{P_1}{T_1} = \frac{P_3}{T_3} \Rightarrow \frac{0.0313}{298} = \frac{P_3}{343} \Rightarrow P_3 = 0.0360\ \text{atm}$$

$$0.0360\ \text{atm} \times \frac{760\ \text{mm Hg}}{1\ \text{atm}} = 27.4\ \text{mm Hg}$$

$$\frac{P_1}{T_1} = \frac{P_4}{T_4} \Rightarrow \frac{0.0313}{298} = \frac{P_4}{373} \Rightarrow P_4 = 0.0392\ \text{atm}$$

$$0.0392\ \text{atm} \times \frac{760\ \text{mm Hg}}{1\ \text{atm}} = 29.8\ \text{mm Hg}$$

Temperature	P (ideal gas)	P (water vapor)
$T_1 = 25°C$	23.76 mm Hg	23.76 mm Hg
$T_2 = 40°C$	25.0 mm Hg	55.3 mm Hg
$T_3 = 70°C$	27.4 mm Hg	233.7 mm Hg
$T_4 = 100°C$	29.8 mm Hg	760.0 mm Hg

These numbers are so different because the number of moles (n) is not constant for the water vapor. As the temperature increases, the amount of water that is vaporized increases. Since pressure is proportional to the moles of gas, the pressure also increases with increasing temperature.

71. Refer to Section 5.6.

Use the equation for average speed of a gas to calculate the molar mass. Compare that value to the molar masses of H_2, He and Ar.

$$u^2 = \frac{3RT}{M} \Rightarrow M = \frac{3RT}{u^2}$$

$$M = \frac{3RT}{u^2} = \frac{(3)(8.31 \times 10^3 \text{ g} \cdot \text{m}^2/\text{s}^2 \cdot \text{mol.} \cdot \text{K})(288 \text{ K})}{(1.12 \times 10^3 \text{ m/s})^2} = 5.72 \text{ g/mol.}$$

Note that molar mass and speed are inversely proportional, thus heavier atoms and molecules will move more slowly. Those atoms and molecules that will have sufficient speed to escape gravity are those with a molar mass of less than 5.72 g/mol., such as H_2 and He. Ar's molar mass is greater than 5.72 g/mol., thus it's speed will be insufficient.

72. Refer to Section 5.6.

Set up an algebraic equation, using x for the distance ammonia travels and $(5 - x)$ for the distance HCl travels, correlating distance traveled with molar mass.

$$\frac{\text{distance}_{NH_3}}{\text{distance}_{HCl}} = \left(\frac{M_{HCl}}{M_{NH_3}} \right)^{\frac{1}{2}} \Rightarrow \frac{x}{5-x} = \left(\frac{35.5}{17} \right)^{\frac{1}{2}}$$

$$\frac{x}{5-x} = 1.45 \Rightarrow x = 7.25 - 1.45x$$

$$2.45x = 7.25$$

$$x = 3.0 \text{ ft}$$

Kelvins and °C are the same size and °R and °F are also the same size. Thus if 1.8°C = 1°F, then 1.8K = 1°R.

273.15 K = 491.67°R

$$R = \frac{PV}{nT} = \frac{(1.00 \, \text{atm})(22.4 \, \text{L})}{(1.00 \, \text{mol.})(491.67 \, °\text{R})} = 0.0456 \, \text{L} \cdot \text{atm/mol.} \cdot °\text{R}$$

Calculate the total amount of H_2 produced, then calculate the moles of H_2 produced by each metal using x grams Zn and $(0.2500 - x)$ grams Al. Solve for x to get the grams of Zn, then calculate the mass percent.

$$P = 755 \, \text{mm Hg} \times \frac{1 \, \text{atm}}{760 \, \text{mm Hg}} = 0.993 \, \text{atm}$$

$$n_{H_2} = \frac{PV}{RT} = \frac{(0.993 \, \text{atm})(0.147 \, \text{L})}{(0.0821 \, \text{L} \cdot \text{atm/mol.} \cdot \text{K})(298 \, \text{K})} = 0.00597 \, \text{moles } H_2$$

$$x \, \text{g Zn} \times \frac{1 \, \text{mol. Zn}}{65.39 \, \text{g Zn}} \times \frac{1 \, \text{mol. } H_2}{1 \, \text{mol. Zn}} = 0.0153x \, \text{mol. } H_2$$

$$(0.2500 - x) \, \text{g Al} \times \frac{1 \, \text{mol. Al}}{26.98 \, \text{g Al}} \times \frac{\frac{3}{2} \, \text{mol. } H_2}{1 \, \text{mol. Al}} = (0.0139 - 0.0556x) \, \text{mol. } H_2$$

$0.00597 = 0.0153x + 0.0139 - 0.0556x$

$0.0403x = 0.00793$

$x = 0.197$ g Zn

$$\text{mass \% Zn} = \frac{0.197 \, \text{g Zn}}{0.2500 \, \text{g total}} \times 100\% = 78.7\%$$

To get off the ground, the buoyant force of the balloon must exceed the gravitational force holding it down. We can find the volume of the balloon by equating the two forces, converting mass to $(d \times V)$ and solving for V. Although we were not given densities, we can solve the ideal gas law for d and substitute. Now we can calculate a numerical value for V and calculate the radius and diameter of the balloon.

$$\text{mass}_{(\text{air})} = \text{mass}_{(\text{H}_2)} + \text{mass}_{(\text{man + balloon})} = \text{mass}_{(\text{H}_2)} + 1.68 \times 10^5 \text{ g}$$

$$d_{(\text{air})} \times V = d_{(\text{H}_2)} \times V + 1.68 \times 10^5 \text{ g}$$

$$V = \frac{1.68 \times 10^5 \text{ g}}{d_{\text{air}} - d_{\text{H}_2}} = \frac{1.68 \times 10^5 \text{ g}}{\dfrac{P\,M_{\text{air}}}{RT} - \dfrac{P\,M_{\text{H}_2}}{RT}} = \frac{(1.68 \times 10^5 \text{ g})\,RT}{P(M_{\text{air}} - M_{\text{H}_2})}$$

$$V = \frac{(1.68 \times 10^5 \text{ g})(0.0821\,\text{L} \cdot \text{atm/mol.} \cdot \text{K})(295\,\text{K})}{\left(758\,\text{mm Hg} \times \dfrac{1\,\text{atm}}{760\,\text{mm Hg}}\right)(29.0\,\text{g/mol.} - 2.01\,\text{g/mol.})} = 1.51 \times 10^5 \text{ L}$$

$$V = 1.51 \times 10^5 \text{ L} \times \frac{1\,\text{m}^3}{1000\,\text{L}} = 1.51 \times 10^2 \text{ m}^3$$

$$V = 1.51 \times 10^2 \text{ m}^3 = \frac{4}{3}\pi r^3$$

$$r = 3.30 \text{ m}$$

$$\text{diameter} = 2r = 6.60 \text{ m}$$

Since R is a constant, we can solve this problem using the ideal gas law, by solving for R, and setting the equations equal to one another. (Remember to convert temperature to K.)

$$\frac{PV}{nT} = R, \text{ thus } \frac{P_i V_i}{n_i T_i} = R = \frac{P_f V_f}{n_f T_f}. \quad \text{Since } V \text{ is constant, we get: } \frac{P_i}{n_i T_i} = \frac{P_f}{n_f T_f}.$$

Solving for P_f, we get: $P_f = \dfrac{P_i\, n_f\, T_f}{n_i\, T_i}$

n_f = unreacted moles of H_2 and O_2 + moles of product.
   (Remember that 3 moles of reactants yields only 2 moles of product.)

Thus, $n_f = (0.120)(n_i) + (0.880)\left(\dfrac{2}{3}n_i\right) = 0.707\,n_i$

$$P_f = \frac{(0.950\,\text{atm})(0.707 n_i)(398\,\text{K})}{(n_i)(298\,\text{K})} = 0.897\,\text{atm}$$

---

77. Refer to Sections 5.2 and 5.5.

Use the ideal gas law, solving for V_A and V, and substitute into the equation for volume fraction. Since R, T and P are constant, they cancel out.

$$\text{volume fraction A} = \frac{V_A}{V} = \frac{\dfrac{n_A\,RT}{P}}{\dfrac{nRT}{P}} = \frac{n_A}{n} = \text{mole fraction A}$$

Volume fraction (and mole fraction) differ from mass fraction because different gases have different molar masses.

Chapter 6: Electronic Structure and the Periodic Table

---

2. *Refer to Section 6.1 and Examples 6.1 and. 2.*

a. $c = \lambda v$

$$423 \text{ nm} \times \frac{1 \text{ m}}{1 \times 10^9 \text{ nm}} = 4.23 \times 10^{-7} \text{ m}$$

$$v = \frac{2.998 \times 10^8 \text{ m/s}}{4.23 \times 10^{-7} \text{ m}} = 7.09 \times 10^{14} \text{ s}^{-1} = 7.09 \times 10^{14} \text{ Hz}$$

b. $E = hv = (6.626 \times 10^{-34} \text{ J·s})(7.09 \times 10^{14} \text{ s}^{-1}) = 4.70 \times 10^{-19} \text{ J}$

Note that this is for one photon, thus units are J/photon.

c. $\dfrac{4.07 \times 10^{-19} \text{ J}}{1 \text{ photon}} \times \dfrac{6.02 \times 10^{23} \text{ photons}}{1 \text{ mol.}} = 2.83 \times 10^5 \text{ J/mol.} = 283 \text{ kJ/mol.}$

---

4. *Refer to Section 6.1, Examples 6.1 and Figure 6.2.*

a. $1498 \text{ nm} \times \dfrac{1 \text{ m}}{1 \times 10^9 \text{ nm}} = 1.498 \times 10^{-6} \text{ m}$

1.498×10^{-6} m is just outside the visible region in the **infrared** region.

b. $c = \lambda v$

$$v = \frac{2.998 \times 10^8 \text{ m/s}}{1498 \times 10^{-6} \text{ m}} = 2.001 \times 10^{14} \text{ s}^{-1} = 2.001 \times 10^{14} \text{ Hz}$$

c. $E = hv = (6.626 \times 10^{-34} \text{ J·s})(2.001 \times 10^{14} \text{ s}^{-1}) = 1.326 \times 10^{-19} \text{ J}$

---

6. *Refer to Example 6.1.*

Calculate the amount of energy in one photon. Use that and the total amount of energy to calculate the number of photons emitted.

$$\lambda = 643 \text{ nm} \times \frac{1 \text{ m}}{1 \times 10^9 \text{ nm}} = 6.43 \times 10^{-7} \text{ m}$$

$$E = \frac{hc}{\lambda} = \frac{(6.626 \times 10^{-34} \text{ J} \cdot \text{s})(2.998 \times 10^8 \text{ m/s})}{6.43 \times 10^{-7} \text{ m}} = 3.09 \times 10^{-19} \text{ J}$$

$$59 \text{ J} \times \frac{1 \text{ photon}}{3.09 \times 10^{-19} \text{ J}} = 1.9 \times 10^{20} \text{ photons}$$

---

8. Refer to Section 6.1 and Example 6.2.

microwave: $\lambda = 5.00 \times 10^6 \text{ nm} \times \dfrac{1 \text{ m}}{1 \times 10^9 \text{ nm}} = 5.00 \times 10^{-3} \text{ m}$

$$E = \frac{hc}{\lambda} = \frac{(6.626 \times 10^{-34} \text{ J} \cdot \text{s})(2.998 \times 10^8 \text{ m/s})}{5.00 \times 10^{-3} \text{ m}} = 3.97 \times 10^{-23} \text{ J}$$

$$\frac{3.97 \times 10^{-23} \text{ J}}{1 \text{ photon}} \times \frac{6.02 \times 10^{23} \text{ photons}}{1 \text{ mol.}} \times \frac{1 \text{kJ}}{1000 \text{ J}} = 0.0239 \text{ kJ/mol.}$$

sun: $\quad\quad \lambda = 100 \text{ nm} \times \dfrac{1 \text{ m}}{1 \times 10^9 \text{ nm}} = 1.00 \times 10^{-7} \text{ m}$

$$E = \frac{hc}{\lambda} = \frac{(6.626 \times 10^{-34} \text{ J} \cdot \text{s})(2.998 \times 10^8 \text{ m/s})}{1.00 \times 10^{-7} \text{ m}} = 1.99 \times 10^{-18} \text{ J}$$

$$\frac{1.99 \times 10^{-18} \text{ J}}{1 \text{ photon}} \times \frac{6.02 \times 10^{23} \text{ photons}}{1 \text{ mol.}} \times \frac{1 \text{ kJ}}{1000 \text{ J}} = 1.20 \times 10^3 \text{ kJ/mol.}$$

The light from the sun has approximately 50,000 times more energy.

---

10. Refer to Section 6.2, Example 6.3 and Figure 6.2.

Use the Rydberg equation to calculate the frequency. Then calculate the wavelength to determine the region of the spectrum associated with that frequency.

a. $v = \dfrac{R_H}{h}\left[\dfrac{1}{(n_{lo})^2} - \dfrac{1}{(n_{hi})^2}\right] = \dfrac{2.180 \times 10^{-18} \text{ J}}{6.626 \times 10^{-34} \text{ J} \cdot \text{s}}\left[\dfrac{1}{2^2} - \dfrac{1}{4^2}\right] = 6.169 \times 10^{14} \text{ s}^{-1}$

b. $\lambda = \dfrac{c}{v} = \dfrac{2.998 \times 10^8 \text{ m/s}}{6.169 \times 10^{14} \text{ s}^{-1}} \times \dfrac{1 \times 10^9 \text{ nm}}{1 \text{ m}} = 486.0 \text{ nm} ; \textbf{ visible.}$

c.   Since the transition is from a high level to a lower one, energy is released, not absorbed.

12. Refer to Section 6.2.

$$E_n = \frac{-R_H}{n^2}$$

$$E_1 = \frac{-R_H}{n^2} = \frac{2.180 \times 10^{-18}\ J}{1^2} = -2.180 \times 10^{-18}\ J$$

$$E_2 = \frac{-R_H}{n^2} = \frac{2.180 \times 10^{-18}\ J}{2^2} = -5.450 \times 10^{-19}\ J$$

$$E_3 = \frac{-R_H}{n^2} = \frac{2.180 \times 10^{-18}\ J}{3^2} = -2.422 \times 10^{-19}\ J$$

$$E_4 = \frac{-R_H}{n^2} = \frac{2.180 \times 10^{-18}\ J}{4^2} = -1.363 \times 10^{-19}\ J$$

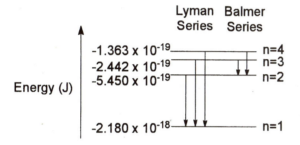

Obviously, this graph assumes that transitions are from E_{hi} to E_{lo}. If one assumes the transitions are in the opposite direction, the only change would be in the directions of the arrows.

14. Refer to Section 6.2, Example 6.3 and Figure 6.2.

a. $$v = \frac{R_H}{h}\left[\frac{1}{(n_{lo})^2} - \frac{1}{(n_{hi})^2}\right] = \frac{2.180 \times 10^{-18}\ J}{6.626 \times 10^{-34}\ J \cdot s}\left[\frac{1}{5^2} - \frac{1}{6^2}\right] = 4.021 \times 10^{13}\ s^{-1}$$

$$\lambda = \frac{c}{v} = \frac{2.998 \times 10^8\ m/s}{4.021 \times 10^{13}\ s^{-1}} \times \frac{1 \times 10^9\ nm}{1\,m} = 7455\ nm$$

b. This wavelength of light occurs in the **infrared** region of the spectrum.

First convert the wavelength to frequency, then apply the Rydberg equation and solve for n_{hi}. Remember that n_{hi} must be a whole number.

$$\nu = \frac{c}{\lambda} = 97.23 \text{ nm} \times \frac{1 \text{ m}}{1 \times 10^9 \text{ nm}} = 9.723 \times 10^{-8} \text{ m}$$

$$\nu = \frac{c}{\lambda} = \frac{2.998 \times 10^8 \text{ m/s}}{9.723 \times 10^{-8} \text{ m}} = 3.083 \times 10^{15} \text{ s}^{-1}$$

$$\nu = \frac{R_H}{h} \left[\frac{1}{(n_{lo})^2} - \frac{1}{(n_{hi})^2} \right] \Rightarrow 3.083 \times 10^{15} \text{ s}^{-1} = \frac{2.180 \times 10^{-18} \text{ J}}{6.626 \times 10^{-34} \text{ J} \cdot \text{s}} \left[\frac{1}{1^2} - \frac{1}{n_{hi}^2} \right]$$

$$3.083 \times 10^{15} \text{ s}^{-1} = (3.290 \times 10^{15} \text{ s}^{-1}) \left[1 - \frac{1}{n_{hi}^2} \right]$$

$$0.9371 = 1 - \frac{1}{n_{hi}^2}$$

$$n_{hi}^2 = \frac{1}{0.0629} = 15.89$$

$$n_{hi} = 4$$

Values of m_ℓ vary from $-\ell$ to $+\ell$ for any given ℓ value.

a. p-sublevel: $\ell = 1$
    $m_\ell = -1, 0, +1.$

b. f-sublevel: $\ell = 3$
    $m_\ell = -3, -2, -1, 0, +1, +2, +3.$

c. For the n=3 shell, there are 3 values for $\ell = 0, 1, 2$.
    $\ell = 0$: s-sublevel: $m_\ell = 0.$
    $\ell = 1$: p-sublevel: $m_\ell = -1, 0, +1.$
    $\ell = 2$: d-sublevel: $m_\ell = -2, -1, 0, +1, +2.$

20. Refer to Section 6.4 and Figures 6.7 and 6.8.

Look at Figures 6.7 (parts a-c) and 6.8 (part d) to determine the order of filling. The orbital that fills first is lower in energy.

a. 3p

b. 4p

c. 1s

d. 4d (*The 4d fills before the 5f.*)

22. Refer to Section 6.3.

The type of orbital is determined by the ℓ value.

a. $\ell = 2$, therefore: d-orbital.

b. $\ell = 3$, therefore: f-orbital.

c. $\ell = 3$, therefore: f-orbital

24. Refer to Section 6.3.

The number of orbitals is equal to $2\ell + 1$. If there is more than one sublevel, add up the orbitals for each sublevel.

a. **n** = 3 has 3 sublevels, $\ell = 0$, $\ell = 1$ and $\ell = 2$
   $\ell = 0$: 1 orbital
   $\ell = 1$: 3 orbitals
   $\ell = 2$: 5 orbitals, thus the **n** = 3 shell has **9 orbitals**.

b. 4p: $\ell = 1$: **3 orbitals**

c. f: $\ell = 3$: **7 orbitals**

d. d: $\ell = 2$: **5 orbitals**

26. Refer to Section 6.3.

a. $s \Rightarrow \ell = 0$ the minimum **n** is $\ell + 1$, therefore **n = 1**.

b. letter designators for sublevels increase alphabetically from f onward
   $\ell = 3$ is f; $\ell = 4$ **is g**; $\ell = 5$ is h.

c. $d \Rightarrow \ell = 2$, $\mathbf{m}_\ell = -2, -1, 0, +1, +2$. Thus there are **5 orbitals.**

d. For $\mathbf{n} = 1$, ℓ can only equal 0, thus **one sublevel**.

---

28. *Refer to Section 6.3.*

Refer to the rules for quantum numbers. If a rule is violated, the set *cannot* occur.

a. $\mathbf{n} = 3 \Rightarrow \ell = 0, 1, 2$
$\ell = 0 \Rightarrow \mathbf{m}_\ell = 0$
$\mathbf{m}_\ell = 0 \Rightarrow \mathbf{m}_s = -\frac{1}{2}, +\frac{1}{2}$
None of the rules are violated, so this set **can occur.**

b. $\mathbf{n} = 2 \Rightarrow \ell = 0, 1$
$\ell = 2$ This is not possible given the allowed values, this set **cannot occur.**

c. $\mathbf{n} = 3 \Rightarrow \ell = 0, 1, 2$
$\ell = 2 \Rightarrow \mathbf{m}_\ell = -2, -1, 0, +1, +2$
$\mathbf{m}_\ell = 1 \Rightarrow \mathbf{m}_s = -\frac{1}{2}, +\frac{1}{2}$
None of the rules are violated, so this set **can occur.**

d. $\mathbf{n} = 3 \Rightarrow \ell = 0, 1, 2$
$\ell = 2 \Rightarrow \mathbf{m}_\ell = -2, -1, 0, +1, +2$
$\mathbf{m}_\ell = 2 \Rightarrow \mathbf{m}_s = -\frac{1}{2}, +\frac{1}{2}$
None of the rules are violated, so this set **can occur.**

e. $\mathbf{n} = 4 \Rightarrow \ell = 0, 1, 2, 3$
$\ell = 2 \Rightarrow \mathbf{m}_\ell = -2, -1, 0, +1, +2$
$\mathbf{m}_\ell = -2 \Rightarrow \mathbf{m}_s = -\frac{1}{2}, +\frac{1}{2}$
$\mathbf{m}_s = 0$ This is not possible given the allowed values, this set **cannot occur.**

---

30. *Refer to Section 6.4 and Examples 6.5 and 6.6.*

Determine the number of electrons. Then write down the filling order and start filling the orbitals until you run out of electrons.

a. B ($5e^-$) : $1s^2 \, 2s^2 \, 2p^1$.

b. Ba ($56e^-$): $1s^2 \, 2s^2 \, 2p^6 \, 3s^2 \, 3p^6 \, 4s^2 \, 3d^{10} \, 4p^6 \, 5s^2 \, 4d^{10} \, 5p^6 \, 6s^2$

c. Be ($4e^-$): $1s^2 \, 2s^2$

d. Bi ($83e^-$): $1s^2 \, 2s^2 \, 2p^6 \, 3s^2 \, 3p^6 \, 4s^2 \, 3d^{10} \, 4p^6 \, 5s^2 \, 4d^{10} \, 5p^6 \, 6s^2 \, 4f^{14} \, 5d^{10} \, 6p^3$

e. Br ($35e^-$): $1s^2 \, 2s^2 \, 2p^6 \, 3s^2 \, 3p^6 \, 4s^2 \, 3d^{10} \, 4p^5$

Locate the element of interest on the periodic table. Move up one period and then move to the extreme right to find the appropriate noble gas. Which period is the element of interest in? That is the s-orbital from which you are to start filling. Then fill until you run out of electrons.

a. Hg: $[Xe]\, 6s^2\, 4f^{14}\, 5d^{10}$

b. Al: $[Ne]\, 3s^2\, 3p^1$

c. As: $[Ar]\, 4s^2\, 3d^{10}\, 4p^3$

d. W: $[Xe]\, 6s^2\, 4f^{14}\, 5d^4$

e. At: $[Xe]\, 6s^2\, 4f^{14}\, 5d^{10}\, 6p^5$

Start writing an electron configuration, filling up the orbitals until you meet the criterion given. Then determine the element that configuration corresponds to.

a. $1s^2\, 2s^2\, 2p^1$
This corresponds to **B.**

b. $1s^2\, 2s^2\, 2p^6\, 3s^2\, 3p^6\, 4s^2\, 3d^{10}\, 4p^6\, 5s^2\, 4d^{10}\, 5p^6\, 6s^2\, 4f^4$
This corresponds to **Nd.**

c. $1s^2\, 2s^2\, 2p^6\, 3s^2\, 3p^6\, 4s^2\, 3d^{10}$
This corresponds to **Zn.**

d. $1s^2\, 2s^2\, 2p^6\, 3s^2$
This corresponds to **Mg.**

Write out the electron configuration, count up the number of electrons in p subshells and divide that number by the total number of electrons.

a. Mg: $1s^2\, 2s^2\, 2p^6\, 3s^2$
(6 p electrons) / (12 total electrons) = **0.5** (or one half)

b. Mn: $1s^2\, 2s^2\, 2p^6\, 3s^2\, 3p^6\, 4s^2\, 3d^5$
(12 p electrons) / (25 total electrons) = **0.48** (or almost one half)

c. Mo: $1s^2 2s^2 2p^6 3s^2 3p^6 4s^2 3d^{10} 4p^6 5s^2 4d^4$
(18 p electrons) / (42 total electrons) = **0.43** (little more than two fifths)

---

38. *Refer to Section 6.4.*

First determine if the given configuration violates one of the rules for quantum numbers. If it does not, determine the ground state configuration for an atom with the given number of electrons and compare that with the configuration given.

a. 4 electrons, ground state: $1s^2 2s^2$
Thus the given configuration is an **excited state**.

b. 4 electrons, ground state: $1s^2 2s^2$
Thus the given configuration is an **excited state**.

c. 7 electrons, ground state: $1s^2 2s^2 2p^3$
Thus the given configuration is the **ground state**.

d. There is no 1p orbital. This would require $n = 1$, $\ell = 1$.
Impossible configuration.

e. 19 electrons, ground state: $1s^2 2s^2 2p^6 3s^2 3p^6 4s^1$
Thus the given configuration is an **excited state**.

f. There is no 3f orbital. This would require $n = 3$, $\ell = 3$.
Impossible configuration.

---

40. *Refer to Section 6.5 and Example 6.7*

a. Li: 1s   2s
($\uparrow\downarrow$) ($\uparrow$)

b. P: 1s   2s   2p   3s   3p
($\uparrow\downarrow$) ($\uparrow\downarrow$) ($\uparrow\downarrow$)($\uparrow\downarrow$)($\uparrow\downarrow$) ($\uparrow\downarrow$) ($\uparrow$)($\uparrow$)($\uparrow$)
Note the application of Hund's rule in filling the 3p orbitals.

c. F: 1s   2s   2p
($\uparrow\downarrow$) ($\uparrow\downarrow$) ($\uparrow\downarrow$)($\uparrow\downarrow$)($\uparrow$)

d. Fe: 1s   2s   2p   3s   3p   4s   3d
($\uparrow\downarrow$) ($\uparrow\downarrow$) ($\uparrow\downarrow$)($\uparrow\downarrow$)($\uparrow\downarrow$) ($\uparrow\downarrow$) ($\uparrow\downarrow$)($\uparrow\downarrow$)($\uparrow\downarrow$) ($\uparrow\downarrow$) ($\uparrow\downarrow$)($\uparrow$)($\uparrow$)($\uparrow$)($\uparrow$)

Count the number of electrons. Since the number of electrons must equal the number of protons (atomic number) in a neutral atom, you can then identify the element from the periodic table.

a.   12 electrons, $Z = 12$, therefore: **Mg**.

b.   15 electrons, $Z = 15$, therefore: **P**.

c.   8 electrons, $Z = 8$, therefore: **O**.

---

44. Refer to Section 6.5.

a.   The outer most shell of the group 17 elements all have the configuration $ns^2\,np^5$. Thus only those group 17 elements in and above the 5th period have filled 4p sublevels, **I** and **At**.

b.   The fourth period non-metals are Se, Br, and Kr. Only **Kr** has a filled 4p sublevel, and thus has no unpaired electrons.

c.   All the metalloids after $Z = 12$ have filled 3s orbitals, thus: **Si**, **Ge**, **As**, **Sb**, and **Te**.

d.   **Li:**   1s   2s   2p
                (↑↓)   (↑)   ( )( )( )

     B: 1s   2s   2p
                (↑↓)   (↑↓)   (↑)( )( )

     F: 1s   2s   2p
                (↑↓)   (↑↓)   (↑↓)(↑↓)(↑)

---

46. Refer to Section 6.5.

All the inner sublevels will be filled, so only the outermost sublevel will contain unpaired electrons.

a.   P: [Ne]$3s^2 3p^3$ $3p^3$: (↑)(↑)(↑) **3 unpaired electrons.**

b.   K: [Ar]$4s^1$ $4s^1$: (↑)   **1 unpaired electron.**

c.   Pu: [Rn]$7s^2 5f^6$ $5f^6$: (↑)(↑)(↑)(↑)(↑)(↑)( ) **6 unpaired electrons.**
     Recall that the f-sublevel ($\ell = 3$) has **7 m_ℓ** values and therefore 7 orbitals.

48. Refer to Section 6.5.

The only main group metals in the 4th period are K, Ca and Ga, the remainder of the elements are transition metals, non metals or metalloids.

```
            4s          4p
    K      (↑ )    (  )(  )(  )
    Ca     (↑↓)    (  )(  )(  )
    Ga     (↑↓)    (↑ )(  )(  )
```

a. Ca

b. K and Ga

c. none

d. none

---

50. Refer to Section 6.6 and Example 6.8.

Write the ground state electron configuration for the atom, then remove (for cations) or add (for anions) the number of electrons indicated by the charge.

a. P: $1s^2 2s^2 2p^6 3s^2 3p^3$
    P^{-3}: $1s^2 2s^2 2p^6 3s^2 3p^6$

b. Ca: $1s^2 2s^2 2p^6 3s^2 3p^6 4s^2$
    Ca^{+2}: $1s^2 2s^2 2p^6 3s^2 3p^6$

c. Ti: $1s^2 2s^2 2p^6 3s^2 3p^6 4s^2 3d^2$
    Ti^{+4}: $1s^2 2s^2 2p^6 3s^2 3p^6$

d. Mn: $1s^2 2s^2 2p^6 3s^2 3p^6 4s^2 3d^5$
    Mn^{+2}: $1s^2 2s^2 2p^6 3s^2 3p^6 4s^0 3d^5 \Rightarrow 1s^2 2s^2 2p^6 3s^2 3p^6 3d^5$
        For transition metals, electrons are lost first from the outermost s sublevel.
    Mn^{+4}: $1s^2 2s^2 2p^6 3s^2 3p^6 4s^0 3d^3 \Rightarrow 1s^2 2s^2 2p^6 3s^2 3p^6 3d^3$

52. Refer to Section 6.6.

Write the ground state electron configuration for the atom, then remove (for cations) or add (for anions) the number of electrons indicated by the charge. Then write the orbital diagram for the ion.

a. Hg: [Xe] $6s^2 4f^{14} 5d^{10}$
    Hg^{+2}:   [Xe] $6s^0 4f^{14} 5d^{10}$
    Hg^{+2}:   [Xe]   6s   4f   5d
              ( )   (↑↓)(↑↓)(↑↓)(↑↓)(↑↓)(↑↓)(↑↓)   (↑↓)(↑↓)(↑↓)(↑↓)(↑↓)
          no unpaired electrons

b. F: [He] $2s^2 2p^5$
    F⁻:   [He] $2s^2 2p^6$
    F⁻:   [He]   2s   2p
              (↑↓)   (↑↓)(↑↓)(↑↓)
          no unpaired electrons

c. Sb: [Kr] $5s^2 4d^{10} 5p^3$
    Sb^{+3}:   [Kr] $5s^2 4d^{10}$
    Sb^{+3}:   [Kr]   5s   4d
              (↑↓)   (↑↓)(↑↓)(↑↓)(↑↓)(↑↓)
          no unpaired electrons

d. Fe: [Ar] $4s^2 3d^6$
    Fe^{+3}:   [Ar] $4s^0 3d^5$
    Fe^{+3}:   [Ar]   4s   3d
              ( )   (↑)(↑)(↑)(↑)(↑)
          5 unpaired electrons

54. Refer to Section 6.7 and Example 6.9.

a. Atomic radius decreases from left to right across a period, therefore:
    Cl < S < Mg

b. Ionization energy increases from left to right across a period, therefore:
    Mg < S < Cl

c. Electronegativity increases from left to right across a period, therefore:
    Mg < S < Cl

56. Refer to Section 6.7 and Example 6.9.

    a.   Atomic radius decreases from left to right across a period and increases down a group, thus the smallest atom would be the upper rightmost: **Sb**.

    b.   Ionization energy increases from left to right across a period and decreases down a group, thus the atom with the lowest ionization energy would be the lower leftmost: **Cs**.

    c.   Electronegativity increases from left to right across a period and decreases down a group, thus the atom with the least electronegativity would be the lower leftmost: **Cs**.

58. Refer to Section 6.7 and Example 6.9.

Cations are smaller than the corresponding atoms, anions are larger.

    a.   N

    b. Ba^{+2}

    c.   Se

    d. Co^{+3}

60. Refer to Section 6.7 and Example 6.9.

For a given atom, as the positive charge increases, the size decreases. Conversely, as the negative charge increases, the size also increases. Size also increases going down a group.

    a. $Co^{+3} < Co^{+2} < Co$

    b. $Cl < Cl^- < Br^-$

62. Refer to Section 6.1.

Calculate 8.5% of the energy used to power a 75 watt bulb to get the energy emitted as visible light. Calculate the energy of one photon of 565 nm light. Then use the energy per photon to calculate the number of photons needed to give the energy emitted as visible light.

75 J/s x 0.085 = 6.4 J/s

$$565 \text{ nm} \times \frac{1 \text{ m}}{1 \times 10^9 \text{ nm}} = 5.65 \times 10^{-7} \text{ m}$$

$$E = \frac{hc}{\lambda} = \frac{(6.626 \times 10^{-34} \text{ J} \cdot \text{s})(2.998 \times 10^{8} \text{ m/s})}{5.65 \times 10^{-7} \text{ m}} = 3.52 \times 10^{-19} \text{ J}$$

$$6.4 \text{ J} / \text{s} \times \frac{1 \text{ photon}}{3.52 \times 10^{-19} \text{ J}} = 1.8 \times 10^{19} \text{ photons/s}$$

---

64. *Refer to Sections 6.4, 6.6 and 6.7.*

a. 17 electrons, therefore 17 protons, thus the element is **Cl**.

b. Since ionization energy decreases going down a group, the element is **Pb**.

c. The +2 ion has 23 electrons, therefore the parent atom has 25 and $Z = 25$, which is **Mn**.

d. Since atomic radii increase going down a group, the smallest would be **Li**.

e. Ionization energy increases from left to right across a period, so the element with the greatest ionization energy is **Kr**.

---

66. *Refer to Section 6.2 and Problem 12 above.*

a. Energy is absorbed when the electrons are excited from lower levels to a higher ones, thus energy is absorbed for **transitions 2 and 4**.

b. Energy is emitted when the electron relaxes from higher level to a lower one, thus energy is emitted for **transitions 1 and 3**.

c. We cannot answer this question without knowing what the element is since one must know the number of electrons to determine the ground state configuration. If we assume the atom to be hydrogen, then **transition 1** involves the ground state.

d. The transition with largest energy difference (see part (a) and problem 12 above) will absorb the most energy, thus **transition 2**.

e. The transition with largest energy difference (see part (b) and problem 12 above) will emit the most energy, thus **transition 1**.

---

68. *Refer to Section 6.3.*

a. The number of orbitals is equal to n^2.

b. The number of orbitals is equal to $2\ell + 1$.

c. The number of spin states (m_s) in **unrelated** to the number of orbitals, it is always 2.

111

a. No two electrons in an atom can have the same four quantum numbers. Also, no two electrons in the same orbital can have the same spin.

b. It takes more energy to put two electrons in the same orbital, than it does to put each electron in a separate orbital (within a given sublevel). Also, electrons enter orbitals of a sublevel singly (until each orbital has one electron) before electrons are paired.

c. A line in an atomic spectrum represents the transition of an electron from a higher energy level to a lower energy level.

d. The principal quantum number gives the relative position and energy of electrons, with smaller n representing electrons closer to the nucleus and with lower energy.

a. This statement is **true**. Photons with very short wavelengths are high energy.

b. This statements is **false**. The energy of an electron is inversely proportional to $\mathbf{n}^2$, not ℓ.

$$E_\mathbf{n} = -\frac{R_H}{\mathbf{n}^2}$$

c. This statements is **false**. Electrons start entering the 5th principal level **before** the fourth is filled. The order of filling is [Kr] 5s 4d 5p …

a. Ionic size depends on the ratio of electrons to protons. The protons pull the electrons toward the nucleus, but the electrons also repel each other. In neutral atoms, these two forces balance one another. In cations, the excess of protons relative to electrons causes the attraction to dominate and the cation is smaller than the atom. In anions, the excess of electrons relative to protons causes the repulsion to dominate and the anion is larger than the atom.

b. Sc has an electron configuration of [Ar] $4s^2$ $3d^1$. Thus only 3 electrons are lost to form the Sc^{+3} ion with a noble gas configuration. By contrast, a later transition metal such as Fe must lose 8 electrons to form an ion with a noble gas configuration, a more formidable task.

c. Electronegativity is a measure of the ability of an atom in a molecule to attract electrons to itself. This attraction is due to the protons of the nucleus. Thus for larger atoms (those down a group) in a molecule, the electron will be further from the nucleus and will experience less attraction.

76. *Refer to Section 6.2.*

For a one electron species, the ground state is **n** = 1, so the first excited state is **n** = 2.

$$E = \frac{-BZ^2}{n^2} = \frac{-(2.180 \times 10^{-18} \text{ J})(3)^2}{2^2} = -4.905 \times 10^{-18} \text{ J}$$

$$E = -4.905 \times 10^{-18} \text{ J} \times \frac{1 \text{ kJ}}{1000 \text{ J}} \times \frac{6.022 \times 10^{23}}{1 \text{ mol.}} = -2.954 \times 10^3 \text{ kJ/mol.}$$

E is the energy of the electron. The energy needed to ionize (or remove) an electron is $+2.954 \times 10^3$ kJ/mol.

77. *Refer to Sections 6.1 and 6.2.*

$$\lambda = \frac{hc}{\Delta E} \quad \text{and} \quad \Delta E = -R_H \left[\frac{1}{n_{hi}^2} - \frac{1}{n_{lo}^2} \right]$$

Substituting for ΔE and $n_{lo} = 2$, we get: $\lambda = \dfrac{hc}{-R_H \left[\dfrac{1}{n_{hi}^2} - \dfrac{1}{2^2} \right]} = -\dfrac{hc}{R_H} \dfrac{1}{\left[\dfrac{1}{n_{hi}^2} - \dfrac{1}{4} \right]}$

$$\lambda = -\frac{(6.626 \times 10^{-34} \text{ J} \cdot \text{s})(2.998 \times 10^8 \text{ m/s})}{2.180 \times 10^{-18} \text{ J}} \frac{1}{\left[\dfrac{1}{n_{hi}^2} - \dfrac{1}{4} \right]} = -9.112 \times 10^{-8} \text{ m} \frac{1}{\left[\dfrac{1}{n_{hi}^2} - \dfrac{1}{4} \right]}$$

$$\lambda = 9.112 \times 10^{-8} \text{ m} \frac{1}{\left[\dfrac{1}{4} - \dfrac{1}{n_{hi}^2} \right]} = 9.112 \times 10^{-8} \text{ m} \frac{4n_{hi}^2}{\dfrac{4n_{hi}^2}{4} - \dfrac{4n_{hi}^2}{n_{hi}^2}}$$

$$\lambda = 9.112 \times 10^{-8} \text{ m} \frac{4n_{hi}^2}{n_{hi}^2 - 4} = \frac{(3.645 \times 10^{-7} \text{ m})n_{hi}^2}{n_{hi}^2 - 4}$$

Converting to nanometers: $\lambda = \dfrac{(3.645 \times 10^2 \text{ nm})n_{hi}^2}{n_{hi}^2 - 4}$

78. *Refer to Section 6.3 and Table 6.3.*

78. *Refer to Section 6.3 and Table 6.3.*

n	1			2		
ℓ	0	1	0	1	2	
sublevel	1s	1p	2s	2p	2d	
m_ℓ	0 1	0 1 2	0 1	0 1 2	0 1 2 3	

electron configuration with eight electrons: $1s^4 1p^4$

79. *Refer to Section 6.3 and Table 6.3.*

a. If there were 3 values for m_s, then each orbital could hold 3 electrons.
s-sublevel: one orbital ($m_\ell = 0$), 3 electrons.
p-sublevel: 3 orbitals ($m_\ell = -1, 0, 1$), 9 electrons.
d-sublevel: 5 orbitals ($m_\ell = -2, -1, 0, 1, 2$), 15 electrons.

b. The $n = 3$ level could hold 27 electrons; 3 in the s-sublevel, 9 in the p-sublevel and 15 in the d-sublevel.

c. AN = 8: $1s^3 2s^3 2p^2$
AN = 17: $1s^3 2s^3 2p^9 3s^2$

80. *Refer to Section 6.1.*

Calculate the energy of the light. The difference between that energy and the kinetic energy is the energy needed to eject the electron (E_{min}).

a. $E = \dfrac{hc}{\lambda} = \dfrac{(6.626 \times 10^{-34} \text{ J} \cdot \text{s})(2.998 \times 10^8 \text{ m/s})}{5.40 \times 10^{-7} \text{ m}} = 3.68 \times 10^{-19}$ J

$E_{min} = 3.68 \times 10^{-19}$ J - 2.60×10^{-20} J = 3.42×10^{-19} J

While the calculations above used the longer wavelength, we also could have used the shorter wavelength and gotten the same answer (within significance).

$E = \dfrac{hc}{\lambda} = \dfrac{(6.626 \times 10^{-34} \text{ J} \cdot \text{s})(2.998 \times 10^8 \text{ m/s})}{4.00 \times 10^{-7} \text{ m}} = 4.97 \times 10^{-19}$ J

$E_{min} = 4.97 \times 10^{-19}$ J - 1.54×10^{-19} J = 3.43×10^{-19} J

b. Since longer wavelengths correspond to lower energies, E_{min} would correspond to the longest wavelength.

$$\lambda = \frac{hc}{E_{min}} = \frac{(6.626 \times 10^{-34} \text{ J} \cdot \text{s})(2.998 \times 10^{8} \text{ m/s})}{3.42 \times 10^{-19} \text{ J}} = 5.81 \times 10^{-7} \text{ m} = 581 \text{ nm}$$

Chapter 7: Covalent Bonding

---

2. *Refer to Section 7.1 and Examples 7.1 and 7.2.*

Add up the total number of valence electrons. Draw the skeletal structure, then add the electrons (remember that each bond represents two electrons).

a. C: 4 valence electrons
    4Cl: 4 x 7 valence electrons
    total: 32 electrons

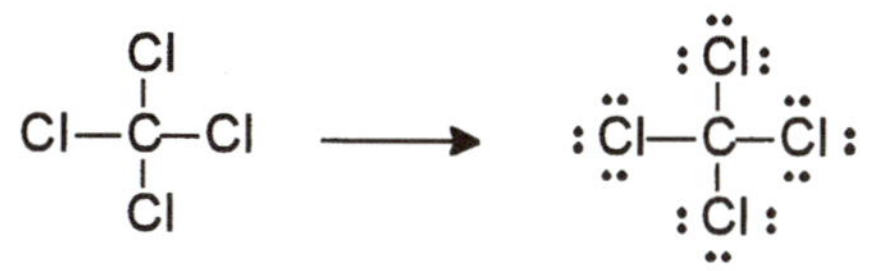

b. N: 5 valence electrons
    3Cl: 3 x 7 valence electrons
    total: 26 electrons

c. C: 4 valence electrons
    O: 6 valence electrons
    2Cl: 2 x 7 valence electrons
    total: 24 electrons

d. S: 6 valence electrons
    3O: 3 x 6 valence electrons
    -2: 2 x 1 valence electrons
    total: 26 electrons

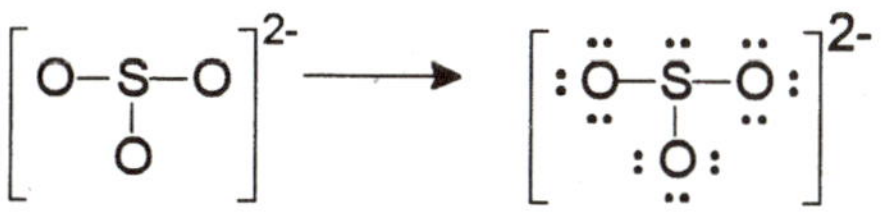

4. Refer to Section 7.1 and Examples 7.1, 7.2 and 7.4.

Add up the total number of valence electrons. Draw the skeletal structure, then add the electrons (remember that each bond represents two electrons).

a. P: 5 valence electrons
    4Cl: 4 x 7 valence electrons
    +1: -1 valence electron
    total: 32 electrons

$$\begin{bmatrix} \text{Cl} \\ | \\ \text{Cl--P--Cl} \\ | \\ \text{Cl} \end{bmatrix}^+ \longrightarrow \begin{bmatrix} :\overset{..}{\text{Cl}}: \\ | \\ :\overset{..}{\text{Cl}}\text{--P--}\overset{..}{\text{Cl}}: \\ | \\ :\overset{..}{\text{Cl}}: \end{bmatrix}^+$$

b. Br: 7 valence electrons
    2F: 2 x 7 valence electrons
    +1: -1 valence electron
    total: 20 electrons

$$\begin{bmatrix} \text{F--Br--F} \end{bmatrix}^+ \longrightarrow \begin{bmatrix} :\overset{..}{\underset{..}{\text{F}}}\text{--}\overset{..}{\underset{..}{\text{Br}}}\text{--}\overset{..}{\underset{..}{\text{F}}}: \end{bmatrix}^+$$

c. 3I: 3 x 7 valence electrons
    -1: 1 valence electron
    total: 22 electrons

$$\begin{bmatrix} \text{I--I--I} \end{bmatrix}^- \longrightarrow \begin{bmatrix} :\overset{..}{\underset{..}{\text{I}}}\text{--}\overset{.}{\underset{..}{\text{I}}}\text{--}\overset{..}{\underset{..}{\text{I}}}: \end{bmatrix}^-$$

d. Se: 6 valence electrons
    6Br: 6 x 7 valence electrons
    total: 48 electrons

$$\begin{matrix} & \text{Br} & \\ \text{Br} & | & \text{Br} \\ & \diagdown \text{Se} \diagup & \\ \text{Br} & \diagup | \diagdown & \text{Br} \\ & \text{Br} & \end{matrix} \longrightarrow \begin{matrix} & :\overset{..}{\text{Br}}: & \\ :\overset{..}{\underset{..}{\text{Br}}} & | & \overset{..}{\underset{..}{\text{Br}}}: \\ & \diagdown \text{Se} \diagup & \\ :\overset{..}{\underset{..}{\text{Br}}} & | & \overset{..}{\underset{..}{\text{Br}}}: \\ & :\overset{..}{\text{Br}}: & \end{matrix}$$

> *6. Refer to Section 7.1 and Examples 7.1, 7.2 and 7.4.*

Add up the total number of valence electrons. Draw the skeletal structure, then add the electrons (remember that each bond represents two electrons). Recall the exceptions to the octet rule (for H, I and B).

a. P: 5 valence electrons
3H: 3 x 1 valence electrons
total: 8 electrons

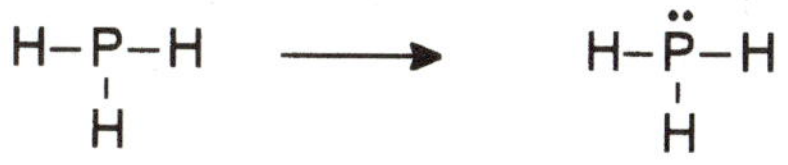

b. S: 6 valence electrons
4O: 4 x 6 valence electrons
-2: 2 x 1 valence electrons
total: 32 electrons

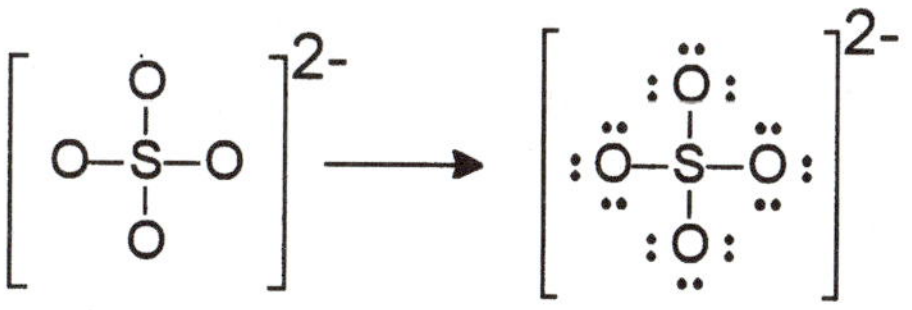

c. I: 7 valence electrons
4Cl: 4 x 7 valence electrons
-1: +1 valence electron
total: 36 electrons

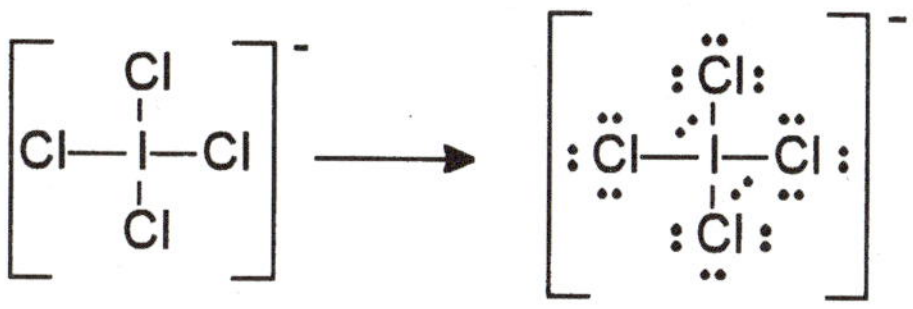

d. B: 3 valence electrons
3F: 3 x 7 valence electrons
total: 24 electrons

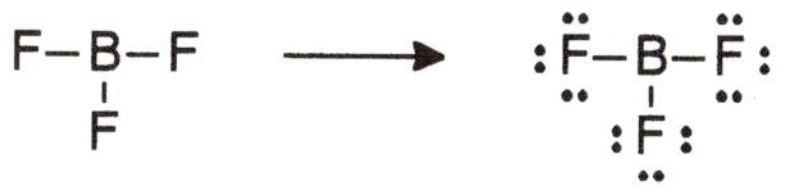

Add up the total number of valence electrons. Draw the skeletal structure, then add the electrons (remember that each bond represents two electrons). Recall that the best structure is one in which there are no formal charges.

2C: 2 x 4 valence electrons
4O: 4 x 6 valence electrons
2H: 2 x 1 valence electron
total: 34 electrons

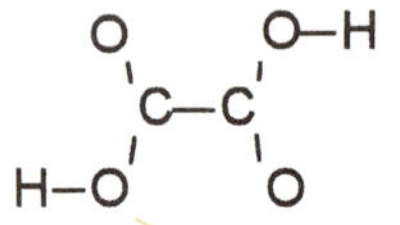

Two structures are possible in which there is one -OH attached to each carbon. Either structure is acceptable and both lead to very similar final structures. A structure with two -OH's on one carbon is also possible, but leads to a final structure with formal charges.

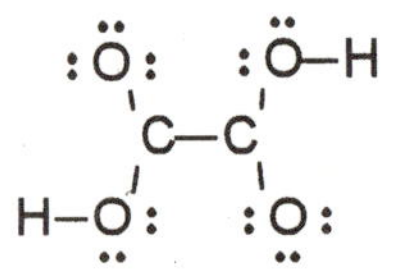

Note that this structure does not have complete octets around the carbon atoms. To complete the octets, move a pair of electrons from the oxygens to form (double) bonds between the carbon and oxygen.

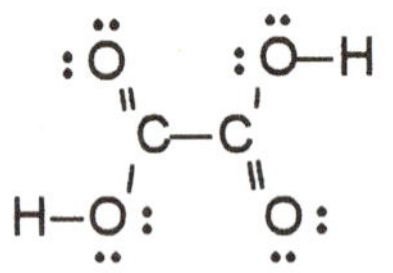

a. 2C: 2 x 4 valence electrons
4H: 4 x 1 valence electrons
O: 6 valence electron
total: 18 electrons

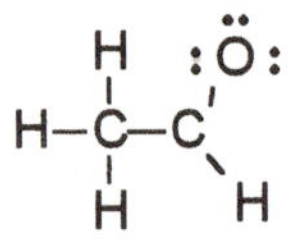

Note that this structure does not have complete octets around the carbon atoms. To complete the octets, move a pair of electrons from the oxygen to form a (double) bond between the carbon and oxygen.

120

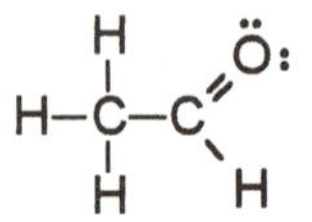

b. S: 6 valence electrons
    2H: 2 x 1 valence electrons
    3O: 3 x 6 valence electrons
    total: 26 electrons

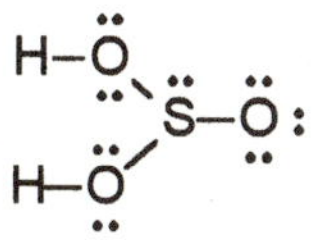

c. 2C: 2 x 4 valence electrons
    2F: 2 x 7 valence electrons
    2Cl: 2 x 7 valence electrons
    total: 36 electrons

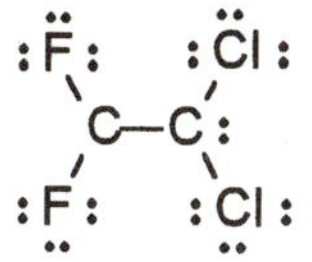

Note that this structure does not have complete octets around both carbon atoms. To complete the octets, move a pair of electrons from one carbon to form a (double) bond between the two carbon atoms.

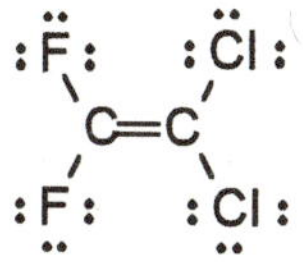

---

12. Refer to Section 7.1 and Problem 8 above.

2C: 2 x 4 valence electrons
3H: 3 x 1 valence electrons
5O: 5 x 6 valence electrons
N: 5 valence electrons
total: 46 electrons

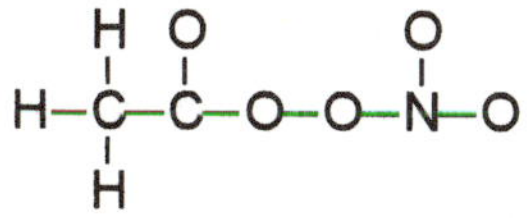

From the information given, the structure must be that on the left.

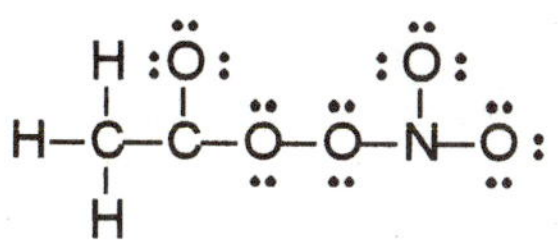

Note that this structure does not have complete octets around one of carbon atoms, nor the nitrogen atom. To complete the octets, move a pair of electrons from the oxygens to form C=O and N=O double bonds.

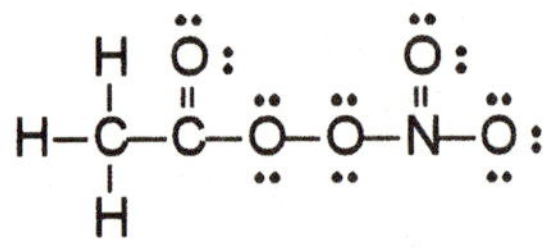

H–C–C–O–O–N–O: (with H, Ö:, Ö: shown above) This structure provides octets for all the atoms with a minimum of formal charges (N is +1, O is –1).

---

14. Refer to Section 7.1 and Problem 8 above.

2C: 2 x 4 valence electrons
2H: 2 x 1 valence electrons
2Cl: 2 x 7 valence electrons
total: 24 electrons

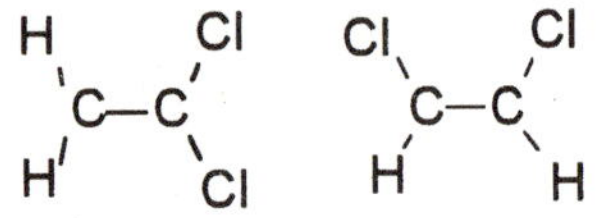

From the information given, the two structures must be those given to the left.

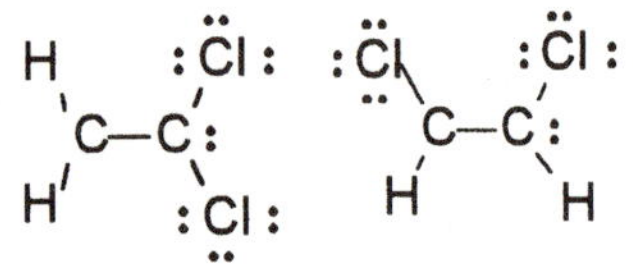

In each of the structures, one of the C's does not have a full octet. Move the pair of electrons from one carbon to form a (double) bond between the carbons.

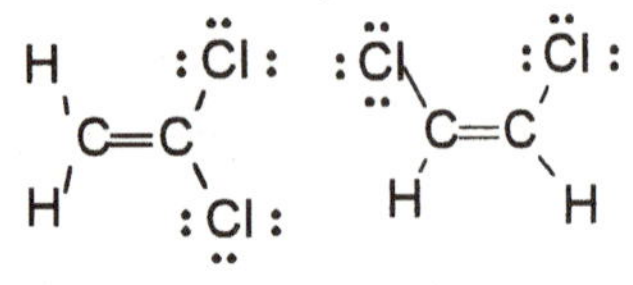

A third structure, drawn to the right, is also possible. This structure is an isomer of the right hand one given above.

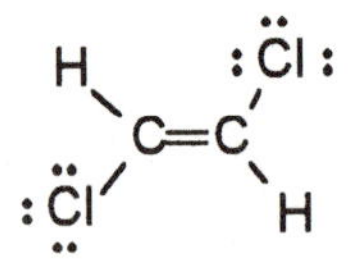

---

16. Refer to Section 7.1.

Draw the Lewis structure of the ion, then draw an atom with the same Lewis structure. This can be facilitated by increasing the atomic number of the atom(s) by the same amount as the negative charge (recall that the negative charge results from extra electrons, and that molecules have equal numbers of electrons and protons.

a. O: 6 valence electrons
    H: 1 valence electron
    -1: 1 valence electron
    total: 8 electrons

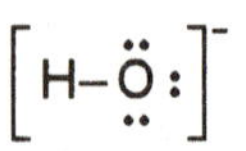

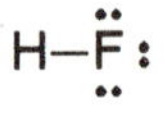

$$\left[H-\ddot{O}: \right]^-$$

$$H-\ddot{F}:$$

122

b. 2O: 2 × 6 valence electrons
   -2: 2 × 1 valence electrons
   total: 14 electrons

$$\left[:\ddot{O}-\ddot{O}: \right]^{2-} \qquad :\ddot{F}-\ddot{F}:$$

c. C: 4 valence electrons
   N: 5 valence electrons
   -1: 1 valence electron
   total: 10 electrons

$$\left[:C\equiv N: \right]^{-} \qquad :N\equiv N:$$

d. S: 6 valence electrons
   4O: 4 × 6 valence electrons
   -2: 2 × 1 valence electrons
   total: 32 electrons

$$\left[\begin{array}{c} :\ddot{O}: \\ :\ddot{O}-S-\ddot{O}: \\ :\ddot{O}: \end{array} \right]^{2-} \qquad \begin{array}{c} :\ddot{O}: \\ :\ddot{F}-S-\ddot{F}: \\ :\ddot{O}: \end{array}$$

---

18. Refer to Section 7.1.

a. B: 3 valence electrons
   4Cl: 4 × 7 valence electrons
   -1: 1 valence electron
   total: 32 electrons

$$\left[\begin{array}{c} Cl \\ | \\ Cl-B-Cl \\ | \\ Cl \end{array} \right]^{-} \qquad \left[\begin{array}{c} :\ddot{Cl}: \\ | \\ :\ddot{Cl}-B-\ddot{Cl}: \\ | \\ :\ddot{Cl}: \end{array} \right]^{-}$$

b. H: 1 valence electron
   4O: 4 × 6 valence electrons
   P: 5 valence electrons
   -2: 2 valence electrons
   total: 32 electrons

$$\left[\begin{array}{c} O \\ || \\ O-P-O-H \\ | \\ O \end{array} \right]^{2-} \qquad \left[\begin{array}{c} :\ddot{O}: \\ | \\ :\ddot{O}-P-\ddot{O}-H \\ | \\ :\ddot{O}: \end{array} \right]^{2-}$$

c. 4O: 4 × 6 valence electrons
   I: 7 valence electrons
   -1: 1 valence electron
   total: 32 electrons

$$\left[\begin{array}{c} O \\ | \\ O-I-O \\ | \\ O \end{array} \right]^{-} \qquad \left[\begin{array}{c} :\ddot{O}: \\ | \\ :\ddot{O}-I-\ddot{O}: \\ | \\ :\ddot{O}: \end{array} \right]^{-}$$

d. 3O: 3 × 6 valence electrons
   2S: 2 × 6 valence electrons
   -2: 2 valence electrons
   total: 32 electrons

$$\left[\begin{array}{c} S \\ | \\ O-S-O \\ | \\ O \end{array} \right]^{2-} \qquad \left[\begin{array}{c} :\ddot{S}: \\ | \\ :\ddot{O}-S-\ddot{O}: \\ | \\ :\ddot{O}: \end{array} \right]^{2-}$$

Parts b-d have odd numbers of electrons. Consequently, the final structures will have an unpaired electron.

a. Be: 2 valence electrons
    2H: 2 x 1 valence electrons
    total: 4 electrons

    H-Be-H

    The skeletal structure is the final Lewis structure.

b. C: 4 valence electron
    O: 6 valence electrons
    -: 1 valence electrons
    total: 11 electrons

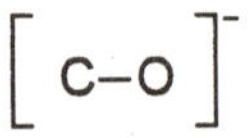

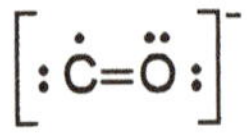

    $\left[\text{C–O} \right]^{-}$ $\left[:\dot{C}=\ddot{O}: \right]^{-}$

c. S: 6 valence electrons
    O: 2 x 6 valence electrons
    -1: 1 valence electron
    total: 19 electrons

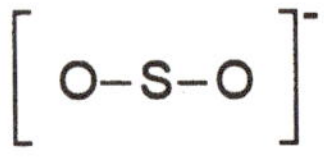

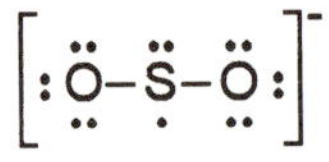

    $\left[\text{O–S–O} \right]^{-}$ $\left[:\ddot{O}-\dot{S}-\ddot{O}: \right]^{-}$

d. C: 4 valence electrons
    3H: 3 x 1 valence electrons
    total: 7 electrons

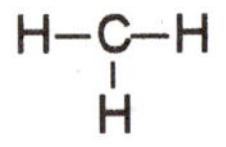

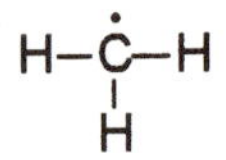

    H–C–H
      |
      H

    H–Ċ–H
      |
      H

Write the Lewis structure for the compound, then draw resonance forms by changing the positions of electron pairs. Remember that the skeletal structure cannot change, and the new structures must also abide by the rules for Lewis structures.

a.  $\left[:\ddot{O}-\ddot{N}=\ddot{O}: \right]^{-}$ ⟷ $\left[:\ddot{O}=\ddot{N}-\ddot{O}: \right]^{-}$

b. 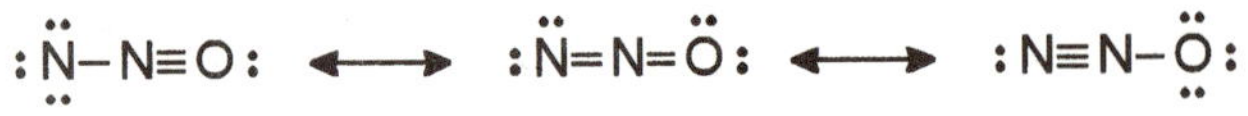 $:\ddot{N}-N≡O:$ ⟷ $:\ddot{N}=N=\ddot{O}:$ ⟷ $:N≡N-\ddot{O}:$

c. 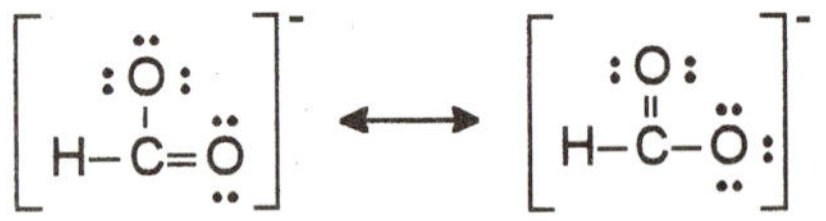 $\left[\begin{array}{c} :\ddot{O}: \\ H-C=\ddot{O} \end{array} \right]^{-}$ ⟷ $\left[\begin{array}{c} :O: \\ \| \\ H-C-\ddot{O}: \end{array} \right]^{-}$

24. Refer to Section 7.1 and Example 7.3.

Write the Lewis structure for hydrazoic acid, then draw resonance forms by changing the positions of electron pairs. Remember that the skeletal structure cannot change, and the new structures must also abide by the rules for Lewis structures.

a.   H—N̈=N=N̈ ⟷ H—N≡N—N̈: ⟷ H—N̈—N≡N̈

b.   This is not a valid resonance structure since the structure (connectivity) has changed. This structure is an isomer.

---

26. Refer to Section 7.1 and Example 7.3.

Write the Lewis structure for borazine, then draw resonance forms by changing the positions of electron pairs. Remember that the skeletal structure cannot change, and the new structures must also abide by the rules for Lewis structures.

a.   3N: 3 x 5 valence electrons
       3B: 3 x 3 valence electrons
       6H: 6 x 1 valence electrons
       total: 30 electrons

---

28. Refer to Section 7.1 and Table 7.2.

Draw the Lewis structure. Then apply the formula: $C_f = e_{valence} - (e_{unshared} + \frac{1}{2}(e_{bonding}))$ [formal charge = valence electrons - (unshared electrons + ½(bonding electrons))] Note that this is the formula given in the text as: $C_f = X - (Y + Z/2)$.

a.   3N: 3 x 5 valence electrons
       -1: +1 valence electron
       total: 16 electrons

       $[:N̈=N=N̈:]^-$

       $C_f = 5 - (0 + \frac{1}{2}(8)) = +1$

b.   Xe: 8 valence electrons
       6F: 6 x 7 valence electrons
       total: 50 electrons

       $C_f = 8 - (2 + \frac{1}{2}(12)) = 0$

c. Br: 7 valence electrons
   3Cl: 3 x 7 valence electrons
   total: 28 electrons

$C_f = 7 - (4 + \frac{1}{2}(6)) = 0$

$$:\ddot{C}l - \ddot{Br} - \ddot{C}l:$$
$$\underset{:\ddot{C}l:}{|}$$

---

30. Refer to Section 7.2, Examples 7.5 and 7.6, Figure 7.5 and Table 7.3.

Draw the Lewis structure of the compound and determine the number of bonded groups and the number of electron pairs. Then use Table 7.3 to assign the geometry.

a. $SO_2 \Rightarrow$ $\quad :\ddot{O} - \ddot{S} = \ddot{O}:$

2 bonded groups, one electron pair, thus AX_2E and **bent**.

b. $BeCl_2 \Rightarrow$ $\quad :\ddot{C}l - Be - \ddot{C}l:$

2 bonded groups, no electron pairs, thus AX_2 and **linear**.

c. $SeCl_4 \Rightarrow$
$$:\ddot{C}l - \ddot{Se} - \ddot{C}l:$$

4 bonded groups, one electron pair, thus AX_4E and **seesaw**.

d. $PCl_5 \Rightarrow$

5 bonded groups, no electron pairs, thus AX_5 and **triangular bipyramid**.

---

32. Refer to Section 7.2, Examples 7.5 and 7.6, Figures 7.4 and 7.5 and Table 7.3.

Draw the Lewis structure of the compound and determine the number of bonded groups and the number of electron pairs. Then use Table 7.3 to assign the geometry.

a. $NNO \Rightarrow$ $\quad :\ddot{N} = N = \ddot{O}:$

2 bonded groups, no electron pairs, thus AX_2 and **linear**.

b. $ONCl \Rightarrow$ $\quad :\ddot{O} = \ddot{N} - \ddot{C}l:$

2 bonded groups, one electron pair, thus AX_2E and **bent**.

c. NH_4^+ ⇒

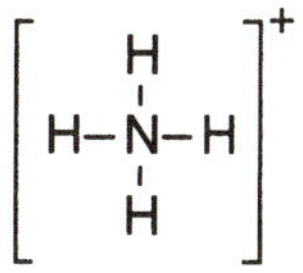

$$\begin{bmatrix} & H & \\ & | & \\ H- & N & -H \\ & | & \\ & H & \end{bmatrix}^+$$

4 bonded groups, no electron pairs, thus AX_4 and **tetrahedron**.

d. O_3 ⇒ :Ö–Ö=Ö:

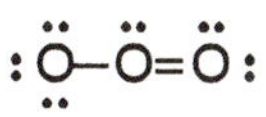

2 bonded groups, one electron pair, thus AX_2E and **bent**.

---

34. Refer to Section 7.2, Examples 7.5 and 7.6, Figures 7.4 and 7.5 and Table 7.3.

Draw the Lewis structure of the compound and determine the number of bonded groups and the number of electron pairs. Then use Table 7.3 to assign the geometry.

a. ClO_4^- ⇒

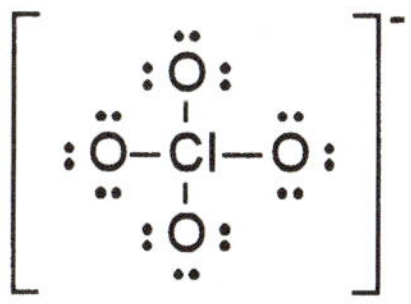

$$\begin{bmatrix} & :\ddot{O}: & \\ & | & \\ :\ddot{O}- & Cl & -\ddot{O}: \\ & | & \\ & :\ddot{O}: & \end{bmatrix}^-$$

4 bonded groups, no electron pairs, thus AX_4 and **tetrahedron**.

b. $TeBr_4$ ⇒

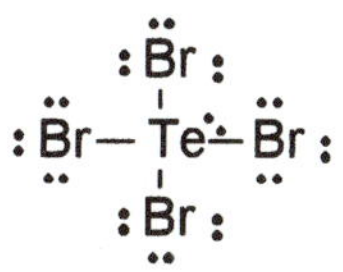

$$\begin{matrix} & :\ddot{Br}: & \\ & | & \\ :\ddot{Br}- & Te\ddot{\cdot}- & \ddot{Br}: \\ & | & \\ & :\ddot{Br}: & \end{matrix}$$

4 bonded groups, one electron pair, thus AX_4E and **seesaw**.

c. IF_3 ⇒

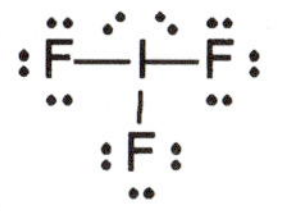

$$\begin{matrix} :\ddot{F}-\overset{\cdot\,\cdot}{I}-\ddot{F}: \\ | \\ :\ddot{F}: \end{matrix}$$

3 bonded groups, two electron pairs, thus AX_3E_2 and **T-shaped**.

d. SeF_6 ⇒

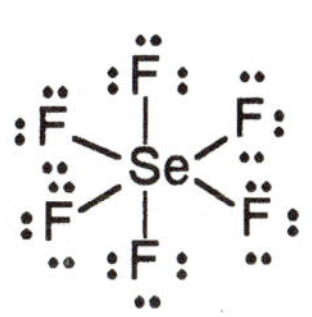

$$\begin{matrix} & :\ddot{F}: & \\ & | & \\ :\ddot{F}\diagdown & Se & \diagup\ddot{F}: \\ :\ddot{F}\diagup & | & \diagdown\ddot{F}: \\ & :\ddot{F}: & \end{matrix}$$

6 bonded groups, no electron pairs, thus AX_6 and **octahedron**.

Draw the Lewis structure, determine the geometry and, from that, the ideal bond angles.

a. $:\ddot{C}l—\ddot{S}—\ddot{C}l:$ AX_2E_2, bent $109.5°$

b. $:\ddot{F}—\overset{..}{X}\overset{.}{e}—\ddot{F}:$ AX_2E_3, linear   The five groups would assume a triangular bipyramid, with the 3 lone pairs in the "equator," leaving the 2 F's **180°** apart.

c. $H—\overset{1}{C}—\overset{2}{C}—\overset{}{N}—H$ (with H, Ö: groups) C^1: AX_4, tetrahedron $109.5°$
     C^2: AX_3, triangular planar $120°$
     N: AX_3E, triangular pyramid $109.5°$

d. $H—\overset{1}{C}=\overset{2}{C}—\overset{3}{C}≡N:$ C^1: AX_3, triangular planar $120°$
     C^2: AX_3, triangular planar $120°$
     C^3: AX_2, linear $180°$

a. $H—\overset{1}{C}—\overset{2}{C}—\overset{3}{O}—\overset{4}{O}—\overset{5}{C}—\overset{6}{C}—H$ b. C^1, C^6: AX_4, tetrahedron $109.5°$
     C^2, C^5: AX_3, triangular planar $120°$
     O^3, O^4: AX_2E_2, bent $109.5°$

Draw the Lewis structure for the molecule. Then determine the geometry of the indicated atom and from that, the bond angles.

1   AX_3 triangular planar $120°$

2   AX_2E_2 bent $109.5°$

3   AX_4 tetrahedral $109.5°$

First determine if the central atom has an octet. Then consider the electronegativity of the atoms to determine if there are dipoles. Then consider the geometry of the molecule to determine if there is a net dipole.

he only molecule with an octet about the central atom is **(a) SO_2**. The molecule is bent, so there is a net dipole.

---

44. Refer to Section 7.2, Example 7.7 and Problem 32 (above).

First determine if the central atom has an octet. Then consider the electronegativity of the atoms to determine if there are dipoles. Then consider the geometry of the molecule to determine if there is a net dipole.

All 4 molecules have octets about the central atom.

a. There is only one dipole, between the central N and the O. Thus there is a net dipole and **the molecule is a dipole**.

b. There are 2 dipoles, between N-O and N-Cl. Since this molecule is bent, there is a net dipole, and **the molecule is a dipole**.

c. There are 4 dipoles, between each of the N-H bonds. Since the molecule is tetrahedron, the 4 dipoles cancel each other out; there is no net dipole, so **the molecule is not a dipole**.

d. Since all the atoms are identical, there are no dipoles and one would thus assume there is no net dipole. However, a dipole depends only on an unsymmetrical distribution of electrons. Since this molecule is bent, that criterion is met, and **the molecule is a dipole**.

---

46. Refer to Section 7.3.

Draw the Lewis structures of the molecules. Then consider the electronegativity of the atoms to determine if there are dipoles. Then consider the geometry of the molecule to determine if there is a net dipole.

cis

The *cis* structure has a dipole along each N-F bond. The two dipoles don't cancel, so **this molecule is a dipole**.

trans

The *trans* structure also has a dipole along each N-F bond, but these two dipoles do cancel, so **this molecule is not a dipole**.

48. *Refer to Section 7.4, Example 7.8 and Problem 30 (above).*

Recall that the total number of groups (bonded atoms and electron pairs) around the central atom is equal to the number of orbitals that hybridized. Furthermore, the sum of the superscripts in the hybrid orbital notation gives the total number of hybrid orbitals.

a. SO_2 AX_2E 3 groups sp^2

b. $BeCl_2$ AX_2 2 groups sp

c. $SeCl_4$ AX_4E 5 groups sp^3d

d. PCl_5 AX_5 5 groups sp^3d

---

50. *Refer to Section 7.4, Example 7.8 and Problem 32 (above).*

Recall that the total number of groups (bonded atoms and electron pairs) around the central atom is equal to the number of orbitals that hybridized. Furthermore, the sum of the superscripts in the hybrid orbital notation gives the total number of hybrid orbitals.

a. NNO AX_2 2 groups sp

b. ONCl AX_2E 3 groups sp^2

c. NH_4^+ AX_4 4 groups sp^3

d. O_3 AX_2E 3 groups sp^2

---

52. *Refer to Section 7.4, Example 7.8 and Problem 34 (above).*

Recall that the total number of groups (bonded atoms and electron pairs) around the central atom is equal to the number of orbitals that hybridized. Furthermore, the sum of the superscripts in the hybrid orbital notation gives the total number of hybrid orbitals.

a. ClO_4^- AX_4 4 groups sp^3

b. $TeBr_4$ AX_4E 5 groups sp^3d

c. IF_3 AX_3E_2 5 groups sp^3d

d. SeF_6 AX_6 6 groups sp^3d^2

54. Refer to Section 7.4, Table 7.2 and Problem 32 above.

Draw the Lewis structure for each of the molecules. Then determine the number of groups (bonded atoms and electrons) and the hybridization.

a.

$\left[:\ddot{F}-\overset{..}{X}\overset{..}{e}-\ddot{F}: \atop :\ddot{F}: \right]^{-}$ AX_3E_3 6 groups sp^3d^2

b.

$\left[\begin{array}{c} :\ddot{F}: \\ :\ddot{F}-Si-\ddot{F}: \\ :\ddot{F}: \\ :\ddot{F}: \end{array} \right]^{2-}$ AX_6 6 groups sp^3d^2

c.

$\left[\begin{array}{c} :\ddot{Cl}: \\ :\ddot{Cl}-P-\ddot{Cl}: \\ :\ddot{Cl}: \end{array} \right]^{-}$ AX_4E 5 groups sp^3d

56. Refer to Section 7.4 and Problem 26 (above).

The Lewis structure for borazine was determined in problem 26. Each B and N is bonded to three other atoms and each has no lone pairs (AX_3). Thus the hybridization is sp^2.

58. Refer to Section 7.4 and Example 7.8.

Draw the Lewis structure for each of the molecules. Then determine the number of groups (bonded atoms and electrons) and the hybridization.

a.

$\begin{array}{c} :\ddot{Cl}: \\ | \\ H-C-H \\ | \\ H \end{array}$ AX_4 4 groups sp^3

b.

$\left[\begin{array}{c} :O: \\ \| \\ :\ddot{O}-C-\ddot{O}: \end{array} \right]^{2-}$ AX_3 3 groups sp^2

c. $\ddot{O}=C=\ddot{O}$ AX$_2$ 2 groups   sp

d.   H–C–Ö–H   AX$_3$ 3 groups   sp^2

---

60. Refer to Section 7.4.

Draw the Lewis structure for each of the molecules. Then determine the number of groups (bonded atoms and electrons) and the hybridization.

a. $H_2\ddot{N}$–C–$\ddot{N}H_2$ AX$_3$ 3 groups   sp^2

b.   :$\ddot{F}$–C–$\ddot{F}$:   AX$_3$ 3 groups   sp^2

c.   H–$\ddot{N}$–H   AX$_3$E   4 groups   sp^3
              H

---

62. Refer to Section 7.4, Example 7.10 and Problem 56 (above).

Recall that each single bond is a sigma (σ) bond and each double bond is composed of a sigma (σ) and a pi (π) bond.

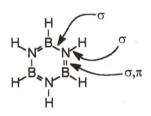

9 single bonds (9 σ) and 3 double bonds (3 σ, and 3 π) adds up to 12 σ-bonds and 3 π-bonds.

---

64. Refer to Section 7.4, Example 7.10 and Problems 58 and 62 (above).

a.   [structure of H–C–H with Cl and H, σ bonds indicated]   4 σ-bonds b.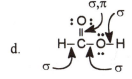

3 σ-bonds, 1 π-bond

c.

2 σ-bonds, 2 π-bonds

d.

4 σ-bonds, 1 π-bond

---

66. Refer to Section 7.1.

Determine the formal charges on each atom (refer to Problem 28, above).

$\ddot{C}l{=}Be{=}\ddot{C}l$ Be: $C_f = 2 - (0 - \frac{1}{2}(8)) = -2$
Cl: $C_f = 7 - (4 + \frac{1}{2}(4)) = +1$

There are two problems with this structure. First, the formal charges on each atom are not zero. Second, the negative formal charge is not on the most electronegative atom. Compare this to the structure with two single bonds:

$:\ddot{C}l{-}Be{-}\ddot{C}l:$ Be: $C_f = 2 - (0 - \frac{1}{2}(4)) = 0$
Cl: $C_f = 7 - (6 + \frac{1}{2}(2)) = 0$

This structure has no formal charges.

---

68. Refer to Section 7.1.

Draw the Lewis structure and determine if there are more than 4 bonds and/or lone pairs around the sulfur atom.

a. SO_2 was solved in problem 30 (above) and does **not** have an expanded octet.

b. SF_4

This sulfur atom has 4 bonds and one lone pair, thus 10 electrons and an **expanded octet**.

133

c. SO_2Cl_2

$$:O:$$
$$\|$$
$$:\ddot{C}l-S-\ddot{C}l:$$
$$\|$$
$$:O:$$

This Lewis structure is the one in which there are no formal charges. The sulfur atom has 6 bonds and thus 12 electrons and an **expanded octet**.

d. SF_6

The sulfur atom has 6 bonds and thus 12 electrons and an **expanded octet**.

---

70. Refer to Sections 7.2, 7.3 and 7.4, Tables 7.3 and 7.4 and Figure 7.8.

For the formula AX_mE_n, m is the number of atoms around Central Atom A and n is the number of electron pairs around A. The geometry is determined by the "Species" formula, as indicated in Table 7.3 and Figure 7.8. The hybridization is also determined by the number of electron pairs (bonding and unshaired, m+n) as shown in Table 7.4. Finally, recall the atoms, X, must be symmetrically disposed around the central atom, A, for the molecule to be free of a dipole.

Species	Atoms Around Central Atom A	Unshared Pairs Around A	Geometry	Hybridization	Polarity
AX_2E_2	2	2	bent	sp^3	polar
AX_3	3	0	trigonal planar	sp^2	non-polar
AX_4E_2	4	2	square planar	sp^3d^2	non-polar
AX_5	5	0	triangular bipyramid	sp^3d	non-polar

---

72. Refer to Section 7.1.

As a rule of thumb, the least electronegative atom will be central atom. This can be confirmed by drawing Lewis structures and determining which is the best (obeys all the rules and has the least formal charges).

$$\left[\ddot{N}=C=\ddot{S}\right]^-$$

The only formal charge is on the N, the most electronegative atom.

$$\left[\ddot{N}=S\equiv\ddot{C}\right]^-$$

The N still bears a negative formal charge, but now the S and C also bear formal charges, so this is a step in the wrong direction.

$$\left[:C\equiv N-\ddot{S}:\right]^-$$

The nitrogen now bears a positive charge, and the S and C still bear formal charges, so this structure is even worse than the one above.

Thus C is the central atom.

Draw Lewis structures of each molecule. Electron pairs in unshared pairs exert greater repulsion than electron pairs in bonds. Thus the structures with unshared pairs will have the bond angles smaller than 109.5°.

$$H-\overset{\displaystyle H}{\underset{\displaystyle H}{Si}}-H \qquad H-\overset{\displaystyle \cdot\cdot}{\underset{\displaystyle H}{P}}-H \qquad H-\overset{\displaystyle \cdot\cdot}{\underset{\displaystyle H}{S}}:$$

PH_3 and H_2S have at least one unshared pair and will have bond angles less than 109.5°.

a. $H-\overset{\displaystyle \cdot\cdot}{\underset{\displaystyle H}{N}}-H$ The sp^3 hybridization indicates that 4 orbitals (1 s and 3p) are involved. If there are 3 bonds, then there must also be one unshared pair of electrons.

b. $:N\equiv N:$ Two pi and one sigma bond indicates the presence of a triple bond.

c. $\overset{\cdot\cdot}{O}=\overset{\cdot\cdot}{O}$ One pi and one sigma bond indicates a double bond is present.

d. $\underset{H}{\overset{H}{\diagdown}}C=C\underset{H}{\overset{H}{\diagup}}$ The three sp^2 hybrid orbitals indicate that the carbon is bonded to 3 other atoms. The remaining p-orbital must be involved in a pi bond. Thus the carbon atom will have 2 single bonds, and 1 double bond.

e. $:\overset{\cdot\cdot}{F}-Xe-\overset{\cdot\cdot}{F}:$ The sp^{3}d^2 hybrid indicates that Xe has 6 electron pairs around it. Since 2 are bonding pairs, the other 4 must be unshared pairs.

The examples given in each of the problems above are but one of many possibilities.

By calculating the moles of Cl and the total moles of F in the products, one can determine the mole ratio of Cl to F and thus the formula of ClF$_x$. Then one can proceed to address the questions of geometry and polarity.

$$n = \frac{PV}{RT} = \frac{(3.00\,\text{atm})(0.457\,\text{L})}{(0.0821\,\text{L}\cdot\text{atm/mol}\cdot\text{K})(348\,\text{K})} = 0.0480\,\text{mol. Cl}$$

$$5.60\,\text{g UF}_6 \times \frac{1\,\text{mol. UF}_6}{352\,\text{g UF}_6} \times \frac{6\,\text{mol. F}}{1\,\text{mol. UF}_6} = 0.0960\,\text{mol. F}$$

moles Cl = 0.0480 moles.

moles F = 0.0480 + 0.0960 = 0.1440 moles.

$$\frac{0.1440 \, \text{mol. F}}{0.0480 \, \text{mol. Cl}} = 3 \, \text{mol. F/1 mol. Cl}$$

Thus, $x = 3$ ClF_3

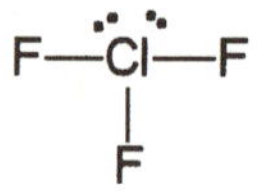

F—Cl—F
|
F

Geometry: AX_3E_2 **T-shaped**.

This molecule is **polar**.

Bond angles are about **90°** and **180°**.

Cl: **sp³d** hybridized

3 σ bonds, **0** π bonds.

---

79. *Refer to Sections 7.1, 7.2 and 7.4.*

Lewis Structure:

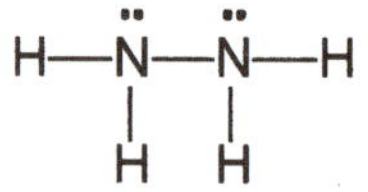

H—N—N—H
   | |
   H   H

Each N is AX_3E, and thus would be **triangular pyramid**. Due to the lone pairs of electrons on the nitrogens, the bond angles will be slightly less than 109.5°.

This molecule is not polar, since the dipoles cancel one another, as can be seen from the perspective drawing at right.

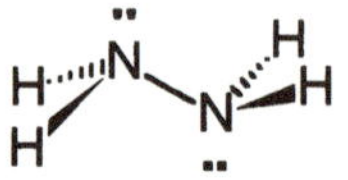

---

80. *Refer to Sections 7.1 - 7.4.*

There are 6 groups around the central iodine atom.

The hybridization sp^3d^2.

The geometry is octahedron.

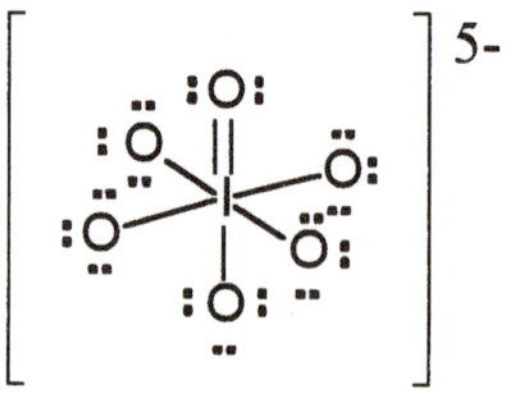

136

Geometry:	AX$_4$: tetrahedron	AX$_4$: tetrahedron
Hybridization:	sp^3	sp^3
C_f(S):	6 - (0 + ½(8)) = 2	6 - (0 + ½(12)) = 0
C_f(O) (with single bond):	6 - (6 + ½(2)) = -1	6 - (6 + ½(2)) = -1
C_f(O) (with double bond):		6 - (4 + ½(4)) = 0

Since the atoms in the second structure have charges closer to zero, it is the better structure.

C_f(P):	5 - (0 + 4) = +1	5 - (0 + 5) = 0
C_f(O):	6 - (6 + 1) = -1	6 - (4 + 2) = 0
C_f(Cl):	7 - (6 + 1) = 0	7 - (6 + 1) = 0

Since the second structure has no formal charges, it is the better structure.

Chapter 8: Thermochemistry

---

2. Refer to Section 8.1 and Example 8.1.

$q = mc\Delta t$

$4.78 \text{ J} = (5.88 \text{ g})(0.523 \text{ J/g°C})(\Delta t)$ ← NoT Right? In this formula $\frac{q}{mc} = t$

$\Delta t = 1.55 \text{ °C}$

---

4. Refer to Section 8.1 and Example 8.1.

Convert the temperature to Celsius, calculate the change in temperature, then calculate q.

$$t(\text{°C}) = \frac{375 - 32}{1.8} = 191\text{°C}$$

$\Delta t = t_f - t_i = 191\text{°C} - 23.00\text{°C} = 168\text{°C}$

$q = mc\Delta t$

$q = (473 \text{ g})(0.902 \text{ J/g°C})(168\text{°C})$

$q = 71700 \text{ J} = 71.7 \text{ kJ}$

---

6. Refer to Sections 8.1 and 8.2 and Example 8.2

a. $KBr(s) \rightarrow K^+(aq) + Br^-(aq)$

b. The temperature of the water drops, therefore heat must be absorbed during the solution process. When heat flows from the surroundings (water) to the system (KBr) the reaction is **endothermic**.

c. $\Delta t = t_f - t_i = 17.279\text{°C} - 18.000\text{°C} = -0.721\text{°C}$

$q = -mc\Delta t = -(74.0 \text{ g})(4.18 \text{ J/g°C})(-0.721\text{°C})$

$q = 223 \text{ J}$

d. Part (c) indicates that 223 J is absorbed per 1.34 g KBr, use this as a conversion factor.

$$1 \text{ mol. KBr} \times \frac{119.002 \text{ g KBr}}{1 \text{ mol. KBr}} \times \frac{223 \text{ J}}{1.34 \text{ g KBr}} = 19800 \text{ J} = 19.8 \text{ kJ}$$

8. **Refer to Section 8.3 and Example 8.3.**

a. $q_{reaction} = -C_{cal} \times \Delta t$
   $q_{reaction} = -(2.115 \times 10^4 \text{ J/°C})(27.71°C - 23.49°C)$
   $q_{reaction} = -8.93 \times 10^4 \text{ J}$
   $q_{cal} = +8.93 \times 10^4 \text{ J}$

b. from part a: $q_{reaction} = -8.93 \times 10^4 \text{ J per 4.50 g fructose}$

c. $1 \text{ mol. fructose} \times \dfrac{180.156 \text{ g fructose}}{1 \text{ mol. fructose}} \times \dfrac{-8.93 \times 10^4 \text{ J}}{4.50 \text{ g fructose}} = -3.58 \times 10^6 \text{ J}$

   $= -3.58 \times 10^3 \text{ kJ}$

10. **Refer to Section 8.3.**

Calculate $q_{reaction}$ for 5.00 g caffeine using 1 mol. caffeine per 4.96×10^3 kJ as a conversion factor. Then use $q_{reaction}$ to calculate C_{cal}.

Since heat is evolved, q is negative. $q_{reaction} = -4.96 \times 10^3 \text{ kJ}$

$5.00 \text{ g caffeine} \times \dfrac{1 \text{ mol. caffeine}}{194.20 \text{ g caffeine}} \times \dfrac{-4.96 \times 10^3 \text{ kJ}}{1 \text{ mol. caffeine}} = -128 \text{ kJ}$

$q_{reaction} = -C_{cal} \times \Delta t$

$-128 \text{ kJ} = -C_{cal} \times 11.37°C$

$C_{cal} = 11.2 \text{ kJ/°C} = 1.12 \times 10^4 \text{ J/°C}$

12. **Refer to Section 8.3.**

Since heat is evolved, q is negative. $q_{reaction} = -1453 \text{ kJ}$

$2.00 \text{ mL CH}_3\text{OH} \times \dfrac{0.791 \text{ g CH}_3\text{OH}}{1 \text{ mL CH}_3\text{OH}} \times \dfrac{1 \text{ mol. CH}_3\text{OH}}{32.04 \text{ g CH}_3\text{OH}} \times \dfrac{-1453 \text{ kJ}}{1 \text{ mol. CH}_3\text{OH}} = -71.7 \text{ kJ}$

$q_{reaction} = -C_{cal} \times \Delta t$

$-71.7 \text{ kJ} = -1.231 \text{ kJ/°C} \times \Delta t$

$\Delta t = 58.2°C = 80.1°C - t_{initial}$

$t_{initial} = 21.9°C$

$$C_{cal} = \frac{q_{cal}}{\Delta t_{cal}} = \frac{9.37\,\text{kJ}}{2.48°C} = 3.78\,\text{kJ/°C}$$

$$q_{reaction} = -C_{cal} \times \Delta t = -(3.78\,\text{kJ/°C})(28.91°C - 23.11°C) = -21.9\,\text{kJ}$$

$$1\,\text{mol. } C_7H_6O_3 \times \frac{138.12\,\text{g } C_7H_6O_3}{1\,\text{mol. } C_7H_6O_3} \times \frac{-21.9\,\text{kJ}}{1.00\,\text{g } C_7H_6O_3} = -3.03 \times 10^3\,\text{kJ}$$

Thus **3.03 x 10³ kJ** of heat are released when one mole of salicylic acid is burned.

a. $CaO(s) + 3C(s) \rightarrow CO(g) + CaC_2(s)$ $\Delta H = +464.8\,\text{kJ}$

b.   Heat is absorbed, therefore the reaction is **endothermic**.

c.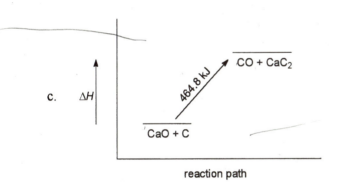

The enthalpy of the products is higher than the enthalpy of the reactants.

d.   The formation of one mole of CaC_2 consumes 464.8 kJ, therefore:

$$1.00\,\text{g } CaC_2 \times \frac{1\,\text{mol. } CaC_2}{64.098\,\text{g } CaC_2} \times \frac{464.8\,\text{kJ}}{1\,\text{mol. } CaC_2} = 7.25\,\text{kJ}$$

e.   From the balanced equation, we know there are 3 mol. C per mole CaC_2, thus:

$$20.00\,\text{kJ} \times \frac{1\,\text{mol. } CaC_2}{464.8\,\text{kJ}} \times \frac{3\,\text{mol. C}}{1\,\text{mol. } CaC_2} \times \frac{12.01\,\text{g C}}{1\,\text{mol. C}} = 1.550\,\text{g C}$$

a.   This is the reverse of the reaction shown, so the change in enthalpy will be the same, but with opposite sign. $\Delta H = -55.8\,\text{kJ}$.

b. One mole of H_2O yields -55.8 kJ, therefore:

$$1.00 \text{ g } H_2O \times \frac{1 \text{ mol. } H_2O}{18.02 \text{ g } H_2O} \times \frac{-55.8 \text{ kJ}}{1 \text{ mol. } H_2O} = -3.10 \text{ kJ}$$

---

20. Refer to Section 8.4 and Chapter 5.

After balancing the reaction, use the ideal gas equation to calculate the moles of oxygen, then calculate ΔH.

a. $Sr(s) + C(s) + \frac{3}{2}O_2(g) \rightarrow SrCO_3(s)$ $\Delta H = -1.220 \times 10^3$ kJ

b. $n = \dfrac{PV}{RT} = \dfrac{(1 \text{ atm})(10.00 \text{ L})}{(0.0821 \text{ L} \cdot \text{atm/mol.} \cdot \text{K})(298 \text{ K})} = 0.409 \text{ mol. } O_2$

$$0.409 \text{ mol. } O_2 \times \frac{1 \text{ mol. } SrCO_3}{\frac{3}{2} \text{ mol. } O_2} \times \frac{-1.220 \times 10^3 \text{ kJ}}{1 \text{ mol. } SrCO_3} = -333 \text{ kJ}$$

---

22. Refer to Sections 8.1 - 8.4.

Calculate the amount of heat needed to raise the temperature of the water 3°C. Recall that the heat absorbed by the water will be equal to the heat released by the fat. Finally, use the enthalpy of the reaction to calculate the amount of fat needed to produce the required heat.

$C_{57}H_{104}O_6(s) + 80O_2(g) \rightarrow 57CO_2(g) + 52H_2O(l)$ $\Delta H = -3.022 \times 10^4$ kJ/mol.

100.0 mL $H_2O = 100.0$ g H_2O

$q_{H_2O} = (100.0 \text{ g})(4.18 \text{ J/g°C})(25.00°C - 22.00°C) = 1254 \text{ J}$

$q_{fat} = -q_{H_2O} = -1254 \text{ J}$

$$-1254 \text{ J} \times \frac{1 \text{ kJ}}{1000 \text{ J}} \times \frac{1 \text{ mol. fat}}{-3.022 \times 10^4 \text{ kJ}} \times \frac{885.4 \text{ g fat}}{1 \text{ mol. fat}} = 3.674 \times 10^{-2} \text{ g fat}$$

---

24. Refer to Section 8.4 and Table 8.2.

Calculate the amount of heat evolved by each process and compare them.

$Br_2(l) \rightarrow Br_2(s)$ $\Delta H = -10.8$ kJ/mol.

$$100.0 \text{ g } Br_2 \times \frac{1 \text{ mol. } Br_2}{159.81 \text{ g } Br_2} \times \frac{-10.8 \text{ kJ}}{1 \text{ mol. } Br_2} = -6.76 \text{ kJ}$$

$$H_2O(g) \rightarrow H_2O(l) \qquad\qquad \Delta H = \text{-40.7 kJ/mol.}$$

$$100.0\,g\,H_2O \times \frac{1\,mol.\,H_2O}{18.02\,g\,H_2O} \times \frac{-40.7\,kJ}{1\,mol.\,H_2O} = -226\,kJ$$

Thus, condensing water evolves more heat than freezing an equal volume of bromine.

---

26. Refer to Section 8.4 and Table 8.2.

Calculate the heat evolved in condensing benzene gas to liquid, then the amount evolved in cooling the liquid to 25.00°C.

$$C_6H_6\,(g,\,80.00°C) \rightarrow C_6H_6\,(l,\,80.00°C) \qquad\qquad \Delta H = \text{-30.8 kJ/mol.}$$

$$\Delta H_{vap} = 100.00\,g \times \frac{1\,mol.\,C_6H_6}{78.1\,g\,C_6H_6} \times \frac{-30.8\,kJ}{1\,mol.} = -39.4\,kJ$$

$$C_6H_6\,(l,\,80.00°C) \rightarrow C_6H_6\,(l,\,25.00°C)$$

$q = mc\Delta t = (100.00\ g)(1.72\ J/g°C)(25.00°C - 80.00°C) = \text{-9460 J = -9.46 kJ}$

$\Delta H = \Delta H\ \text{(condensation)} + q\ \text{(cooling)}$

$\Delta H = (\text{-39.4 kJ}) + (\text{-9.46 kJ}) = \text{-48.9 kJ}$

---

28. Refer to Section 8.4.

Combine the two reactions given to get the desired, overall reaction. The ΔH for the reaction will be the sum of the combined reactions.

$PbS(s) + {}^3/_2O_2(g) \rightarrow PbO(s) + SO_2(g)$	$\Delta H = \text{-415.4 kJ/mol.}$
$\underline{PbO(s) + C(s) \rightarrow Pb(s) + CO(g)}$	$\underline{\Delta H = \text{+108.5 kJ/mol.}}$
$PbS(s) + {}^3/_2O_2(g) + C(s) \rightarrow Pb(s) + CO(g) + SO_2(g)$	$\Delta H = \text{-309.6 kJ/mol.}$

---

30. Refer to Section 8.4.

Combine the reactions given to get the desired, overall reaction. The ΔH for the reaction will be the sum of the combined reactions. Remember that when a reaction is reversed, the sign for ΔH changes and if the reaction is multiplied through by some factor, ΔH must be multiplied by that same factor.

$$2HNO_3(l) \rightarrow N_2O_5(g) + H_2O(l) \qquad\qquad \Delta H = +73.7 \text{ kJ/mol.}$$
$$\underline{2[\tfrac{1}{2}N_2(g) + \tfrac{3}{2}O_2(g) + \tfrac{1}{2}H_2(g) \rightarrow HNO_3(l)] \qquad\qquad \Delta H = 2(-174.1 \text{ kJ/mol.})}$$
$$2HNO_3(l) + N_2(g) + 3O_2(g) + H_2(g) \rightarrow 2HNO_3(l) + N_2O_5(g) + H_2O(l)$$
$$N_2(g) + 3O_2(g) + H_2(g) \rightarrow N_2O_5(g) + H_2O(l) \qquad\qquad \Delta H = -274.5 \text{ kJ/mol.}$$

$$N_2(g) + 3O_2(g) + H_2(g) \rightarrow N_2O_5(g) + H_2O(l) \qquad\qquad \Delta H = -274.5 \text{ kJ/mol.}$$
$$\underline{\tfrac{1}{2}[2H_2O(l) \rightarrow 2H_2(g) + O_2(g)] \qquad\qquad \Delta H = \tfrac{1}{2}(+571.6 \text{ kJ/mol.})}$$
$$N_2(g) + \tfrac{5}{2}O_2(g) \rightarrow N_2O_5(g) \qquad\qquad \Delta H = 11.3 \text{ kJ/mol.}$$

---

32. Refer to Section 8.5 and Table 8.3.

a. $N_2(g) + 2O_2(g) \rightarrow N_2O_4(g)$ $\qquad\qquad \Delta H = \Delta H_f^\circ = +9.2 \text{ kJ/mol.}$

b. $Ca(s) + S(g) + 2O_2(g) \rightarrow CaSO_4(s)$ $\qquad \Delta H = \Delta H_f^\circ = -1434.1 \text{ kJ/mol.}$

c. $Ag(s) + \tfrac{1}{2}Cl_2(g) \rightarrow AgCl(s)$ $\qquad\qquad \Delta H = \Delta H_f^\circ = -127.1 \text{ kJ/mol.}$

d. $\tfrac{1}{2}H_2(g) + \tfrac{1}{2}I_2(s) \rightarrow HI(g)$ $\qquad\qquad \Delta H = \Delta H_f^\circ = +26.5 \text{ kJ/mol.}$

---

34. Refer to Sections 8.4 and 8.5.

The heat of formation is defined for the formation of one mole of a substance from the elements in stable states. The equation given is for the decomposition of two moles of Al_2O_3. Thus, reverse the equation and multiply through by one half ($\tfrac{1}{2}$). Then change the sign of ΔH°, and multiply it also by one half. Finally, calculate moles of Al_2O_3 in 12.5 g, then ΔH°.

a. $2Al(s) + \tfrac{3}{2}O_2(g) \rightarrow Al_2O_3(s)$ $\qquad\qquad \Delta H^\circ = -1675.7 \text{ kJ/mol.}$

b. $\Delta H^\circ = 12.50 \text{ g } Al_2O_3 \times \dfrac{1 \text{ mol. } Al_2O_3}{101.96 \text{ g } Al_2O_3} \times \dfrac{-1675.7 \text{ kJ}}{1 \text{ mol. } Al_2O_3} = -205.4 \text{ kJ}$

---

36. Refer to Section 8.5 and Chapter 5.

Calculate the amount of energy released by one mole of NH_3, then determine the number of moles of NH_3 in one liter at the given conditions. Finally calculate the amount of energy released by the one liter of ammonia.

$$NH_3(g) + \tfrac{5}{4}O_2(g) \rightarrow NO(g) + \tfrac{3}{2}H_2O(g)$$

$$\Delta H_{reaction} = \Sigma \Delta H_f^\circ \text{ (products)} - \Sigma \Delta H_f^\circ \text{ (reactants)}$$
$$= [(1 \text{ mol.})(90.2 \text{ kJ/mol.}) + (\tfrac{3}{2} \text{ mol.})(-241.8 \text{ kJ/mol.})]$$
$$- [(1 \text{ mol.})(-46.1 \text{ kJ/mol.}) + (\tfrac{5}{4} \text{ mol.})(0.0 \text{ kJ/mol.})]$$
$$= -226.4 \text{ kJ}$$

$$PV = nRT \Rightarrow n = \frac{PV}{RT} = \frac{(1.01\,\text{atm})(1.00\,\text{L})}{(0.0821\,\text{L} \cdot \text{atm/mol.} \cdot \text{K})(296\,\text{K})} = 0.0416\,\text{mol.}$$

$$\Delta H = \frac{-226.4\,\text{kJ}}{1\,\text{mol.}} \times 0.0416\,\text{mol.} = -9.42\,\text{kJ}$$

Thus **9.42 kJ** of heat are evolved.

---

38. *Refer to Section 8.5, Example 8.8 and Table 8.3.*

a. $\Delta H_{\text{reaction}} = \Sigma\,\Delta H_f^o\,_{(\text{products})} - \Sigma\,\Delta H_f^o\,_{(\text{reactants})}$
    $\Delta H_{\text{reaction}} = [2\Delta H_f^o\,\text{Cl}^-(aq) + 2\Delta H_f^o\,\text{H}_2\text{O}(l)] - [\Delta H_f^o\,\text{Cl}_2(g) + \Delta H_f^o\,\text{H}_2(g) + 2\Delta H_f^o\,\text{OH}^-(aq)]$
    $\Delta H_{\text{reaction}} = [(2\,\text{mol.})(-167.7\,\text{kJ/mol.}) + (2\,\text{mol.})(-285.8\,\text{kJ/mol.})]$
    $\qquad\qquad - [(1\,\text{mol.})(0.0\,\text{kJ/mol.}) + (1\,\text{mol.})(0.0\,\text{kJ/mol.}) + (2\,\text{mol.})(-230.0\,\text{kJ/mol.})]$
    $\Delta H_{\text{reaction}} = -446.0\,\text{kJ}$

b. $\Delta H_{\text{reaction}} = \Sigma\,\Delta H_f^o\,_{(\text{products})} - \Sigma\,\Delta H_f^o\,_{(\text{reactants})}$
    $\Delta H_{\text{reaction}} = [5\Delta H_f^o\,\text{NO}_3^-\,(aq) + 3\Delta H_f^o\,\text{Mn}^{2+}\,(aq) + 2\Delta H_f^o\,\text{H}_2\text{O}\,(l)] -$
    $\qquad\qquad [5\Delta H_f^o\,\text{NO}(g) + 3\Delta H_f^o\,\text{MnO}_4^-(aq) + 4\Delta H_f^o\,\text{H}^+(aq)]$
    $\Delta H_{\text{reaction}} = [(5\,\text{mol.})(-205.0\,\text{kJ/mol.}) + (3\,\text{mol.})(-220.8\,\text{kJ/mol.})$
    $\qquad\qquad + (2\,\text{mol.})(-285.8\,\text{kJ/mol.})]$
    $\qquad\qquad - [(5\,\text{mol.})(+90.2\,\text{kJ/mol.}) + (3\,\text{mol.})(-541.4\,\text{kJ/mol.}) + (4\,\text{mol.})(0.0\,\text{kJ/mol.})]$
    $\Delta H_{\text{reaction}} = -1085.8\,\text{kJ}$

c. $\Delta H_{\text{reaction}} = \Sigma\,\Delta H_f^o\,_{(\text{products})} - \Sigma\,\Delta H_f^o\,_{(\text{reactants})}$
    $\Delta H_{\text{reaction}} = [\Delta H_f^o\,\text{Cr}(s) + 6\Delta H_f^o\,\text{Fe}^{+3}(aq) + 4\Delta H_f^o\,\text{H}_2\text{O}(l)] -$
    $\qquad\qquad [6\Delta H_f^o\,\text{Fe}^{+2}(aq) + \Delta H_f^o\,\text{CrO}_4^{-2}(aq) + 8\Delta H_f^o\,\text{H}^+(aq)]$
    $\Delta H_{\text{reaction}} = [(1\,\text{mol.})(0.0\,\text{kJ/mol.}) + (6\,\text{mol.})(-48.5\,\text{kJ/mol.}) + (4\,\text{mol.})(-285.5\,\text{kJ/mol.})]$
    $\qquad\qquad - [(6\,\text{mol.})(-89.1\,\text{kJ/mol.}) + (1\,\text{mol.})(-881.2\,\text{kJ/mol.}) + (8\,\text{mol.})(0.0\,\text{kJ/mol.})]$
    $\Delta H_{\text{reaction}} = -18.4\,\text{kJ}$

---

40. *Refer to Section 8.5.*

a. $\text{CuO}(s) + \text{CO}(g) \rightarrow \text{Cu}(s) + \text{CO}_2(g)$
    $\Delta H_{\text{reaction}} = \Sigma\,\Delta H_f^o\,_{(\text{products})} - \Sigma\,\Delta H_f^o\,_{(\text{reactants})}$
    $\Delta H_{\text{reaction}} = [\Delta H_f^o\,\text{Cu}(s) + \Delta H_f^o\,\text{CO}_2(g)] - [\Delta H_f^o\,\text{CuO}(s) + \Delta H_f^o\,\text{CO}(g)]$
    $\Delta H_{\text{reaction}} = [(1\,\text{mol.})(0.0\,\text{kJ/mol.}) + (1\,\text{mol.})(-393.5\,\text{kJ/mol.})]$
    $\qquad\qquad - [(1\,\text{mol.})(-157.3\,\text{kJ/mol.}) + (1\,\text{mol.})(-110.5\,\text{kJ/mol.})]$
    $\Delta H_{\text{reaction}} = -125.7\,\text{kJ}$

b. $CH_3OH(l) \rightarrow CH_4(g) + \frac{1}{2}O_2(g)$

$\Delta H_{reaction} = \Sigma\ \Delta H_f^\circ\ _{(products)} - \Sigma\ \Delta H_f^\circ\ _{(reactants)}$

$\Delta H_{reaction} = [\Delta H_f^\circ\ CH_4(g) + \frac{1}{2}\Delta H_f^\circ\ O_2(g)] - [\Delta H_f^\circ\ CH_3OH(l)]$

$\Delta H_{reaction} = [(1\ mol.)(-74.8\ kJ/mol.) + (\frac{1}{2}\ mol.)(0.0\ kJ/mol.)] - [(1\ mol.)(-238.7\ kJ/mol.)]$

$\Delta H_{reaction} = +163.9\ kJ$

---

42. Refer to Section 8.5 and Example 8.9.

a. $CaCO_3(s) + 2NH_3(g) \rightarrow CaCN_2(s) + 3H_2O(l)$ $\Delta H = 90.1\ kJ/mol.$

b. $\Delta H_{reaction} = \Sigma\ \Delta H_f^\circ\ _{(products)} - \Sigma\ \Delta H_f^\circ\ _{(reactants)}$

$\Delta H_{reaction} = [\Delta H_f^\circ\ CaCN_2(s) + 3\Delta H_f^\circ\ H_2O(l)] - [\Delta H_f^\circ\ CaCO_3(s) + 2\Delta H_f^\circ\ NH_3(g)]$

$90.1\ kJ = [(1\ mol.)(\Delta H_f^\circ\ CaCN_2\ (s)) + (3\ mol.)(-285.8\ kJ/mol.)]$

$\qquad\qquad - [(1\ mol.)(-1206.9\ kJ/mol.) + (2\ mol.)(-46.1\ kJ/mol.)]]$

$\Delta H_f^\circ\ CaCN_2(s) = -351.6\ kJ/mol.$

---

44. Refer to Section 8.5 and Example 8.9.

$\Delta H_{reaction} = \Sigma\ \Delta H_f^\circ\ _{(products)} - \Sigma\ \Delta H_f^\circ\ _{(reactants)}$

$\Delta H_{reaction} = [2\Delta H_f^\circ\ Al^{+3}(aq) + \Delta H_f^\circ\ Cr_2O_3(s) + 4\Delta H_f^\circ\ H_2O(l)] -$

$\qquad\qquad [8\Delta H_f^\circ\ H^+(aq) + \Delta H_f^\circ\ Cr_2O_7^{-2}(aq) + 2\Delta H_f^\circ\ Al(s)]$

$-1854.9\ kJ = [(2\ mol.)(-531.0\ kJ/mol.) + (1\ mol.)(\Delta H_f^\circ\ Cr_2O_3(s)) + (4\ mol.)(-285.8\ kJ/mol.)]$

$\qquad - [(8\ mol.)(0.0\ kJ/mol.) + (1\ mol.)(-1490.3\ kJ/mol.) + (2\ mol.)(0.0\ kJ/mol.)]$

$\Delta H_f^\circ\ Cr_2O_3(s) = -1140.0\ kJ/mol.$

---

46. Refer to Section 8.5.

Write a balanced thermochemical equation. Then calculate the amount of heat liberated by one mole of glucose. Calculate the amount of ethyl alcohol in 750.0 mL of wine, and the amount of heat liberated in producing that quantity of ethyl alcohol.

$C_6H_{12}O_6(s) \rightarrow 2C_2H_5OH(l) + 2CO_2\ (g)$

$\Delta H_{reaction} = \Sigma\ \Delta H_f^\circ\ _{(products)} - \Sigma\ \Delta H_f^\circ\ _{(reactants)}$

$\Delta H_{reaction} = [2\Delta H_f^\circ\ C_2H_5OH(l) + 2\Delta H_f^\circ\ CO_2(g)] - [\Delta H_f^\circ\ C_6H_{12}O_6(s)]$

$\Delta H_{reaction} = [(2\ mol.)(-277.7\ kJ/mol.) + (2\ mol.)(-393.5\ kJ/mol.)] - [(1\ mol.)(-1275.2\ kJ/mol.)]$

$\Delta H_{reaction} = -67.2\ kJ$

$$750.0\ mL\ wine \times \frac{12.0\ mL\ C_2H_5OH}{100\ mL\ wine} \times \frac{1\ cm^3}{1\ mL} \times \frac{0.789\ g\ C_2H_5OH}{1\ cm^3\ C_2H_5OH} = 71.01\ g\ C_2H_5OH$$

$$71.01 \text{ g } C_2H_5OH \times \frac{1 \text{ mol. } C_2H_5OH}{46.07 \text{ g } C_2H_5OH} \times \frac{-67.2 \text{ kJ}}{2 \text{ mol. } C_2H_5OH} = -51.8 \text{ kJ}$$

Thus **51.8 kJ** of heat are evolved.

---

48. Refer to Section 8.7 and Example 8.11.

$$12.2 \text{ kJ} \times \frac{1 \text{ L} \cdot \text{atm}}{0.1013 \text{ kJ}} = 120 \text{ L} \cdot \text{atm}$$

---

50. Refer to Section 8.7 and Example 8.11.

a. $\Delta E = q + w$
   $\Delta E = 18 \text{ J} + 13 \text{ J} = 31 \text{ J}$

b. $\Delta E = q + w$
   $+61 \text{ J} = q + 72 \text{ J}$
   $q = -11 \text{ J}$

---

52. Refer to Sections 8.5 and 8.7 and Chapter 5.

$H_2O(l) \rightarrow H_2O(g)$

a. $\Delta H_{vap} = 40.7 \text{ kJ}$ *(from Table 8.2)*

b. $\Delta PV = PV_{products} - PV_{reactants}$ (assume $P = 1.00$ atm)

$$V_{reactants} = 1 \text{ mol. } H_2O \times \frac{18.02 \text{ g } H_2O}{1 \text{ mol. } H_2O} \times \frac{1 \text{ mL } H_2O}{1.00 \text{ g } H_2O} \times \frac{1 \text{ L}}{1000 \text{ mL}} = 0.01802 \text{ L } H_2O$$

$PV_{reactants} = 1.00 \text{ atm} \times 0.01802 \text{ L} = 0.01802 \text{ L} \cdot \text{atm}$

$PV_{products} = nRT = (1.00 \text{ mol.})(0.0821 \text{ L} \cdot \text{atm/mol.} \cdot \text{K})(373 \text{ K}) = 30.6 \text{ L} \cdot \text{atm}$

$\Delta PV = 30.6 \text{ L} \cdot \text{atm} - 0.0821 \text{ L} \cdot \text{atm} = 30.6 \text{ L} \cdot \text{atm}$
   (Note that PV for the liquid is so small relative to that of the gas it could be ignored.)

$$30.6 \text{ L} \cdot \text{atm} \times \frac{0.1013 \text{ kJ}}{1 \text{ L} \cdot \text{atm}} = 3.10 \text{ kJ}$$

c. $\Delta H = \Delta E + \Delta(PV)$
   $40.7 \text{ kJ} = \Delta E + 3.10 \text{ kJ}$
   $\Delta E = 37.6 \text{ kJ}$

54. Refer to Section 8.5 and 8.7.

a. $C_3H_8(g) + 5O_2(g) \rightarrow 3CO_2(g) + 4H_2O(l)$

$\Delta H^\circ = [3\Delta H_f^\circ\ CO_2(g) + 4\Delta H_f^\circ\ H_2O(l)] - [\Delta H_f^\circ\ C_3H_8(g) + 5\Delta H_f^\circ\ O_2(g)]$

$\Delta H^\circ = [(3\ \text{mol.})(-393.5\ \text{kJ/mol.}) + (4\ \text{mol.})(-285.5\ \text{kJ/mol.})]$
$\quad\quad - [(1\ \text{mol.})(-103.8\ \text{kJ/mol.}) + (5\ \text{mol.})(0.0\ \text{kJ/mol.})]$

$\Delta H^\circ = -2219.9\ \text{kJ}$

b. $\Delta H = \Delta E + \Delta(PV)$

$\Delta(PV) = \Delta nRT$

$\quad\quad\quad \Delta n = n_{products} - n_{reactants} = 3 - (5 + 1) = -3$ *(this applies to the gases only.)*

$\Delta(PV) = -3(0.0821\ \text{L·atm/mol.·K})(298\ \text{K}) = -73.4\ \text{L·atm}$

$\Delta PV = -73.4\ \text{L·atm} \times \dfrac{0.1013\ \text{kJ}}{1\ \text{L·atm}} = -7.44\ \text{kJ}$

$-2219.9\ \text{kJ} = \Delta E + (-7.44\ \text{kJ})$

$\Delta E = -2212.5\ \text{kJ}$

56. Refer to Section 8.1.

a. $3\ \text{hr} \times \dfrac{60\ \text{min}}{1\ \text{hr}} \times \dfrac{60\ \text{sec}}{1\ \text{min}} = 10800\ \text{s}$

$q_{halogen} = 75\ \text{W} \times 10800\ \text{s} = 8.1 \times 10^5\ \text{W·s} = 8.1 \times 10^5\ \text{J}$
$q_{fluorescent} = 20\ \text{W} \times 10800\ \text{s} = 2.2 \times 10^5\ \text{W·s} = 2.2 \times 10^5\ \text{J}$

b. Calculate the volume and then mass of air in the room. Then use the numbers calculated above to determine the change in temperature of the room.

$13 \times 15 \times 8.0 = 1560\ \text{ft}^3$

$1560\ \text{ft}^3 \times \dfrac{(12\ \text{in})^3}{(1\ \text{ft})^3} \times \dfrac{(2.54\ \text{cm})^3}{(1\ \text{in})^3} \times \dfrac{1\ \text{mL}}{1\ \text{cm}^3} \times \dfrac{1\ \text{L}}{1000\ \text{mL}} = 4.4 \times 10^4\ \text{L}$

$4.4 \times 10^4\ \text{L} \times \dfrac{1.20\ \text{g}}{1\ \text{L}} = 5.3 \times 10^4\ \text{g}$

$q = mc\Delta t$

$q_{halogen} = 8.1 \times 10^5\ \text{J} = (5.3 \times 10^4\ \text{g})(1.007\ \text{J/g°C})\Delta t$
$\Delta t = 15°\text{C}$

$q_{fluorescent} = 2.2 \times 10^5\ \text{J} = (5.3 \times 10^4\ \text{g})(1.007\ \text{J/g°C})\Delta t$
$\Delta t = 4.1°\text{C}$

58. Refer to Section 8.1 and Chemistry Beyond the Classroom..

$$120 \text{ kcal} \times \frac{1 \text{ hr}}{250 \text{ kcal}} \times \frac{60 \text{ min}}{1 \text{ hr}} = 29 \text{ min}$$

60. Refer to Section 8.1, Table 1.3 and Chemistry Beyond the Classroom.

$$1 \text{ lb fat} \times \frac{453.6 \text{ g}}{1 \text{ lb}} \times \frac{32 \text{ kJ}}{1.00 \text{ g fat}} \times \frac{1 \text{ kcal}}{4.184 \text{ kJ}} \times \frac{0.5 \text{ hr}}{225 \text{ kcal}} = 7.7 \text{ hr}$$

62. Refer to Sections 8.1 and 8.2.

Calculate the volume and mass of the brass cube. The final temperature of the water will also be the final temperature of the brass cube. Also, the heat lost by the cube will equal the heat gained by the water. Set up the equations for each, set them equal to one another, and solve for t_f.

$$(22.00 \text{ mm})^3 \times \frac{(1 \text{ cm})^3}{(10 \text{ mm})^3} \times \frac{8.25 \text{ g}}{1 \text{ cm}^3} = 87.9 \text{ g brass}$$

$q_{brass} = -q_{water}$

87.9 g $\times$ 0.362 J/g°C $\times$ (t_f - 95.0°C) = -[20.0 g $\times$ 4.18 J/g°C $\times$ (t_f - 22.0°C)]

(31.8 J/°C)(t_f - 95.0°C) = -(83.6 J/°C)(t_f - 22.0°C)

31.8 J/°C(t_f) - 3021 J = -83.6 J/°C(t_f) - 1839 J

115.4 J/°C(t_f) = 4860 J

t_f = 42.1°C

64. Refer to Chapter 6 and Section 8.3.

$q_{reaction} = -mc\Delta t$

$q_{reaction} = -(350 \text{ g})(4.18 \text{ J/g·°C})(99.0°C - 23.0°C)$

$q_{reaction} = -1.11 \times 10^5 \text{ J}$

$$12.5 \text{ cm} \times \frac{1 \text{ m}}{100 \text{ cm}} = 0.125 \text{ m}$$

$$E_{(per \; photon)} = \frac{hc}{\lambda} = \frac{6.626 \times 10^{-34} \text{ J·s})(2.998 \times 10^8 \text{ m/s})}{0.125 \text{ m}} = 1.59 \times 10^{-24} \text{ J}$$

$$E_{(\text{per mole})} = \frac{1.59 \times 10^{-24} \text{ J}}{1 \text{ photon}} \times \frac{6.022 \times 10^{23} \text{ photons}}{1 \text{ mol.}} = 0.957 \text{ J/mol.}$$

$$1.11 \times 10^5 \text{ J} \times \frac{1 \text{ mol.}}{0.957 \text{ J}} = 1.16 \times 10^5 \text{ moles of photons}$$

---

66. Refer to Section 8.4.

Write the equations. Remember that changes which require the input of energy (such as melting and vaporization) will have a positive (+) ΔH, while those that release energy (such as freezing and condensing) will have a negative (-) ΔH.

a. $Hg(s) \rightarrow Hg(l)$ $\Delta H = +2.33$ kJ/mol.

b. $Br_2(l) \rightarrow Br_2(g)$ $\Delta H = +29.6$ kJ/mol.

c. $C_6H_6(l) \rightarrow C_6H_6(s)$ $\Delta H = -9.84$ kJ/mol.

d. $Hg(g) \rightarrow Hg(l)$ $\Delta H = -59.4$ kJ/mol

e. This phase change is not in the table and must be calculated with Hess's Law.

$C_{10}H_8(s) \rightarrow C_{10}H_8(l)$	$\Delta H = +43.3$ kJ/mol.
$C_{10}H_8(l) \rightarrow C_{10}H_8(g)$	$\Delta H = +19.3$ kJ/mol.
$C_{10}H_8(s) \rightarrow C_{10}H_8(g)$	$\Delta H = +62.6$ kJ/mol.

---

68. Refer to Section 8.7.

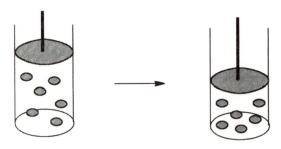

---

70. Refer to Section 8.6.

a. **True**. Heat is always absorbed when chemical bonds are broken.

b. **True**. Heat is always given off when bonds are formed.

c. **True**. Bond enthalpies are defined for atoms/molecules in the gaseous state.

d. **False**. Bond type does affect the geometry of a molecule. Furthermore, enthalpies do vary with bond type. ΔH for a π-bond is not the same for ΔH of a σ-bond.

e. **False**. As shown in Table 8.4, the bond enthalpy for a double bond (for example C=C) is not twice that of a single bond (for example C-C).

---

71. Refer to Section 8.3.

---

a. Mass of soda $= 6 \text{ cans} \times \dfrac{12.0 \text{ oz}}{1 \text{ can}} \times \dfrac{1 \text{ lb}}{16 \text{ oz}} \times \dfrac{454 \text{ g}}{1 \text{ lb}} = 2.04 \times 10^3 \text{ g}$

$q = -mc\Delta t$

$q_{\text{soda}} = -(2.04 \times 10^3 \text{ g})(4.10 \text{ J/g·°C})(5.0 \text{ °C} - 25.0 \text{ °C})$
$q_{\text{soda}} = 1.68 \times 10^5 \text{ J}$

$q_{\text{Al}} = -(6 \text{ cans} \times 12.5 \text{ g/can})(0.902 \text{ J/g·°C})(5.0 \text{ °C} - 25.0 \text{ °C})$
$q_{\text{Al}} = 1.35 \times 10^3 \text{ J}$

$q_{\text{total}} = q_{\text{soda}} + q_{\text{Al}} = 1.68 \times 10^5 \text{ J} + 1.35 \times 10^3 \text{ J} = 1.69 \times 10^5 \text{ J}$

b. $\Delta H_{\text{fus}} = 6.00 \text{ kJ/mol.} = 6.00 \times 10^3 \text{ J/mol.}$

$1.69 \times 10^5 \text{ J} \times \dfrac{1 \text{ mol. ice}}{6.00 \times 10^3 \text{ J}} \times \dfrac{18.02 \text{ g ice}}{1 \text{ mol. ice}} = 508 \text{ g ice}$

---

72. Refer to Section 8.3.

Some reasonable assumptions to make are:
    density of tea = density of H_2O = 1.00 g/mL
    density of ice = 1.0 g/mL (although, since ice floats, the density is actually less than 1)
    room temperature = 25°C
    volume of the glass = 480 mL (16 fl. oz.)
Then, let x = mass of ice and $(480 - x)$ = mass of tea.

q_{ice} = heat to melt ice + heat to warm ice to 25°C

$q_{\text{melt}} = x \text{ g } H_2O \times \dfrac{1 \text{ mol. } H_2O}{18.02 \text{ g } H_2O} \times \dfrac{6.00 \text{ kJ}}{1 \text{ mol } H_2O} \times \dfrac{1000 \text{ J}}{1 \text{ kJ}} = 333x \text{ J}$

$q_{\text{warming}} = mc\Delta t$

(note that the sign is positive since the water is warming, and thus gaining heat)
    $q_{\text{warming}} = (x \text{ g})(4.18 \text{ J/g·°C})(25 \text{ °C} - 0 \text{ °C})$

$q_{warming} = 105x$ J

$q_{ice} = 333x$ J $+ 105x$ J $= 438x$ J

$q_{tea} = -(480$ g $- x$ g$)(4.18$ J/g$\cdot$°C$)(0$ °C $- 25$ °C$)$
$q_{tea} = 50200 - 105x$

$q_{ice} = q_{tea}$
$438x$ J $= 50200$ J $- 105x$ J
$543x$ J $= 50200$ J
$x = 92.4$ g (of ice)
$(480 - x = 480 - 92.4 = 388$ g tea$)$

Since the densities of the tea and water (from the ice) equal 1.00 g/mL, density = mass and:

$$\text{fraction to be left empty} = \frac{92.4 \text{ mL}}{480 \text{ mL}} \times 100\% = 19.3\%$$

---

73. Refer to Section 8.5.

a. $\Delta H_{reaction} = \Sigma \, \Delta H_f^\circ \, _{(products)} - \Sigma \, \Delta H_f^\circ \, _{(reactants)}$
$= [(1 \text{ mol.})(-1675.7 \text{ kJ/mol.}) + (2 \text{ mol.})(0.0 \text{ kJ/mol.})]$
$- [(2 \text{ mol.})(0.0 \text{ kJ/mol.}) + (1 \text{ mol.})(-824.2 \text{ kJ/mol.})]$
$= -851.5$ kJ

b. First calculate the mass of Al_2O_3 and Fe for the reaction of 1 mole of Al_2O_3. Then calculate t_f, using $\Delta H_{reaction} = q_{Al_2O_3} + q_{Fe}$ and $q = -mc(\Delta t)$

$$1 \text{ mol. Al}_2O_3 \times \frac{101.96 \text{ g Al}_2O_3}{1 \text{ mol. Al}_2O_3} = 101.96 \text{ g Al}_2O_3$$

$$2 \text{ mol. Fe} \times \frac{55.85 \text{ g Fe}}{1 \text{ mol. Fe}} = 111.7 \text{ g Fe}$$

$\Delta H_{reaction} = -851.5$ kJ $= -8.515 \times 10^5$ J
$= -(101.96$ g $Al_2O_3)(0.77$ J/g$\cdot$°C$)(t_f - 25$ °C$) - (111.7$ g Fe$)(0.45$ J/g$\cdot$°C$)(t_f - 25$ °C$)$
$= -78.5t_f + 1963 - 50.3t_f + 1257$
-8.515×10^5 J $= -128.8t_f + 3220$
-8.547×10^5 J $= -128.8t_f$
$t_f = 6636$ °C

c. $111.7 \text{ g Fe} \times \dfrac{270 \text{ J}}{1 \text{ g}} = 30200 \text{ J} = 30.2 \text{ kJ}$

The 6636°C is much higher than the melting point of Fe, and the 851.5 kJ produced in the reaction (part a) is more than the 30.2 kJ needed to melt the Fe, so the reaction will definitely produce molten iron.

---

74. *Refer to Section 7.3.*

$q_{reaction} = -C_{cal}\Delta T$
$\qquad = -(22.51 \text{ kJ/°C})(1.67\text{°C})$
$\qquad = -37.59 \text{ kJ}$

$$-37.59\,\text{kJ} \times \frac{1\,\text{mol. sucrose}}{-5.64 \times 10^3\,\text{kJ}} \times \frac{342.30\,\text{g sucrose}}{1\,\text{mol. sucrose}} = 2.28\,\text{g sucrose}$$

$$\text{mass \% sucrose} = \frac{2.28\,\text{g}}{3.000\,\text{g}} \times 100\% = 76.0\%$$

Chapter 9: Liquids and Solids.

2. Refer to Section 9.1, Example 9.1 and Chapter 5.

$$P = 254 \text{ mm Hg} \times \frac{1 \text{ atm}}{760 \text{ mm Hg}} = 0.334 \text{ atm}$$

$$T = 57°C + 273 = 330 \text{ K}$$

a. $\dfrac{P_1 V_1}{n_1 T_1} = \dfrac{P_2 V_2}{n_2 T_2}$

Since neither volume nor moles change, V and n are constant and the equation becomes:

$$\frac{P_1}{T_1} = \frac{P_2}{T_2} \quad \text{and} \quad P_2 = \frac{P_1 T_2}{T_1}$$

at 35°C: $\qquad P_2 = \dfrac{(0.334 \text{ atm})(35 + 273)}{330 \text{ K}} = 0.312 \text{ atm}$

$$0.312 \text{ atm} \times \frac{760 \text{ mm Hg}}{1 \text{ atm}} = 237 \text{ mm Hg}$$

at 45°C: $\qquad P_2 = \dfrac{(0.334 \text{ atm})(45 + 273)}{330 \text{ K}} = 0.322 \text{ atm}$

$$0.322 \text{ atm} \times \frac{760 \text{ mm Hg}}{1 \text{ atm}} = 245 \text{ mm Hg}$$

b. At 35°C, the calculated vapor pressure is **greater than** the equilibrium vapor pressure of methyl alcohol.
At 45°C, the calculated vapor pressure is **less than** the equilibrium vapor pressure of methyl alcohol.

c. The pressure exerted by the methyl alcohol vapor will never exceed the vapor pressure, therefore, at 35°C, $P = 203$ mm Hg, and at 45°C, $P = 245$ mm Hg.

d. **At 35°C** the equilibrium vapor pressure is less than the calculated vapor pressure, thus **both liquid and vapor** will exist.
At 45°C the equilibrium vapor pressure is greater than the calculated vapor pressure, thus **only vapor** will exist.

4. *Refer to Section 9.1, Example 9.1 and Chapter 5.*

a. $P = 0.466 \text{ mm Hg} \times \dfrac{1 \text{ atm}}{760 \text{ mm Hg}} = 6.13 \times 10^{-4} \text{ atm}$

$30°C + 273 = 303 \text{ K}$

$PV = nRT$
$(6.13 \times 10^{-4} \text{ atm})(1.00 \text{ L}) = (n)(0.0821 \text{ L·atm/mol·K})(303 \text{ K})$
$n = 2.46 \times 10^{-5} \text{ mol. } I_2$

$2.46 \times 10^{-5} \text{ mol. } I_2 \times \dfrac{253.8 \text{ g}}{1 \text{ mol. } I_2} = 6.26 \times 10^{-3} \text{ g } I_2 = 6.26 \text{ mg } I_2$

b. $2.0 \text{ mg} \times \dfrac{1 \text{ g}}{1000 \text{ mg}} \times \dfrac{1 \text{ mol. } I_2}{253.8 \text{ g}} = 7.9 \times 10^{-6} \text{ mol. } I_2$

$PV = nRT$
$(P)(1.00 \text{ L}) = (7.9 \times 10^{-6} \text{ mol.})(0.0821 \text{ L·atm/mol·K})(303 \text{ K})$
$P = 2.0 \times 10^{-4} \text{ atm}$

$P = 2.0 \times 10^{-4} \text{ atm} \times \dfrac{760 \text{ mm Hg}}{1 \text{ atm}} = 0.15 \text{ mm Hg}$

c. There is more I_2 (10 mg) than will sublime (6.26 mg), thus there will be some unsublimed I_2, and P will equal the vapor pressure of I_2, 0.466 mm Hg.

6. *Refer to Section 9.1, Example 9.1 and Chapter 5.*

a. $10.00 \text{ mL} \times \dfrac{0.692 \text{ g}}{1 \text{ mL}} \times \dfrac{1 \text{ mol.}}{114.22 \text{ g}} = 0.0606 \text{ mol.}$

$P = 45.2 \text{ mm Hg} \times \dfrac{1 \text{ atm}}{760 \text{ mm Hg}} = 0.0595 \text{ atm}$

$PV = nRT$
$(0.0595 \text{ atm})(15.00 \text{ L}) = (n)(0.0821 \text{ L·atm/mol·K})(298 \text{ K})$
$n = 0.0365 \text{ mol.}$

There is more isooctane (0.0606 mol.) than will vaporize (0.0365 mol.), thus there will be left over liquid.

b. $PV = nRT$
$(0.0595 \text{ atm})(V) = (0.0606 \text{ mol.})(0.0821 \text{ L·atm/mol·K})(298 \text{ K})$
$V = 24.9 \text{ L}$

c. $PV = nRT$

$(P)(35.00 \text{ L}) = (0.0606 \text{ mol.})(0.0821 \text{ L·atm/mol.·K})(298 \text{ K})$

$P = 0.0424 \text{ atm}$

$$P = 0.0424 \text{ atm} \times \frac{760 \text{ mm Hg}}{1 \text{ atm}} = 32.2 \text{ mm Hg}$$

---

8. Refer to Section 9.1 and Example 9.2.

a. $$\ln\left(\frac{P_2}{P_1}\right) = \frac{+\Delta H_{vap}}{R}\left(\frac{1}{T_1} - \frac{1}{T_2}\right)$$

$$\ln\left(\frac{448 \text{ atm}}{197 \text{ atm}}\right) = \frac{+\Delta H_{vap}}{8.31 \text{ J/mol.·K}}\left(\frac{1}{296 \text{ K}} - \frac{1}{318 \text{ K}}\right)$$

$$0.822 = \frac{+\Delta H_{vap}}{8.31 \text{ J/mol.·K}}\left(2.34 \times 10^{-4} \text{ K}^{-1}\right)$$

$\Delta H_{vap} = 2.92 \times 10^4 \text{ J/mol.}$

$\Delta H_{vap} = 29.2 \text{ kJ/mol.}$

b. $$\ln\left(\frac{P_2}{P_1}\right) = \frac{+\Delta H_{vap}}{R}\left(\frac{1}{T_1} - \frac{1}{T_2}\right)$$

$$\ln\left(\frac{760 \text{ atm}}{197 \text{ atm}}\right) = \frac{2.92 \times 10^4 \text{ J/mol.}}{8.31 \text{ J/mol.·K}}\left(\frac{1}{296 \text{ K}} - \frac{1}{T_2}\right)$$

$$3.84 \times 10^{-4} \text{ K}^{-1} = \left(3.38 \times 10^{-3} \text{ K}^{-1} - \frac{1}{T_2}\right)$$

$T_2 = 334 \text{ K} = 61°C$

---

10. Refer to Section 9.1 and Example 9.2.

$$P = 681 \text{ mm Hg} \times \frac{1 \text{ atm}}{760 \text{ mm Hg}} = 0.896 \text{ atm}$$

$$\ln\left(\frac{P_2}{P_1}\right) = \frac{+\Delta H_{vap}}{R}\left(\frac{1}{T_1} - \frac{1}{T_2}\right)$$

$$\ln\left(\frac{0.896 \text{ atm}}{1.00 \text{ atm}}\right) = \frac{4.07 \times 10^4 \text{ J/mol.}}{8.31 \text{ J/mol.·K}}\left(\frac{1}{373 \text{ K}} - \frac{1}{T_2}\right)$$

$$-0.110 = (4.90 \times 10^3 \text{ K})\left(0.00268 \text{ K}^{-1} - \frac{1}{T_2}\right)$$

$$-2.25 \times 10^{-5} \text{ K}^{-1} = 0.00268 \text{ K}^{-1} - \frac{1}{T_2}$$

$$\frac{1}{T_2} = 0.00270$$

$$T_2 = 370 \text{ K} = 97°C$$

---

12. Refer to Section 9.1.

At sea level, $P = 760$ mm Hg and the boiling point of water is 100°C (373 K). From problem 11, the pressure in the pressure cooker is 1.500×10^3 mm Hg.

$$\ln\left(\frac{P_2}{P_1}\right) = \frac{+\Delta H_{vap}}{R}\left(\frac{1}{T_1} - \frac{1}{T_2}\right)$$

$$\ln\left(\frac{760 \text{ mm Hg}}{1500 \text{ mm Hg}}\right) = \frac{4.07 \times 10^4 \text{ J/mol.}}{8.31 \text{ J/mol.} \cdot \text{K}}\left(\frac{1}{T_1} - \frac{1}{373 \text{ K}}\right)$$

$$-0.680 = (4.90 \times 10^3 \text{ K})\left(\frac{1}{T_1} - 0.00268 \text{ K}^{-1}\right)$$

$$\frac{1}{T_1} = -0.000139 \text{ K}^{-1} + 0.00268 \text{ K}^{-1}$$

$$\frac{1}{T_1} = 0.00254 \text{ K}^{-1}$$

$$T_1 = 394 \text{ K} = 121°C$$

$$\Delta T = 121°C - 100°C = 21°C$$

---

14. Refer to Section 9.2 and Figure 9.5.

a. **Solid**

b. 0.5 atm = 380 mm Hg. **Liquid**

c. **Vapor**

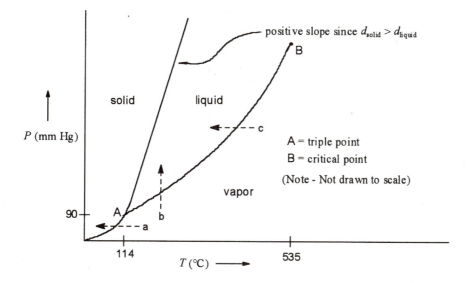

a. Iodine vapor at 80 mm Hg condenses to the **solid** when cooled sufficiently. Since this is below the triple point of 90 mm Hg, the liquid cannot form.

b. Iodine vapor at 125°C condenses to the **liquid** when enough pressure is applied. Since the temperature is above that of the triple point, condensation will be to the liquid state.

c. Iodine vapor at 700 mm Hg condenses to the **liquid** when cooled above the triple point temperature. The pressure is above that of the triple point, so condensation will be to the liquid state.

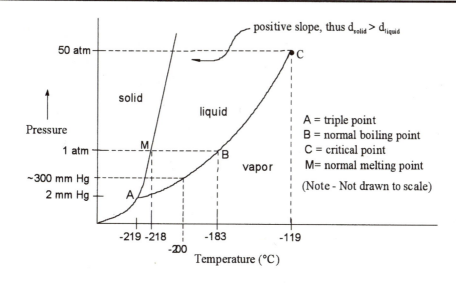

a. The phase diagram is shown above.

b. The slope of the line representing the boundary between solid and liquid is positive, thus $O_2(s)$ is more dense than $O_2(g)$.

c. As shown in the phase diagram above, the pressure will be about 300 mm Hg. The value one gets will depend on the curve draw for the vapor-liquid boundary, but must be between 2 mm Hg (triple point) and 760 mm Hg (normal boiling point).

---

20. Refer to Section 9.2.

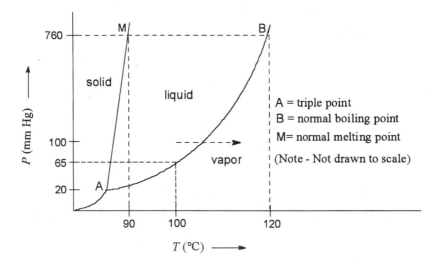

a, b. The phase diagram is drawn above with the appropriate labels. Note that the slope is positive since mp increases slightly with increasing pressure.

c. The phase change will be from liquid to vapor, as noted with dashed arrow in the phase diagram.

---

22. Refer to Section 9.3 and Example 9.4.

The only intermolecular forces present for each of the elements listed are dispersion forces. Thus boiling point will correspond with the strength of the dispersion forces, which depends on two factors: (1) the number of electrons and (2) ease of electron dispersion. The second factor is the same for each of the listed elements, thus the determining factor is the number of electrons. Xe has the most, He has the least.

He < Ne < Ar < Kr < Xe

24. Refer to Section 9.3, Examples 9.4 and 9.6 and Chapter 7.

All molecules have dispersion forces. One can consider each molecule to be roughly spherical. Those that have an even distribution of charge over the surface (b and c) will not have a dipole, while those that have an uneven distribution (a and d) will have a dipole.

a. $CHCl_3$: **Dispersion and dipole.** The electrons are not symmetrically distributed, so this molecule has a dipole. (Each Cl bears a slight negative charge and the H a slight positive charge.)

b. Cl_2: **Dispersion.** The electrons are symmetrically distributed, so this molecule does not have a dipole. (Each Cl has identical, zero, charge).

c. CCl_4: **Dispersion.** The electrons are symmetrically distributed, so this molecule does not have a dipole. (Each Cl bears an identical, slight negative charge, thus the surface of the "sphere" has an even distribution of charge.)

d. H_2O: **Dispersion and dipole.** The electrons are not symmetrically distributed, so this molecule has a dipole. (The O bears a slight negative charge, while the H's each bear a slight positive charge.) Recall from Chapter 7 that water is bent.

26. Refer to Section 9.3, Example 9.5 and Chapter 7.

In order for a molecule to have H-bonding, a hydrogen must be bonded to one of the following molecules: N, O or F. Draw the molecule and look for the appropriate bond.

a. This molecule has an O-H bond, thus there is **H-bonding.**

b. This molecule does not have an N-H bond (all H's are on C's), thus there is **no H-bonding.**

c. This molecule has an N-H bond, thus there is **H-bonding.**

d. This molecule has an F-H bond, thus there is **H-bonding.**

28. Refer to Sections 9.3 and 9.4 and Example 9.4.

    a.   NaBr is an ionic compound, which typically have high melting points.
           Br_2 has dispersion forces only, which results in Br_2 having the lower melting point.

    b. C_2H_5OH has dispersion forces, dipole forces and H-bonding, while
           C_4H_{10} has only dispersion forces. Thus C_4H_{10} will have the lower boiling point.

    c. H_2O has dispersion forces, dipole forces and H-bonding, while
           H_2Te has only dispersion and dipole forces, resulting a lower boiling point for H_2Te.

    d. CH_3CO_2H and $C_6H_5CO_2H$ both have dispersion forces, dipole forces and H-bonding.
           However, $C_6H_5CO_2H$ is a larger molecule with more electrons, consequently it has greater
           dispersion forces and a higher boiling point.

30. Refer to Section 9.3.

    a.   **IMF's**. This is a phase change from solid to liquid, phase changes only involve
           overcoming intermolecular forces.

    b.   **Breaking Covalent Bonds.** HBr dissociates into H^+ and Br^-, thus the covalent bond must
           be broken.

    c.   **IMF's**. This is a phase change from liquid to gas, phase changes only involve overcoming
           intermolecular forces.

    d.   **Breaking Covalent Bonds.** This is a chemical reaction. Chemical reactions involve the
           making and/or breaking of bonds.

32. Refer to Section 9.3 and Examples 9.4 - 9.6.

    a. AsH_3 and **PH_3**. Both of these compounds have both dispersion and dipole forces. PH_3,
           however, has fewer electrons and thus weaker dispersion forces and a lower boiling point.

    b.   **C_6H_6** and $C_{10}H_8$. Both of these molecules have only dispersion forces. C_6H_6, however,
           has fewer electrons and thus weaker dispersion forces and a lower boiling point.

    c. NH_3 and **PH_3**. Both of these compounds have both dispersion and dipole forces. NH_3,
           however, also has hydrogen bonding, and would thus have a higher boiling point.

    d.   LiCl and **C_3H_8**. LiCl is an ionic compound, which typically have high melting and boiling
           points, while C_3H_8 has only dispersion forces, and thus the lower boiling point.

34. Refer to Section 9.3.

a. Melting H_2O requires breaking IMF's, the strongest of which is **H-bonding**.

b. Subliming Br_2 requires breaking IMF's. The only intermolecular force between Br_2 molecules is **dispersion forces**.

c. Boiling $CHCl_3$ requires breaking IMF's, the strongest of which is **dipole forces**. (see problem 24a)

d. Vaporizing benzene (C_6H_6) requires breaking **dispersion forces**, the only IMF's present between benzene molecules (see 32b).

36. Refer to Section 9.4, Example 9.7 and Table 9.5.

a. The water insolubility points to network covalent or metallic. The lack of conductivity indicates the solid is not metallic. The high melting point also suggests network covalent. Thus the solid is **network covalent**.

b. The lack of conductivity in solution or as a melt indicates the solid is not ionic. The lack of conductivity in the solid state indicates it is also not metallic. The water solubility indicates the molecule is not network covalent. Thus it is a **molecular** solid.

c. Water solubility and conduction of electricity in aqueous solution indicates the compound is **ionic**.

38. Refer to Section 9.4, Example 9.7 and Table 9.5.

a. Three types of solids are generally insoluble in water: **metallic, network covalent and non-polar molecular solids**.

b. **Network covalent and ionic compounds** generally have high melting points. Many of the metallic solids have high melting points, but several (notably Hg) have quite low melting points.

c. **Metals and some network covalent** compounds conduct electricity as solids (most network covalent compounds do not conduct electricity, but a few (such as graphite) do. Ionic compounds only do so in solution or in a melt.

163

40. Refer to Section 9.4, Example 9.7 and Table 9.5.

 a. **Metallic**, note position in periodic table.

 b. **Molecular**. Compounds of two non-metals will be molecular.

 c. **Network Covalent**. Sand has an empirical formula of SiO_2, but is a 3-dimensional network of -O-Si-O-Si- bonds.

 d. **Ionic**. This molecule is composed of two ions, Cr^{3+} and SO_4^{2-}.

 e. **Molecular**. Cl_2 is not a metal (note position on periodic table), not a network covalent molecule (very low b.p.) and cannot be ionic since it is composed of a single element.

42. Refer to Section 9.4.

Note that many answers are possible given the large number of molecules that contain C.

 a. $C_{12}H_{22}O_{11}$, sugar.

 b. $CaCO_3$ (found in TUMS®)

 c. C(graphite) or C(diamond)

 d. Steel. Steel is an alloy of Fe and C containing up to 1.7% carbon.

44. Refer to Section 9.4.

 a. Graphite is a network covalent compound, so the C atoms are held together in an extended network of covalent bonds.

 b. Silicon carbide is a network covalent compound, so the Si and C atoms are held together in an extended network of covalent bonds.

 c. $FeCl_2$ is an ionic solid. The structural units are composed of individual Fe^{2+} and Cl^- ions.

 d. Acetylene (C_2H_2) is a molecular solid, composed of individual C_2H_2 molecules held together by dispersion forces.

46. *Refer to Section 9.5 and Example 9.8.*

Determine which of the equations relating the sides of the unit cell to the given atomic radius is valid. The geometry corresponding to the equation is the geometry of the unit cell.

$2(0.186) \neq 0.430$ (not simple cubic)

$4(0.186) \neq 0.430\sqrt{2}$ (not face centered cubic)

$4(0.186) = 0.430\sqrt{3}$ Thus the unit cell is **body centered cubic**.

48. *Refer to Section 9.5 and Problem 46 (above).*

$4r = s\sqrt{2}$

$4r = 0.620\sqrt{2} = 0.877$ nm

$r = 0.219$ nm

50. *Refer to Section 9.5 and Figure 9.7.*

$s = 0.513$ nm
$r = 0.181$ nm

a. $s - 2r$ = distance between Cl^- ions
0.513 nm - 2(0.181 nm) = 0.151 nm

b. Neither Na^+ and K^+ ions are small enough to fit into this space. (Remember, the **diameter** of the ion must be smaller than the distance calculated above.)

52. *Refer to Section 9.5.*

Cs^+ : $r = 0.169$ nm
Cl^- : $r = 0.181$ nm
Figure 9.18 indicates that CsCl forms a BCC structure.

a.   Body diagonal $= 4r = 2(r_{Cr^+}) + 2(r_{Cl^-})$
$= 2(0.169$ nm$) + 2(0.181$ nm$)$
$= 0.338$ nm $+ 0.362$ nm
$= 0.700$ nm

b. $4r = s\sqrt{3}$
0.700 nm $= s(1.73)$
$s = 0.404$ nm

165

54. Refer to Section 9.5 and Figure 9.18.

Note that you are seeing only two dimensions. Visualize the unit cell in three dimensions, with ions coming up towards you, and going back away from you.

a. **4**. The sodium ion is on an edge of the cube, which is common to 4 cubes.

b. **8**, for chloride ions at the corners of the cube. Each corner is shared by 8 cubes.
    2, for chloride ions in the face of the cube. Each face is shared by 2 cubes.

56. Refer to Section 9.4 and Table 9.5.

The properties describe a **network covalent** molecule.

58. Refer to Sections 9.1 and Chapter 5.

$V = 15 \text{ ft} \times 12 \text{ ft} \times 8.0 \text{ ft} = 1440 \text{ ft}^3$

$$1440 \text{ ft}^3 \times \frac{(12 \text{ in})^3}{(1 \text{ ft})^3} \times \frac{(2.54 \text{ cm})^3}{(1 \text{ in})^3} \times \frac{1 \text{ mL}}{1 \text{ cm}^3} \times \frac{1 \text{ L}}{1000 \text{ mL}} = 4.1 \times 10^4 \text{ L}$$

$T = 25°C + 273 = 298 \text{ K}$

$$P = 1.63 \times 10^{-3} \text{ mm Hg} \times \frac{1 \text{ atm}}{760 \text{ mm Hg}} = 2.14 \times 10^{-6} \text{ atm}$$

$PV = nRT$

$(2.14 \times 10^{-6} \text{ atm})(4.1 \times 10^4 \text{ L}) = (n)(0.0821 \text{ L·atm/mol·K})(298 \text{ K})$

$n = 3.6 \times 10^{-3} \text{ mol.}$

$$3.6 \times 10^{-3} \text{ mol.} \times \frac{200.6 \text{ g}}{1 \text{ mol. Hg}} \times \frac{1 \text{ mL}}{13 \text{ g Hg}} = 0.055 \text{ mL}$$

60. Refer to Sections 9.1, 9.2, 9.3 and 9.5.

a. **False**.

b. **False**. (True only if $d_{solid} > d_{liquid}$). Consider the phase diagram of water, where $d_{solid} < d_{liquid}$. The melting point of water at 760 mm Hg = 0°C, this is lower than the temperature of the triple point.

c. **False**. CHF_3 does not have hydrogen bonding since the H is attached to the carbon and not to an electronegative atom.

d. **True**. If $V_{bc} = V_{fc}$, then $s_{bc} = s_{fc}$

$$4r_{bc} = s\sqrt{2} \quad \text{and} \quad 4r_{fc} = s\sqrt{3}$$

$$\frac{4r_{bc}}{\sqrt{2}} = s \quad \text{and} \quad \frac{4r_{fc}}{\sqrt{3}} = s$$

thus, $\quad \dfrac{4r_{bc}}{\sqrt{2}} = \dfrac{4r_{fc}}{\sqrt{3}} \quad$ and $\quad \dfrac{4\sqrt{3}}{4\sqrt{2}} = \dfrac{r_{fc}}{r_{bc}}$

which gives: $\quad \dfrac{r_{fc}}{r_{bc}} = \dfrac{\sqrt{3}}{\sqrt{2}} = \sqrt{\dfrac{3}{2}} = \sqrt{1.5}$

---

62. Refer to Sections 9.1, 9.2 and 9.3.

---

a. A covalent bond is the sharing of 2 electrons between two atoms within a molecule (an **intra**molecular interaction) while hydrogen bonding is the sharing of 2 electrons between a hydrogen (on an electronegative atom) and a second electronegative atom (usually on another molecule). Hydrogen bonding is an **inter**molecular interaction.

b. Boiling point is the temperature at which the vapor pressure equals ambient pressure, while normal boiling point is the temperature at which the vapor pressure equals 1 atm.

c. The triple point is the temperature and pressure at which all three phases (solid, liquid and gas) coexist in equilibrium. The critical point is the condition beyond which a liquid can no longer exist.

d. The vapor pressure curve is a plot of temperature versus pressure at which vapor and liquid are in equilibrium. The phase diagram also includes plots for solid-liquid and solid-gas equilibria. Thus a vapor pressure curve is a subset of a phase diagram.

e. Volume changes have no effect on vapor pressure for a given liquid. Temperature changes, however, do affect vapor pressure.

---

64. Refer to Section 9.5.

First calculate the volume of one cell. Using density as a conversion factor, calculate the mass per cell and mass per atom of Fe. Using the mass of a single Fe atom and the mass of a mole of Fe (atomic mass), calculate the number of atoms in a mole.

For a body centered cube: $4r = s\sqrt{3}$. Thus

$$s = \frac{4(0.124 \text{ nm})}{\sqrt{3}} = 0.286 \text{ nm}$$

$$s = 0.286 \text{ nm} \times \frac{1 \text{ m}}{1 \times 10^9 \text{ nm}} \times \frac{100 \text{ cm}}{1 \text{ m}} = 2.86 \times 10^{-8} \text{ cm}$$

$$V = s^3 = (2.86 \times 10^{-8} \text{ cm})^3 = 2.34 \times 10^{-23} \text{ cm}^3$$

$$\frac{2.34 \times 10^{-23} \text{ cm}^3}{1 \text{ cell}} \times \frac{7.86 \text{ g}}{1 \text{ cm}^3} \times \frac{1 \text{ cell}}{2 \text{ atoms}} = 9.19 \times 10^{-23} \text{ g / atom}$$

$$\frac{55.847 \text{ g}}{1 \text{ mole}} \times \frac{1 \text{ atom}}{9.19 \times 10^{-23} \text{ g}} = 6.08 \times 10^{23} \text{ atoms / mole}$$

---

65. Refer to Section 9.1, Chapters 3 and 5 and Appendix 1.

First, determine the limiting reagent in the reaction. Then calculate the moles of water that would be produced, and the pressure that would result if all the water existed as vapor. Compare that to the vapor pressure of water at 27°C.

a. $$0.400 \text{ g H}_2 \times \frac{1 \text{ mol. H}_2}{2.016 \text{ g}} \times \frac{2 \text{ mol H}_2\text{O}}{2 \text{ mol. H}_2} = 0.198 \text{ mol. H}_2\text{O}$$

$$3.20 \text{ g O}_2 \times \frac{1 \text{ mol. O}_2}{32.00 \text{ g}} \times \frac{2 \text{ mol H}_2\text{O}}{1 \text{ mol. O}_2} = 0.200 \text{ mol. H}_2\text{O}$$

Thus O_2 and H_2 are present in nearly stoichiometric quantities. Consequently, the only gas in the flask will be H_2O. Our calculations will be based on H_2.

$$P = \frac{nRT}{V} = \frac{(0.198 \text{ mol.})(0.0821 \text{ L} \cdot \text{atm/mol.} \cdot \text{K})(300 \text{ K})}{10.0 \text{ L}} = 0.488 \text{ atm}$$

$$P_{\text{H}_2\text{O}\,(27^\circ\text{C})} = 26.74 \text{ mm Hg} \times \frac{1 \text{ atm}}{760 \text{ mm Hg}} = 0.0352 \text{ atm}$$

$P > P_{\text{H}_2\text{O}}$ Since the total pressure cannot exceed the vapor pressure of the water at 27°C, much of the water will exist as liquid, and **both the liquid and vapor phases** will be present.

b. The final pressure in the flask will be 0.0352 atm (26.7 mm Hg), the vapor pressure of the H_2O at 27°C.

c. $3.2 \,\text{g H}_2 \times \dfrac{1\,\text{mol. H}_2}{2.016\,\text{g}} \times \dfrac{2\,\text{mol H}_2\text{O}}{2\,\text{mol. H}_2} = 1.6\,\text{mol. H}_2\text{O}$

$3.2 \,\text{g O}_2 \times \dfrac{1\,\text{mol. O}_2}{32.00\,\text{g}} \times \dfrac{2\,\text{mol H}_2\text{O}}{1\,\text{mol. O}_2} = 0.20\,\text{mol. H}_2\text{O}$

Thus O_2 is the limiting reactant and 1.4 mol. H_2 will be left over. The total pressure will be P_{H_2O} (0.0352 atm, see part a) plus P_{H_2} .

$P = \dfrac{nRT}{V} = \dfrac{(1.4\,\text{mol.})(0.0821\,\text{L}\cdot\text{atm/mol.}\cdot\text{K})(300\,\text{K})}{10.0\,\text{L}} = 3.4\,\text{atm}$

$P_{\text{total}} = 3.4 \text{ atm} + 0.0352 \text{ atm} = 3.4 \text{ atm}$

---

66. *Refer to Section 9.1 and Chapter 5.*

This problem is similar to the one above. Calculate the moles of trichloroethane and the pressure that would result if all the material vaporized.

$1 \,\text{cup} \times \dfrac{1\,\text{qt.}}{4\,\text{cups}} \times \dfrac{1\,\text{L}}{1.057\,\text{qt.}} \times \dfrac{1000\,\text{mL}}{1\,\text{L}} \times \dfrac{1.325\,\text{g}}{1\,\text{mL}} \times \dfrac{1\,\text{mol.}}{133.39\,\text{g}} = 2.35\,\text{mol. C}_2\text{H}_3\text{Cl}_3$

$18\,\text{ft}^3 \times \dfrac{28.32\,\text{L}}{1\,\text{ft}^3} = 510\,\text{L}$

$(39°\text{F} - 32) \times \dfrac{5}{9} = 3.9°\text{C} = 280\,\text{K}$

$P = \dfrac{nRT}{V} = \dfrac{(2.35\,\text{mol.})(0.0821\,\text{L}\cdot\text{atm/mol.}\cdot\text{K})(280\,\text{K})}{510\,\text{L}} = 0.11\,\text{atm}$

Now we must calculate the vapor pressure of trichloroethane at 3.9°C and compare the vapor pressure to the pressure calculated above. Calculate ΔH_{vap} and from that, the vapor pressure at 3.9°C.

$\ln\left(\dfrac{P_2}{P_1}\right) = \dfrac{+\Delta H_{\text{vap}}}{R}\left(\dfrac{1}{T_1} - \dfrac{1}{T_2}\right)$

$\ln\left(\dfrac{760}{100}\right) = \dfrac{+\Delta H_{\text{vap}}}{8.31\,\text{J/mol.}\cdot\text{K}}\left(\dfrac{1}{293\,\text{K}} - \dfrac{1}{347.3\,\text{K}}\right)$

$2.03 = \dfrac{+\Delta H_{\text{vap}}}{8.31\,\text{J/mol.}\cdot\text{K}}\left(5.34 \times 10^{-4}\,\text{K}^{-1}\right)$

$\Delta H_{\text{vap}} = 3.17 \times 10^4$ J/mol. (Now calculate vapor pressure at 3.9°C, (277 K))

$$\ln\left(\frac{760}{P_1}\right) = \frac{3.17 \times 10^4 \text{ J/mol.}}{8.31 \text{ J/mol.} \cdot \text{K}}\left(\frac{1}{277\,\text{K}} - \frac{1}{347.3\,\text{K}}\right) = 2.79$$

$$\frac{760 \text{ mm Hg}}{P_1} = e^{2.79} = 16.3$$

$$P_1 = 46.6 \text{ mm Hg} \times \frac{1 \text{ atm}}{760 \text{ mm Hg}} = 0.0613 \text{ atm}$$

Since the pressure that would result from complete vaporization is greater than the vapor pressure of trichloroethane, not all the $C_2H_3Cl_3$ would evaporate. To calculate the percentage that remains as liquid, we need to determine the amount that would evaporate to produce a pressure of 0.0613 atm. The rest would remain as liquid.

$$n = \frac{PV}{RT} = \frac{(0.0613 \text{ atm})(510 \text{ L})}{(0.0821 \text{ L} \cdot \text{atm/mol.} \cdot \text{K})(277 \text{ K})} = 1.37 \text{ mol. (would vaporize)}$$

2.35 mol. - 1.37 mol. = 0.98 mol. (would remain)

$$\text{Percent remaining} = \frac{0.98}{2.35} \times 100\% = 42\%$$

---

67. Refer to Section 9.2.

The pressure exerted by a skater is:

$$\text{Pressure} = \frac{120 \text{ lbs}}{0.10 \text{ in}^2} \times \frac{1 \text{ atm}}{15 \text{ lbs/in}^2} = 80 \text{ atm}$$

Statement #2 under the heading melting point states "An increase in pressure of 134 atm is required to lower the melting point of ice by 1°C." Thus, unless the ice is only a half degree below freezing, the concept that pressure melts the ice is implausible.

Another possible scenario is heat conduction, given that metals are good conductors of heat. This explanation also falls short, however, since skating is frequently enjoyed when ambient temperatures are well below the freezing point of water.

Friction also fails as an explanation since skating is relatively frictionless.

The most likely explanation for the ease at which a skater glides over ice is a low coefficient of friction between ice and steel.

LiCl is a face centered cube, thus
$$4r = s\sqrt{2}$$

The length of one side $= s = 2(r_{cation} + r_{anion})$

The diagonal $= 4r_{anion} = s\sqrt{2}$

Substituting: $4r_{anion} = 2(r_{cation} + r_{anion})\sqrt{2}$

$$4r_{anion} = 2\sqrt{2}\,r_{cation} + 2\sqrt{2}\,r_{anion}$$

$$\frac{4r_{anion}}{2\sqrt{2}\,r_{anion}} = \frac{2\sqrt{2}\,r_{cation}}{2\sqrt{2}\,r_{anion}} + \frac{2\sqrt{2}\,r_{anion}}{2\sqrt{2}\,r_{anion}} = \frac{r_{cation}}{r_{anion}} + 1$$

$$\frac{r_{cation}}{r_{anion}} = \frac{4}{2\sqrt{2}} - 1 = 0.414$$

69. Refer to Section 9.1, Table 9.1 and Chapter 5.

The critical temperature of N_2 is -147°C, thus N_2 exists entirely as a gas at both 20°C and 10°C. Propane, on the other hand, has a critical temperature of 97°C and thus is present as an equilibrium between the vapor and liquid phases. The change in pressure observed with the N_2 is due to a "contraction" of the gas, or a reduction in the thermal motion of the N_2 molecules, while the pressure change with the propane is due to a change in vapor pressure. Changes in vapor pressure are much more sensitive to temperature changes than are changes in gas pressure.

Chapter 10: Solutions

---

2. *Refer to Section 10.1.*

---

a.   mass percent of $C_2H_6O = \dfrac{mass_{C_2H_6O}}{mass_{C_2H_6O} + mass_{C_3H_8O}} \times 100\%$

The mass of C_2H_6O is given, but the mass of C_3H_8O must be calculated.

$$45.0\,mL\ C_3H_8O \times \dfrac{0.7855\,g\ C_3H_8O}{1\,mL\ C_3H_8O} = 35.3\,g\ C_3H_8O$$

mass percent of $C_2H_6O = \dfrac{19.7\,g}{19.7\,g + 35.3\,g} \times 100\% = 35.8\%$

b.   volume percent of $C_3H_8O = \dfrac{volume_{C_3H_8O}}{volume_{C_3H_8O} + volume_{C_2H_6O}} \times 100\%$

volume percent of $C_3H_8O = \dfrac{45.0\,mL}{45.0\,mL + 25.0\,mL} \times 100\% = 64.3\%$

c.   mole fraction of $C_2H_6O = \dfrac{mole_{C_2H_6O}}{mole_{C_2H_6O} + mole_{C_3H_8O}}$

Moles of C_2H_6O and C_3H_8O must each be calculated.

$$25.0\,mL\ C_2H_6O \times \dfrac{0.789\,g\ C_2H_6O}{1\,mL\ C_2H_6O} \times \dfrac{1\,mol.\ C_2H_6O}{46.07\,g\ C_2H_6O} = 0.428\,mol.\ C_2H_6O$$

$$35.3\,g\ C_3H_8O \times \dfrac{1\,mol.\ C_3H_8O}{60.09\,g\ C_3H_8O} = 0.587\,mol.\ C_3H_8O$$

$$X = \dfrac{0.428}{0.428 + 0.587} = 0.422$$

4. *Refer to Section 10.1 and Example 10.4.*

Recall that 5.00% by mass means 5.00 g CH₃COOH per 100 g solution. Assume 5.00 g acetic acid and 100 g solution. Calculate moles of acetic acid and liters of solution.

$$5.00 \text{ g CH}_3\text{COOH} \times \frac{1 \text{ mol. CH}_3\text{COOH}}{60.05 \text{ g CH}_3\text{COOH}} = 0.0833 \text{ mol. CH}_3\text{COOH}$$

$$100 \text{ g vinegar} \times \frac{1 \text{ mL vinegar}}{1.006 \text{ g vinegar}} = 99.4 \text{ mL vinegar} = 0.0994 \text{ L vinegar}$$

$$M = \frac{\text{moles solute}}{\text{L solution}} = \frac{0.0833 \text{ mol. CH}_3\text{COOH}}{0.0994 \text{ L}} = 0.838 \, M$$

---

6. *Refer to Section 10.1.*

$$\text{ppm} = \frac{\text{mass of solute}}{\text{mass of solution}} \times 10^6$$

$$0.250 \text{ ppm} = \frac{\text{mass (g) of Pb}}{1.00 \text{ g blood}} \times 10^6$$

$$\text{mass of Pb} = 2.50 \times 10^{-7} \text{ g Pb}$$

$$2.50 \times 10^{-7} \text{ g Pb} \times \frac{1 \text{ mol. Pb}}{207.2 \text{ g Pb}} = 1.21 \times 10^{-9} \text{ mol. Pb}$$

---

8. *Refer to Section 10.1 and Examples 10.4 and 10.5.*

	Mass of Solute	Volume of Solution	Molarity
a.	**334.9 g**	750.0 mL	2.757 M
b.	1.500 g	**80.18 mL**	0.1155 M
c.	12.00 g	1.50 L	**0.0494 M**

a. $0.7500 \text{ L} \times \dfrac{2.757 \text{ mol. Na}_2\text{CrO}_4}{1 \text{ L}} \times \dfrac{161.98 \text{ g Na}_2\text{CrO}_4}{1 \text{ mol. Na}_2\text{CrO}_4} = 334.9 \text{ g Na}_2\text{CrO}_4$

b. $1.500 \text{ g Na}_2\text{Cr}_2\text{O}_7 \times \dfrac{1 \text{ mol. Na}_2\text{Cr}_2\text{O}_7}{161.98 \text{ g Na}_2\text{Cr}_2\text{O}_7} \times \dfrac{1 \text{ L}}{0.1155 \text{ mol. Na}_2\text{Cr}_2\text{O}_7} = 0.08018 \text{ L}$

c. $12.00\,g\,Na_2CrO_4 \times \dfrac{1\,mol.\,Na_2CrO_4}{161.98\,g\,Na_2CrO_4} = 0.07408\,mol.\,Na_2CrO_4$

$$\dfrac{0.07408\,mol.\,Na_2CrO_4}{1.50\,L} = 0.0494\,M$$

---

10. Refer to Section 10.1 and Examples 10.4 and 10.5.

	Molality	Mass Percent of Solvent	Ppm Solute	Mole Fraction of Solvent
a.	6.17	45.5%	5.45×10^5	0.900
b.	6.543×10^{-3}	99.8731%	1269	0.999882
c.	0.873	85.5%	1.45×10^5	0.984
d.	0.2560	95.26%	4.737×10^4	0.9954

Since these are aqueous solutions, the solvent is water. The solute, caffeine, has a molar mass of 194.2 g/mol.

a. $X_{solvent} = 0.900$ indicates that there are 0.900 moles H_2O per 1 mole of solvent and solute combined. Consequently, there must be 0.100 mol. $C_8H_{10}O_2N_4$.

$$0.100\,mol.\,C_8H_{10}O_2N_4 \times \dfrac{194.2\,g\,C_8H_{10}O_2N_4}{1\,mol.\,C_8H_{10}O_2N_4} = 19.4\,g\,C_8H_{10}O_2N_4$$

$$0.900\,mol.\,H_2O \times \dfrac{18.02\,g}{1\,mol.} = 16.2\,g\,H_2O$$

$$\dfrac{0.100\,mol.\,C_8H_{10}O_2N_4}{0.0162\,kg\,H_2O} = 6.17\,m \quad \text{(remember to convert mass of solvent to kg)}$$

$$\text{Mass}\% = \dfrac{\text{mass of solute}}{\text{total mass}} \times 100\% = \dfrac{16.2\,g}{(16.2\,g + 19.4\,g)} \times 100\% = 45.5\%$$

$$\text{ppm} = \dfrac{\text{mass of solute}}{\text{mass of solution}} \times 10^6 = \dfrac{19.4\,g\,C_8H_{10}O_2N_4}{(16.2\,g + 19.4\,g)} \times 10^6 = 5.45 \times 10^5$$

b. $\text{ppm} = \dfrac{\text{mass of solute}}{\text{mass of solution}} \times 10^6 = 1269\,g$

If we assume a total mass of 10^6 g, then the mass of solute = 1269 g by definition. Any assumption for mass is valid here, this was chosen for simplicity.

$$1269 \text{ g C}_8\text{H}_{10}\text{O}_2\text{N}_4 \times \frac{1 \text{ mol. C}_8\text{H}_{10}\text{O}_2\text{N}_4}{194.2 \text{ g C}_8\text{H}_{10}\text{O}_2\text{N}_4} = 6.535 \text{ mol. C}_8\text{H}_{10}\text{O}_2\text{N}_4$$

$$10^6 \text{ g} - 1269 \text{ g} = 998731 \text{ g}; \quad 998731 \text{ g H}_2\text{O} \times \frac{1 \text{ mol.}}{18.0152 \text{ g}} = 55438.2 \text{ mol. H}_2\text{O}$$

$$X_{\text{H}_2\text{O}} = \frac{55438.2 \text{ mol.}}{(55438.2 \text{ mol.} + 6.535 \text{ mol.})} = \mathbf{0.999882}$$

$$\text{mass \%} = \frac{9.98731 \times 10^5 \text{ g}}{10^6 \text{ g}} \times 100\% = \mathbf{99.8731\%}$$

$$\text{molality} = \frac{6.535 \text{ mol.}}{998.731 \text{ kg}} = \mathbf{6.543 \times 10^{-3} \, m}$$

c. 85.5 mass % of solvent means that 85.5 g solvent are present for each 100 g solution. Calculations are then based on this ratio.

$$85.5 \text{ g} \times \frac{1 \text{ mol. H}_2\text{O}}{18.02 \text{ g}} = 4.74 \text{ mol. H}_2\text{O}$$

mass of solute = 100 g (total) - 85.5 g (water) = 14.5 g $C_8H_{10}O_2N_4$

$$14.5 \text{ g C}_8\text{H}_{10}\text{O}_2\text{N}_4 \times \frac{1 \text{ mol. C}_8\text{H}_{10}\text{O}_2\text{N}_4}{194.2 \text{ g C}_8\text{H}_{10}\text{O}_2\text{N}_4} = 0.0747 \text{ mol. C}_8\text{H}_{10}\text{O}_2\text{N}_4$$

$$\text{molality} = \frac{0.0747 \text{ mol. C}_8\text{H}_{10}\text{O}_2\text{N}_4}{0.0855 \text{ kg H}_2\text{O}} = \mathbf{0.873 \, m}$$

$$\text{ppm solute} = \frac{14.5 \text{ g C}_8\text{H}_{10}\text{O}_2\text{N}_4}{100 \text{ g}} \times 10^6 = \mathbf{1.45 \times 10^5}$$

$$X_{\text{H}_2\text{O}} = \frac{4.74 \text{ mol.}}{(4.74 \text{ mol.} + 0.0747 \text{ mol.})} = \mathbf{0.984}$$

d. 0.2560 m indicates that there are 0.2560 mol. solute per 1000 g solvent and the calculations are based on that ratio.

$$0.2560 \text{ mol.} \times \frac{194.2 \text{ g C}_8\text{H}_{10}\text{O}_2\text{N}_4}{1 \text{ mol. C}_8\text{H}_{10}\text{O}_2\text{N}_4} = 49.72 \text{ g C}_8\text{H}_{10}\text{O}_2\text{N}_4$$

$$1 \text{ kg} \times \frac{1000 \text{ g}}{1 \text{ kg}} \times \frac{1 \text{ mol. H}_2\text{O}}{18.02 \text{ g H}_2\text{O}} = 55.49 \text{ mol. H}_2\text{O}$$

$$\text{mass \%} = \frac{1000 \text{ g}}{(1000 \text{ g} + 49.72 \text{ g})} \times 100\% = \mathbf{95.26\%}$$

$$\text{ppm solute} = \frac{49.72\,\text{g}}{(1000\,\text{g} + 49.72\,\text{g})} \times 10^6 = \textbf{4.737} \times \textbf{10}^{\textbf{4}}$$

$$X_{\text{H}_2\text{O}} = \frac{55.49\,\text{mol.}}{(55.49\,\text{mol.} + 0.2560\,\text{mol.})} = \textbf{0.9954}$$

---

12. *Refer to Section 10.1 and Example 10.1.*

---

a. A 0.750 M solution requires 0.750 mol. $Ba(OH)_2$ per 1 L of solution. Thus, calculate the mass of 0.750 mol. $Ba(OH)_2$, and dissolve that in sufficient water to make exactly one liter of solution.

$$0.750\,\text{mol. } Ba(OH)_2 \times \frac{171.32\,\text{g } Ba(OH)_2}{1\,\text{mol. } Ba(OH)_2} = 128\,\text{g } Ba(OH)_2$$

b. $[Ba(OH)_2]_c(V)_c = [Ba(OH)_2]_d(V)_d$
    $(6.00\,M)(V)_c = (0.750\,M)(1.00\,\text{L})$
    $(V)_c = 0.125\,\text{L}$

Thus one would obtain 0.125 L of 6.00 M $Ba(OH)_2$ and add sufficient water to make one liter of solution.

---

14. *Refer to Section 10.1 and Example 10.1.*

---

a. $[CoCl_2]_c(V)_c = [CoCl_2]_d(V)_d$
    $(0.773\,M)(0.450\,\text{L}) = [CoCl_2]_d(1.25\,\text{L})$
    $[CoCl_2]_d = 0.278\,M$

To calculate $[Co^{2+}]$ and $[Cl^-]$, recall that $CoCl_2$ is an ionic compound, which dissolves in water according to the equation:

$CoCl_2(aq) \rightarrow Co^{2+}(aq) + 2Cl^-(aq)$

$[Co^{2+}] = [CoCl_2] = 0.278\,M\ Co^{2+}$
$[Cl^-] = 2[CoCl_2] = 0.556\,M\ Cl^-$

b. $0.450\,\text{L} \times \dfrac{0.773\,\text{mol. } CoCl_2}{1\,\text{L}} \times \dfrac{129.83\,\text{g } CoCl_2}{1\,\text{mol. } CoCl_2} = 45.2\,\text{g } CoCl_2$

16. Refer to Section 10.1 and Examples 10.2 - 10.5.

85.0% H_3PO_4 means that there are 85.0 g H_3PO_4 per 100 g solution. Consequently, there must be 15.0 g H_2O.

Molarity: Convert grams H_3PO_4 to moles, and grams of solution to volume.

$$85.0\,g\,H_3PO_4 \times \frac{1\,mol.\,H_3PO_4}{97.994\,g\,H_3PO_4} = 0.867\,mol.\,H_3PO_4$$

$$100\,g\,solution \times \frac{1\,cm^3}{1.689\,g} \times \frac{1\,mL}{1\,cm^3} \times \frac{1\,L}{1000\,mL} = 0.0592\,L\,solution$$

$$M = \frac{0.867\,mol.\,H_3PO_4}{0.0592\,L} = 14.7\,M$$

Molality: Divide moles of solute by *kilograms* of solvent.

$$m = \frac{0.867\,mol.\,H_3PO_4}{0.0150\,kg\,H_2O} = 57.8\,m$$

Mole fraction: Divide moles of H_3PO_4 by total moles of solute and solvent.

$$15.0\,g\,H_2O \times \frac{1\,mol.\,H_2O}{18.02\,g\,H_2O} = 0.832\,mol.\,H_2O$$

$$X_{H_3PO_4} = \frac{0.867\,mol.\,H_3PO_4}{0.867\,mol.\,H_3PO_4 + 0.832\,mol.\,H_2O} = 0.510$$

18. Refer to Section 10.1 and Examples 10.4 and 10.5.

	Density g/mL	Molarity (M)	Molality (m)	Mass % of Solute
a.	1.06	0.886	**0.940**	**11.0%**
b.	1.15	**2.27**	**2.66**	26.0%
c.	1.23	**2.71**	3.11	**29.1%**

a.   0.866 M indicates that there are 0.866 moles $(NH_4)_2SO_4$ per 1 L of solvent. For simplicity, our calculations will be based on a volume of 1 L. Molality requires moles of solute and kg of solvent, thus we must calculate mass of H_2O.

$$0.866 \text{ mol. } (NH_4)_2SO_4 \times \frac{132.15 \text{ g } (NH_4)_2SO_4}{1 \text{ mol. } (NH_4)_2SO_4} = 117 \text{ g } (NH_4)_2SO_4$$

$$1 \text{ L solution} \times \frac{1000 \text{ mL}}{1 \text{ L}} \times \frac{1.06 \text{ g}}{1 \text{ mL}} = 1060 \text{ g solution}$$

1060 g solution - 117 g $(NH_4)_2SO_4$ = 943 g H_2O = 0.943 kg H_2O

$$\frac{0.886 \text{ mol. } (NH_4)_2SO_4}{0.943 \text{ kg } H_2O} = \mathbf{0.940\ m}$$

$$\text{Mass \%} = \frac{\text{mass of solute}}{\text{total mass}} \times 100\% = \frac{117 \text{ g}}{(117 \text{ g} + 943 \text{ g})} \times 100\% = \mathbf{11.0\%}$$

b. 26.0% indicates that there are 26.0 g $(NH_4)_2SO_4$ per 100 g of solution (and 74.0 g H_2O). For simplicity, our calculations will be based on these masses.

$$26.0 \text{ g } (NH_4)_2SO_4 \times \frac{1 \text{ mol. } (NH_4)_2SO_4}{132.15 \text{ g } (NH_4)_2SO_4} = 0.197 \text{ mol. } (NH_4)_2SO_4$$

$$\frac{0.197 \text{ mol. } (NH_4)_2SO_4}{0.0740 \text{ kg } H_2O} = \mathbf{2.66\ m}$$

$$100 \text{ g solution} \times \frac{1 \text{ cm}^3}{1.15 \text{ g}} \times \frac{1 \text{ mL}}{1 \text{ cm}^3} \times \frac{1 \text{ L}}{1000 \text{ mL}} = 0.0870 \text{ L solution}$$

$$M = \frac{0.197 \text{ mol. } (NH_4)_2SO_4}{0.0870 \text{ L}} = \mathbf{2.27\ M}$$

c. 3.11 m indicates that there are 3.11 mol. $(NH_4)_2SO_4$ per 1 kg of solvent. Again, our calculations will be based on these amounts.

$$3.11 \text{ mol. } (NH_4)_2SO_4 \times \frac{132.15 \text{ g } (NH_4)_2SO_4}{1 \text{ mol. } (NH_4)_2SO_4} = 411 \text{ g } (NH_4)_2SO_4$$

$$(1000 + 411) \text{ g solution} \times \frac{1 \text{ cm}^3}{1.23 \text{ g}} \times \frac{1 \text{ mL}}{1 \text{ cm}^3} \times \frac{1 \text{ L}}{1000 \text{ mL}} = 1.15 \text{ L solution}$$

$$M = \frac{3.11 \text{ mol. } (NH_4)_2SO_4}{1.15 \text{ L}} = \mathbf{2.70\ M}$$

$$\text{Mass \%} = \frac{\text{mass of solute}}{\text{total mass}} \times 100\% = \frac{411 \text{ g}}{(1000 \text{ g} + 411 \text{ g})} \times 100\% = \mathbf{29.1\%}$$

20. Refer to Sections 10.2 and 9.3.

The compound which exhibits intermolecular forces most similar to water will be the more soluble in water (like dissolves like). Recall that water has dispersion, dipole and H-bonding forces.

a. $C_{10}H_8$: dispersion forces
   H_2O_2: dispersion, dipole and H-bonding forces
   H_2O_2 would be more soluble since it shares H-bonding with water.

b. SiO_2: network covalent (not intermolecular forces, only intramolecular)
   NaOH: ionic
   NaOH would be more soluble because it is ionic and ionic compounds generally exhibit high solubility in water.

c. $CHCl_3$: dispersion and dipole forces
   HCl: ionic
   HCl would be more soluble because it is ionic and ionic compounds generally exhibit high solubility in water.

d. CH_3OH: dispersion, dipole and H-bonding forces
   CH_3OCH_3: dispersion and dipole forces (since the H's are bonded to the carbons, and not to the oxygen, this molecule doesn't hydrogen bond).
   CH_3OH would be more soluble because of the H-bonding.

---

22. Refer to Section 10.2 and Chapter 8.

a. $\Delta H_{reaction} = \Sigma \Delta H^{\circ}_{f(products)} - \Sigma \Delta H^{\circ}_{f(reactants)}$

   $= [(1 \text{ mol.})(-542.8 \text{ kJ/mol.}) + (1 \text{ mol.})(-677.1 \text{ kJ/mol.})] - [(1 \text{ mol.})(-1206.9 \text{ kJ/mol.})]$

   $= -13.0 \text{ kJ}$

b. Remember that an increase in temperature always shifts the equilibrium so as to favor an endothermic process. This reaction is exothermic, meaning it releases heat into the environment. Therefore a temperature increase will result in a **decreased solubility**.

---

24. Refer to Section 10.2 and Example 10.6.

a. $\dfrac{6.40 \times 10^{-4} M}{1 \text{ atm}} \times \dfrac{1 \text{ atm}}{760 \text{ mm Hg}} = 8.42 \times 10^{-7} M /\text{mm Hg}$

b. The pressure given is the total pressure, the pressure exerted by the nitrogen and the water vapor. Calculate the pressure exerted by the N_2 alone, then calculate the concentration of N_2 in solution and finally the mass of N_2.

$$P_{H_2O} = 23.8 \text{ mm Hg} \times \frac{1 \text{ atm}}{760 \text{ mm Hg}} = 0.0313 \text{ atm}$$

$$P_{total} = P_{N_2} + P_{H_2O} = 1.00 \text{ atm}$$

$$P_{N_2} = 1.00 \text{ atm} - 0.0313 \text{ atm} = 0.969 \text{ atm}$$

$$0.969 \text{ atm} \times \frac{6.40 \times 10^{-4} M}{1 \text{ atm}} = 6.20 \times 10^{-4} M$$

$$1 \text{ L} \times \frac{6.20 \times 10^{-4} \text{ mol. } N_2}{1 \text{ L}} \times \frac{28.02 \text{ g } N_2}{1 \text{ mol. } N_2} = 0.0174 \text{ g } N_2$$

c. This is identical to b, except the volume of water and pressure have changed.

$$P_{total} = P_{N_2} + P_{H_2O} = 0.332 \text{ atm}$$

$$P_{N_2} = 0.332 \text{ atm} - 0.0313 \text{ atm} = 0.301 \text{ atm}$$

$$0.301 \text{ atm} \times \frac{6.40 \times 10^{-4} M}{1 \text{ atm}} = 1.93 \times 10^{-4} M$$

$$0.525 \text{ L} \times \frac{1.93 \times 10^{-4} \text{ mol. } N_2}{1 \text{ L}} \times \frac{28.02 \text{ g } N_2}{1 \text{ mol. } N_2} = 2.83 \times 10^{-3} \text{ g } N_2$$

---

26. *Refer to Section 10.2 and Example 10.6.*

a. 21 mol. % O_2 is identical to $X = 0.21$.

$$P_{O_2} = X_{O_2} \cdot P_{tot} = (0.21)(1.00 \text{ atm}) = 0.21 \text{ atm}$$

$$0.21 \text{ atm} \times \frac{3.30 \times 10^{-4} M}{1 \text{ atm}} = 6.9 \times 10^{-5} M$$

$$1.00 \text{ L} \times \frac{6.9 \times 10^{-5} \text{ mol. } O_2}{1 \text{ L}} \times \frac{32.00 \text{ g } O_2}{1 \text{ mol. } O_2} = 2.2 \times 10^{-3} \text{ g } O_2$$

b. $0.21\,\text{atm} \times \dfrac{2.85 \times 10^{-4}\,M}{1\,\text{atm}} = 6.0 \times 10^{-5}\,M$

$1.00\,\text{L} \times \dfrac{6.0 \times 10^{-5}\,\text{mol. O}_2}{1\,\text{L}} \times \dfrac{32.00\,\text{g O}_2}{1\,\text{mol. O}_2} = 1.9 \times 10^{-3}\,\text{g O}_2$

c. $2.2 \times 10^{-3}\,\text{g} - 1.9 \times 10^{-3}\,\text{g} = 0.3 \times 10^{-3}\,\text{g} = 3 \times 10^{-4}\,\text{g}$

$\dfrac{3 \times 10^{-4}\,\text{g}}{2.2 \times 10^{-3}\,\text{g}} \times 100\% = 10\%$ (note that this answer has only one significant figure)

---

28. Refer to Sections 10.1 and 10.3 and Example 10.7.

---

a. $\Delta P = (X_{\text{C}_2\text{H}_6\text{O}_2})(P^{\circ}_{\text{H}_2\text{O}}) = (0.288)(657.6\,\text{mm Hg}) = 189\,\text{mm Hg}$

$\Delta P = P^{\circ}_{\text{H}_2\text{O}} - P_{\text{H}_2\text{O}} \quad\Rightarrow\quad 189\,\text{mm Hg} = 657.6\,\text{mm Hg} - P_{\text{H}_2\text{O}}$

$P_{\text{H}_2\text{O}} = 469\,\text{mm Hg}$

b. 39.0 mass % means 39.0 g $C_2H_6O_2$ per 100 g solution, while the remainder (61.0 g) must be water.

$39.0\,\text{g C}_2\text{H}_6\text{O}_2 \times \dfrac{1\,\text{mol. C}_2\text{H}_6\text{O}_2}{62.07\,\text{g C}_2\text{H}_6\text{O}_2} = 0.628\,\text{mol. C}_2\text{H}_6\text{O}_2$

$61.0\,\text{g H}_2\text{O} \times \dfrac{1\,\text{mol. H}_2\text{O}}{18.02\,\text{g H}_2\text{O}} = 3.39\,\text{mol. H}_2\text{O}$

$X_{\text{C}_2\text{H}_6\text{O}_2} = \dfrac{0.628\,\text{mol. C}_2\text{H}_6\text{O}_2}{(3.39 + 0.628)\,\text{mol.}} = 0.156$

$\Delta P = (X_{\text{C}_2\text{H}_6\text{O}_2})(P^{\circ}_{\text{H}_2\text{O}}) = (0.156)(657.6\,\text{mm Hg}) = 103\,\text{mm Hg}$

$P_{\text{H}_2\text{O}} = 657.6 - 103 = 555\,\text{mm Hg}$

c. 2.42 m means 2.42 moles $C_2H_6O_2$ are present in 1.00 kg solvent (H_2O). Calculate the moles of water, and from that the mole fraction. Then proceed as above.

$1.00\,\text{kg H}_2\text{O} \times \dfrac{1000\,\text{g}}{1\,\text{kg}} \times \dfrac{1\,\text{mol. H}_2\text{O}}{18.02\,\text{g H}_2\text{O}} = 55.5\,\text{mol. H}_2\text{O}$

$X_{\text{C}_2\text{H}_6\text{O}_2} = \dfrac{2.42\,\text{mol. C}_2\text{H}_6\text{O}_2}{(55.5 + 2.42)\,\text{mol.}} = 0.0418$

$$\Delta P = (X_{C_2H_6O_2})(P^\circ_{H_2O}) = (0.0418)(657.6 \text{ mm Hg}) = 27.5 \text{ mm Hg}$$

$$P_{H_2O} = 657.6 - 27.5 = 630.1 \text{ mm Hg}$$

---

30. Refer to Section 10.3.

The first step is to calculate the mole fraction of oxalic acid.

$$\Delta P = P^\circ_{H_2O} - P_{solution} = 22.38 \text{ mm Hg} - 21.97 \text{ mm Hg} = 0.41 \text{ mm Hg}$$

$$\Delta P = (X_{H_2C_2O_4})(P^\circ_{H_2O}) \implies 0.41 \text{ mm Hg} = (X_{H_2C_2O_4})(22.38 \text{ mm Hg})$$

$$X_{H_2C_2O_4} = 0.018$$

Assuming 1 mole total, this means we have 0.018 mol. $H_2C_2O_4$ and 0.982 mol water (1.00 - 0.018 = 0.982). The next step is to calculate the masses associated with these quantities, and from that the mass of solution and volume of solution.

$$0.982 \text{ mol } H_2O \times \frac{18.02 \text{ g } H_2O}{1 \text{ mol. } H_2O} = 17.7 \text{ g } H_2O$$

$$0.018 \text{ mol } H_2C_2O_4 \times \frac{90.04 \text{ g } H_2C_2O_4}{1 \text{ mol. } H_2C_2O_4} = 1.6 \text{ g } H_2C_2O_4$$

$$(17.7 + 1.6) \text{ g solution} \times \frac{1 \text{ mL}}{1.05 \text{ g}} \times \frac{1 \text{ L}}{1000 \text{ mL}} = 0.0184 \text{ L}$$

Now one can either:

1. Calculate molarity (mol./L) and convert mol./L to grams/L (using molecular mass) or

2. Directly calculate grams of $H_2C_2O_4$ in one liter (as shown below)

$$1.00 \text{ L solution} \times \frac{1.6 \text{ g } H_2C_2O_4}{0.0184 \text{ L soltion}} = 87 \text{ g } H_2C_2O_4$$

Thus, to prepare the prescribed solution, one must dissolve 87 g $H_2C_2O_4$ in enough water to make 1.00 L of solution.

---

32. Refer to Section 10.3.

$$25 \text{ mg} \times \frac{1 \text{ g}}{1000 \text{ mg}} \times \frac{1 \text{ mol.}}{182 \text{ g}} = 1.37 \times 10^{-4} \text{ mol.}$$

$$\frac{1.37 \times 10^{-4} \text{ mol.}}{0.0750 \text{ L}} = 1.83 \times 10^{-3} \, M$$

$\pi = MRT$

$\pi = (1.83 \times 10^{-3} \, M)(0.0821 \text{ L·atm/mol.·K})(298 \text{ K})$

$\pi = 0.0448$ atm

$$0.0448 \text{ atm} \times \frac{760 \text{ mm Hg}}{1 \text{ atm}} = 34.1 \text{ mm Hg}$$

---

34. Refer to Section 10.3 and Example 10.10.

$$2.5 \text{ mm Hg} \times \frac{1 \text{ atm}}{760 \text{ mm Hg}} = 0.0033 \text{ atm}$$

$\pi = MRT$

$0.0033 \text{ atm} = M(0.0821 \text{ L·atm/mol.·K})(298 \text{ K})$

$M = 1.3 \times 10^{-4} \, M$

$$0.125 \text{ L} \times \frac{1.3 \times 10^{-4} \text{ mol.}}{1 \text{ L}} = 1.6 \times 10^{-5} \text{ mol. insulin}$$

$$\text{molar mass} = \frac{0.100 \text{ g insulin}}{1.6 \times 10^{-5} \text{ mol. insulin}} = 6.3 \times 10^{3} \text{ g/mol.}$$

---

36. Refer to Section 10.3.

a. $\Delta T_b = k_b m$

$2.0°C = (2.75°C/m)(m)$

$m = 0.73$

$$0.1000 \text{ kg cyclohexane} \times \frac{0.73 \text{ mol. } C_6H_8O_7}{1 \text{ kg cyclohexane}} \times \frac{192.2 \text{ g } C_6H_8O_7}{1 \text{ mol. } C_6H_8O_7} = 14 \text{ g } C_6H_8O_7$$

$\Delta T_f = k_f m$

$1.0°C = (20.2°C/m)(m)$

$m = 0.050$

$$0.1000 \text{ kg cyclohexane} \times \frac{0.050 \text{ mol. } C_6H_8O_7}{1 \text{ kg cyclohexane}} \times \frac{192.2 \text{ g}}{1 \text{ mol. } C_6H_8O_7} = 0.96 \text{ g } C_6H_8O_7$$

b. $\Delta T_b = k_b m$

$2.0°C = (2.75°C/m)(m)$

$m = 0.73$

$0.1000 \text{ kg cyclohexane} \times \dfrac{0.73 \text{ mol.}}{1 \text{ kg cyclohexane}} \times \dfrac{194.2 \text{ g } C_8H_{10}N_4O_2}{1 \text{ mol.}} = 14 \text{ g } C_8H_{10}N_4O_2$

$\Delta T_f = k_f m$

$1.0°C = (20.2°C/m)(m)$

$m = 0.050$

$0.1000 \text{ kg cyclohexane} \times \dfrac{0.050 \text{ mol.}}{1 \text{ kg cyclohexane}} \times \dfrac{194.2 \text{ g } C_8H_{10}N_4O_2}{1 \text{ mol.}} = 0.97 \text{ g } C_8H_{10}N_4O_2$

---

38. Refer to Section 10.3 and Example 10.8.

Assume 400 mL $C_2H_6O_2$ and 600 mL of H_2O to give 1.00 L total solution. Calculate molality of this solution and the freezing point depression. Then determine if this drop in freezing point is sufficient to prevent freezing at -20°F (be sure to convert to °C).

$400 \text{ mL} \times \dfrac{1 \text{ cm}^3}{1 \text{ mL}} \times \dfrac{1.12 \text{ g } C_2H_6O_2}{1 \text{ cm}^3} \times \dfrac{1 \text{ mol. } C_2H_6O_2}{62.07 \text{ g } C_2H_6O_2} = 7.22 \text{ mol. } C_2H_6O_2$

$m = \dfrac{\text{mol. solute } (C_2H_6O_2)}{\text{kg solvent}} = \dfrac{7.22 \text{ mol. } C_2H_6O_2}{0.600 \text{ kg } H_2O} = 12.0 \text{ } m$

$\Delta T_f = k_f m = (1.86°C/m)(12.0 \text{ } m) = 22.4°C$

$\Delta T_f = T_f° - T_f$
$22.4°C = 0°C - T_f$
$T_f = -22.4°C$

$-20°F = 1.8t_{°C} + 32$
$t_{°C} = -28.9°C$

No, that antifreeze mixture will not reduce the freezing point enough.

---

40. Refer to Section 10.3 and Example 10.8.

Calculate ΔT_f and molality and then solve for k_f.

$\Delta T_f = 69.4°C - 67.2°C = 6.7°C$

$$13.66 \text{ g } C_3H_6O_3 \times \frac{1 \text{ mol. } C_3H_6O_3}{90.108 \text{ g } C_3H_6O_3} = 0.1516 \text{ mol. } C_3H_6O_3$$

$$m = \frac{\text{mol. solute } (C_3H_6O_3)}{\text{kg solvent (stearic acid)}} = \frac{0.1516 \text{ mol. } C_3H_6O_3}{0.115 \text{ kg stearic acid}} = 1.32 \text{ } m$$

$$\Delta T_f = k_f m$$

$$k_f = \frac{\Delta T_f}{m} = \frac{6.7 \text{ °C}}{1.32 \text{ } m} = 5.1 \text{ °C}/m$$

---

42. Refer to Sections 10.3 and 3.3.

Use the freezing point depression to calculate molality, and from that, moles of the compound and the compound's molecular mass. Then calculate empirical formula and molecular formula.

$$\Delta T_f = T_f° - T_f = 6.50°C - 0.0°C = 6.5°C$$

$$\Delta T_f = k_f m$$

$$6.5°C = (20.2°C/m)(m)$$
$$m = 0.322 \text{ } m$$

$$75.0 \text{ mL cyclohexane} \times \frac{1 \text{ cm}^3}{1 \text{ mL}} \times \frac{0.779 \text{ g}}{1 \text{ cm}^3} \times \frac{1 \text{ kg}}{1000 \text{ g}} = 0.0584 \text{ kg cyclohexane}$$

$$m = \frac{\text{mol. solute}}{\text{kg solvent}} \Rightarrow 0.322 \text{ } m = \frac{x \text{ mol.}}{0.0584 \text{ kg cyclohexane}}$$

$$x = 0.0188 \text{ mol.}$$

$$\frac{3.16 \text{ g}}{0.0188 \text{ mol.}} = 168 \text{ g/mol.}$$

Empirical formula:

$$42.9 \text{ g C} \times \frac{1 \text{ mol. C}}{12.01 \text{ g C}} = 3.57 \text{ mol. C}$$

$$2.4 \text{ g H} \times \frac{1 \text{ mol. H}}{1.008 \text{ g H}} = 2.38 \text{ mol. H}$$

$$16.6 \text{ g N} \times \frac{1 \text{ mol. N}}{14.01 \text{ g N}} = 1.18 \text{ mol. N}$$

$$38.1 \text{ g O} \times \frac{1 \text{ mol. O}}{16.00 \text{ g O}} = 2.38 \text{ mol. O}$$

(3.57 mol. C) / (1.18 mol.) = 3
(2.38 mol. H) / (1.18 mol.) = 2
(1.18 mol. N) / (1.18 mol.) = 1
(2.38 mol. O) / (1.18 mol.) = 2

empirical formula = $C_3H_2NO_2$ (emp. mass = 84.06 g/mol.)

$\frac{168}{84} = 2$, thus multiply the subscripts of empirical formula by two ($C_{3x2}H_{2x2}N_{1x2}O_{2x2}$).

empirical formula = $C_6H_4N_2O_4$ (mol. mass = 168 g/mol.)

---

44. Refer to Section 10.3 and Problem 42 (above).

$\Delta T_f = k_f m$

$0.376°C = (20.2°C/m)(m)$
$m = 0.0186\ m$

$m = \frac{\text{mol. solute}}{\text{kg solvent}} \Rightarrow 0.0186\ m = \frac{x\ \text{mol.}}{0.00500\ \text{kg cyclohexane}}$

$x = 9.30 \times 10^{-5}$ mol. β-carotene

$\frac{0.0500\ \text{g}}{9.30 \times 10^{-5}\ \text{mol.}} = 538$ g/mol.

Empirical formula:

If β-carotene is 89.5% C, then it must be 10.5% H since it is composed only of C and H.

$89.5\ \text{g C} \times \frac{1\ \text{mol. C}}{12.01\ \text{g C}} = 7.45\ \text{mol. C}$

$10.5\ \text{g H} \times \frac{1\ \text{mol. H}}{1.008\ \text{g H}} = 10.4\ \text{mol. H}$

(7.45 mol. C) /(7.45 mol.) = 1
(10.4 mol. H) /(7.45 mol.) = 1.4

$C_1H_{1.4}$, to convert to whole numbers, multiply through by 5.

empirical formula = C_5H_7 (emp. mass = 67.11 g/mol.)

$\frac{538}{67} = 8$, thus multiply the subscripts of empirical formula by 8 ($C_{5x8}H_{7x8}$).

molecular formula = $C_{40}H_{56}$ (molar mass = 538 g/mol.)

46. Refer to Section 10.3.

$\pi = MRT$

$7.7 \text{ atm} = (M)(0.0821 \text{ L·atm/mol.·K})(298 \text{ K})$

$M = 0.32 \text{ mol./L}$

48. Refer to Section 10.3 and Example 10.10.

$$4.60 \text{ mm Hg} \times \frac{1 \text{ atm}}{760 \text{ mm Hg}} = 6.05 \times 10^{-3} \text{ atm}$$

$\pi = MRT$

$6.05 \times 10^{-3} \text{ atm} = (M)(0.0821 \text{ L·atm/mol.·K})(293 \text{ K})$

$M = 2.52 \times 10^{-4} \text{ mol./L}$

$$0.200 \text{ L} \times \frac{2.52 \times 10^{-4} \text{ mol.}}{1 \text{ L}} = 5.03 \times 10^{-5} \text{ mol.}$$

$$\frac{3.27 \text{ g hemoglobin}}{5.03 \times 10^{-5} \text{ mol.}} = 6.50 \times 10^{4} \text{ g/mol.}$$

50. Refer to Section 10.3.

All molecules are of the same concentration, so determine which molecule produces the greatest number of species in solution (i). The ones with the larger i's will be the ones with the lowest freezing point and the highest boiling point.

a. $Ni(NO_3)_2(aq) \rightarrow Ni^{2+}(aq) + 2NO_3^-(aq)$ $\quad i = 3$

b. $CH_3OH(aq) \rightarrow CH_3OH(aq)$ $\quad i = 1$

c. $Al_2(SO_4)_3(aq) \rightarrow 2Al^{3+}(aq) + 3SO_4^{2-}(aq)$ $\quad i = 5$

d. $KMnO_4(aq) \rightarrow K^+(aq) + MnO_4^-(aq)$ $\quad i = 2$

freezing point: $CH_3OH > KMnO_4 > Ni(NO_3)_2 > Al_2(SO_4)_3$
boiling point: $Al_2(SO_4)_3 > Ni(NO_3)_2 > KMnO_4 > CH_3OH$

52. Refer to Section 10.1 and Chemistry Beyond the Classroom: Maple Syrup.

Calculate the moles of sucrose in 1 L of maple syrup. This is also the moles of sucrose in the sap since only the water is being evaporated. Calculate the amount of water present after one fourth has been removed (¾ remains), then calculate the molality.

1 kg syrup is 66% sucrose, thus 660 g sucrose, with the remainder (340 g) being the water.

$$660\,g\,C_{12}H_{22}O_{11} \times \frac{1\,mol.\,C_{12}H_{22}O_{11}}{342.3\,g\,C_{12}H_{22}O_{11}} = 1.93\,mol.\,C_{12}H_{22}O_{11}$$

$$30\,L \times \frac{3}{4} = 22.5\,L\,H_2O = 22.5\,kg\,H_2O$$

$$\frac{1.93\,mol.}{22.5\,kg} = 0.086\,m$$

54. Refer to Sections 10.1 and 10.3.

a. $32.48\,g\,FeCl_3 \times \dfrac{1\,mol.\,FeCl_3}{162.2\,g\,FeCl_3} = 0.2002\,mol.\,FeCl_3$

$Molarity = \dfrac{0.2002\,mol.}{0.1000\,L} = 2.002\,M$

b. $100.0\,mL\,solution \times \dfrac{1.249\,g}{1\,mL} = 124.9\,g\,solution$

$124.9\,g\,solution - 32.48\,g\,FeCl_3 = 92.4\,g\,H_2O = 0.0924\,kg\,H_2O$

$Molality = \dfrac{0.2002\,mol.}{0.0924\,kg} = 2.17\,m$

c. $\pi = iMRT$

$\pi = (4)(2.002\,mol./L)(0.0821\,L\cdot atm/mol.\cdot K)(298\,K)$

$\pi = 196\,atm$

d. $\Delta T_f = ik_f m$

$\Delta T_f = (4)(1.86°C/m)(2.17\,m) = 16.1°C$

$T_f = 0°C - 16.1°C = -16.1°C$

56. Refer to Section 10.3.

a. $750.0 \, \text{mL} \times \dfrac{1.933 \, \text{g}}{1 \, \text{mL}} = 1450 \, \text{g} = 1.450 \, \text{kg}$

$\text{Molality} = \dfrac{0.300 \, \text{mol.}}{1.450 \, \text{kg}} = 0.207 \, m$

$\Delta T_b = 73.5°\text{C} - 72.5°\text{C} = 1.0°\text{C}$

$\Delta T_b = k_b m$

$1.0°\text{C} = (k_b)(0.207 \, m)$

$k_b = 4.8°\text{C}/m$

b. $\Delta T_b = 74.9°\text{C} - 72.5°\text{C} = 2.4°\text{C}$

$\Delta T_b = k_b m$

$2.4°\text{C} = (4.8°\text{C}/m)(m)$

$m = 0.50 m$

$100.0 \, \text{mL} \times \dfrac{1.933 \, \text{g}}{1 \, \text{mL}} = 193.3 \, \text{g} = 0.1933 \, \text{kg}$

$\text{Molality} = 0.50 \, \text{mol.}/\text{kg} = \dfrac{x \, \text{mol.}}{0.1933 \, \text{kg}}$

$x = 0.097 \, \text{mol.}$

$\text{molar mass} = \dfrac{12.5 \, \text{g}}{0.097 \, \text{mol.}} = 1.3 \times 10^2 \, \text{g/mol.}$

58. Refer to Sections 3.4, 4.2 and 10.3.

a. $H_2SO_4(aq) + BaCl_2(aq) \rightarrow BaSO_4(s) + 2HCl(aq)$
$BaSO_4$ is the expected precipitate.

b. $25.00 \, \text{mL} \times \dfrac{1.107 \, \text{g}}{1 \, \text{mL}} \times \dfrac{15.25 \, \text{g} \, H_2SO_4}{100 \, \text{g solution}} = 4.220 \, \text{g} \, H_2SO_4$

$4.220 \, \text{g} \, H_2SO_4 \times \dfrac{1 \, \text{mol.} \, H_2SO_4}{98.086 \, \text{g} \, H_2SO_4} = 0.04302 \, \text{mol.} \, H_2SO_4$

$0.04302 \, \text{mol.} \, H_2SO_4 \times \dfrac{1 \, \text{mol.} \, BaSO_4}{1 \, \text{mol.} \, H_2SO_4} \times \dfrac{233.37 \, \text{g} \, BaSO_4}{1 \, \text{mol.} \, BaSO_4} = 10.04 \, \text{g} \, BaSO_4$

$$0.0500 \, \text{L} \times \frac{2.45 \, \text{mol. BaCl}_2}{1 \, \text{L}} = 0.123 \, \text{mol. BaCl}_2$$

$$0.123 \, \text{mol. BaCl}_2 \times \frac{1 \, \text{mol. BaSO}_4}{1 \, \text{mol. BaCl}_2} \times \frac{233.37 \, \text{g BaSO}_4}{1 \, \text{mol. BaSO}_4} = 28.7 \, \text{g BaSO}_4$$

Therefore H_2SO_4 is the limiting reactant, and the theoretical yield is 10.04 g

c. $0.123 \, \text{mol. BaCl}_2 \times \dfrac{2 \, \text{mol. Cl}^-}{1 \, \text{mol. BaCl}_2} = 0.246 \, \text{mol. Cl}^-$

total volume = 25.00 mL + 50.0 mL = 75.0 mL = 0.0750 L

$$\frac{0.246 \, \text{mol. Cl}^-}{0.0750 \, \text{L}} = 3.28 \, M \, \text{Cl}^-$$

---

60. Refer to Section 10.1.

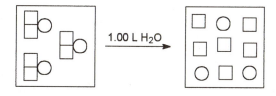

There are 3 moles Na_2S in 1.00 L, thus the molarity is 3.

$$3 \, \text{mol.} \times \frac{1 \, \text{L}}{3 \, \text{mol.}} = 1 \, \text{L}$$

$$Na_2S \rightarrow 2Na^+ + S^{2-}$$

$$\frac{3.0 \, \text{mol. Na}_2\text{S}}{1 \, \text{L}} \times \frac{2 \, \text{mol. Na}^+}{1 \, \text{mol. Na}_2\text{S}} = 6.0 \, M \, \text{Na}^+$$

$$\frac{3.0 \, \text{mol. Na}_2\text{S}}{1 \, \text{L}} \times \frac{1 \, \text{mol. S}^{2-}}{1 \, \text{mol. Na}_2\text{S}} = 3.0 \, M \, \text{S}^{2-}$$

---

62. Refer to Section 10.2.

a. At 40°C, approximately 240 g sucrose will dissolve in 100 g water.

$$80 \, \text{g H}_2\text{O} \times \frac{240 \, \text{g sucrose}}{100 \, \text{g H}_2\text{O}} = 190 \, \text{g sucrose}$$

b. At 25°C, approximately 210 g sucrose dissolves in 100 g water.

$$125 \text{ g H}_2\text{O} \times \frac{210 \text{ g sucrose}}{100 \text{ g H}_2\text{O}} = 260 \text{ g sucrose} \quad \text{(remains in solution)}$$

750 g - 260 g = 490 g sucrose crystallizes out of solution.

---

64. Refer to Section 10.2.

Since heat is evolved during dissolution, $\Delta H_{solution}$ is exothermic. Consequently, adding heat (increasing temperature) would decrease solubility, while removing heat (lowering temperature) would make it more soluble. According to Henry's law, increasing pressure will increase solubility, while decreasing pressure will decrease solubility.

a. $$\frac{C_{g1}}{C_{g2}} = \frac{(k)(P_{g1})}{(k)(P_{g2})} \implies \frac{0.0200}{C_{g2}} = \frac{(4.00 \text{ atm})}{(6.00 \text{ atm})}$$

$C_{g2} = 0.0300$ M, thus solubility is **greater** than 0.0200 M

b. Temperature is higher (thus decreased solubility) and pressure is less (also decreased solubility), thus the combined effect is **less** solubility.

c. Temperature is lower (thus increased solubility) and pressure is the same (thus no change in solubility), thus the combined effect is **greater** solubility.

d. Temperature is the same (thus no change in solubility) and pressure is lower (thus decreased solubility), thus the combined effect is **less** solubility.

---

66. Refer to Section 10.3.

For dilute solution, $M \approx m$

$\Delta T_f = k_f m \, i$
$0.38°C = (1.86°C/m)(0.10 \, m)(i)$
$i = 2$

Thus two moles of ions are formed, indicating that the best equation is **b**.

---

68. Refer to Sections 10.2 and 10.3.

a. Sea water has a lot of salt in it, thus the freezing point should be lower.

b. If it is truly a solution, it would be smooth, not grainy. The graininess indicates that some of the solute (sugar) must have crystallized from solution.

c. If the IV solution is not isotonic, blood cells would experience osmosis, either shrinking or expanding the cells. Neither effect is generally considered beneficial.

d. Warmer water will have less dissolved O_2 than cooler water. Since the cooler water will be deeper, fish get more O_2 there.

e. Champagne has dissolved CO_2. When the bottle is opened the pressure in bottle decreases to equal atmospheric pressure. Consequently, the solubility of the CO_2 decreases and comes out of solution.

---

70. Refer to all of Chapter 10.

---

a. Determine if the solution conducts a current. If so, it is an electrolyte.

b. Beer is a solution of CO_2 (among other things). As it warms, the solubility of the CO_2 decreases (see 68(e) above).

c. Molality is moles of solute per kg of solvent, while mole fraction is moles of solute per total moles (solvent plus solute). In general, the kg of solvent is considerably less than the moles of solvent or even moles of solvent + solute, thus molality is usually greater than mole fraction.

d. The presence of a (non-volatile) solute raises the boiling point of the solvent because it lowers the vapor pressure of the solvent. (Remember that boiling point is the temperature at which the vapor pressure equals the ambient pressure.) This vapor pressure lowering is a consequence of the relative disorders between the pure solvent and solvent vapor and between solution and solvent vapor.

---

72. Refer to Sections 10.1 and 10.3.

From the change in boiling point, one can calculate the molality and from that the mass of solute in 1.00 kg of solvent. This also yields the total mass of the solution. To calculate density, one also needs to know the volume of solution. From osmotic pressure, one can calculate molarity. Using the moles of solute and mass of **solution** calculated from molality, one can now calculate density.

$$KNO_3(s) \rightarrow K^+(aq) + NO_3^-(aq) \qquad (i = 2)$$

$$\Delta T_b = 103.0°C - 100.0°C = 3.0°C$$

$$\Delta T_b = ik_b m$$

$$3.0°C = (2)(0.52°C/m)(m)$$

$$m = 2.9 \ m$$

$$\frac{2.9\,\text{mol. KNO}_3}{1.000\,\text{kg solvent}} \times \frac{101\,\text{g}}{1\,\text{mol. KNO}_3} = 290\,\text{g KNO}_3/1.000\,\text{kg solvent}$$

Thus the total mass of solution is: 290 g solute + 1000 g solvent = 1290 g solution.

$$\pi = MRTi$$
$$122\,\text{atm} = (M)(0.0821\,\text{L·atm/mol.·K})(298\,\text{K})(2)$$
$$M = 2.49\,M$$

$$\frac{2.49\,\text{mol. KNO}_3}{1.000\,\text{L sol'n}} \times \frac{1290\,\text{g sol'n}}{2.9\,\text{mol. KNO}_3} \times \frac{1\,\text{L}}{1000\,\text{mL}} = 1.1\,\text{g/mL}$$

---

73. *Refer to Section 10.1.*

Molality is moles solute per kilogram of solvent, so convert mass of solute to moles, and liters of solution to kg of solvent.

$$158.2\,\text{g KOH} \times \frac{1\,\text{mol. KOH}}{56.11\,\text{g KOH}} = 2.820\,\text{mol. KOH}$$

$$1.000\,\text{L solution} \times \frac{1000\,\text{mL}}{1\,\text{L}} \times \frac{1.13\,\text{g}}{1\,\text{mL}} = 1130\,\text{g solution}$$

1130 g solution - 158.2 g solute = 972 g solvent = 0.972 kg solvent

$$\text{molality} = \frac{2.820\,\text{mol.}}{0.972\,\text{kg}} = 2.90\,m$$

This solution is about 2.9 m, approximately 10 times stronger than the solution the technician needs. Thus he will need to increase the volume approximately 10 fold (to ~1 L).

If there are 2.820 moles KOH in 1.000 L, then there are 0.2820 moles in the 100 mL sample. Likewise, the 100 mL sample also has only 0.0972 kg of solvent. The problem can then be solved by setting up a ratio.

$$\frac{0.250\,\text{mol. KOH}}{1.000\,\text{kg solvent}} = \frac{0.2820\,\text{moles KOH}}{(0.0972\,\text{kg} + x\,\text{kg})\,\text{solvent}}$$

$$0.0972\,\text{kg H}_2\text{O} + x\,\text{kg H}_2\text{O} = 1.128\,\text{kg H}_2\text{O}$$

$$x = 1.03\,\text{kg H}_2\text{O}$$

Thus the technician must add 1.03 kg (1.03 L) H_2O to the 100 mL of KOH solution.

One can prove the generality of this equation in one of two ways:

1. Combine simple relationships (i.e. m = moles solute/ kg solvent) to come up with the complex equation given (as is done in the answer in the back of the text book).

2. Simplify the complex equation to come up with a final simple relationship which we already know to be true (as is done below).

$$m = \cfrac{M}{d - \cfrac{(M)(M)}{1000}} = \cfrac{\cfrac{\text{moles of solute}}{\text{L of solution}}}{\cfrac{\text{g solution}}{\text{mL of solution}} - \cfrac{\left(\cfrac{\text{g solute}}{\text{moles solute}}\right)\left(\cfrac{\text{moles of solute}}{\text{L of solution}}\right)}{1000}}$$

Multiplying through the $(M)(M)$ terms simplifies the equation slightly to:

$$m = \cfrac{\cfrac{\text{moles of solute}}{\text{L of solution}}}{\cfrac{\text{g solution}}{\text{mL of solution}} - \cfrac{\cfrac{\text{g solute}}{\text{L of solution}}}{1000}}$$

Since $\cfrac{\cfrac{\text{g solute}}{\text{L of solution}}}{1000} = \cfrac{\text{g solute}}{(1000)(\text{L of solution})}$ and $(1000)(L) = mL$, we get:

$$m = \cfrac{\cfrac{\text{moles of solute}}{\text{L of solution}}}{\cfrac{\text{g solution}}{\text{mL of solution}} - \cfrac{\text{g solute}}{\text{mL of solution}}} = \cfrac{\cfrac{\text{moles of solute}}{\text{L of solution}}}{\cfrac{\text{g solution - g solute}}{\text{mL of solution}}}$$

Recognizing that (g solution) – (g solute) = (g solvent), one can substitute and further simplify to:

$$m = \cfrac{\cfrac{\text{moles of solute}}{\text{L of solution}}}{\cfrac{\text{g solvent}}{\text{mL of solution}}} = \left(\cfrac{\text{moles of solute}}{\text{L of solution}}\right)\left(\cfrac{\text{mL of solution}}{\text{g solvent}}\right)$$

Recognizing that $(1000)(L) = mL$, and likewise $(1000)(kg) = g$, one can substitute and simplify the equation further to:

$$m = \cfrac{\text{moles of solute}}{\text{L of solution}} \times \cfrac{(1000)(\text{L of solution})}{(1000)(\text{kg solvent})} = \cfrac{\text{moles of solute}}{\text{kg solvent}}$$

This is the definition of molality, so the equation is true and therefore valid for any solution.

The second part of the question asks why $m = M$ for dilute aqueous solutions. This is because such solutions have a density of approximately 1. Consequently, L solution = kg solvent.

Substituting into the first equation above, one gets:

$$m = \frac{M}{1 - \dfrac{\left(\dfrac{\text{g solute}}{\text{moles solute}}\right)\left(\dfrac{\text{moles of solute}}{\text{kg of solvent}}\right)}{1000}} = \frac{M}{1 - \dfrac{\text{g solute}}{\text{g of solvent}}}$$

For dilute solutions, grams of solute is so much smaller than grams of solvent that

$$1 - \frac{\text{g solute}}{\text{g of solvent}} \approx 1 \quad \text{and so} \quad m = \frac{M}{1} = M .$$

---

75. Refer to Section 10.3.

From freezing point depression one can calculate the molality of the solution. Since the mass of the solvent is given, one can also calculate the moles of solute. From the molar masses of the two solutes and the mass of the mixture, one can then determine the mole ratio.

$$\Delta T_f = k_f m$$
$$0.500°C = (1.86°C/m)(m)$$
$$m = 0.269 \ m$$

$$0.269 \ m = \frac{\text{mol. solute}}{1.00 \times 10^{-3} \text{ kg solvent}} \Rightarrow \text{moles of solute} = 2.69 \times 10^{-4} \text{ mol.}$$

Thus the total moles of solute (moles sugar + moles X) = 2.69×10^{-4} mol.

If mass of solute = x g, then mass of sugar = $(0.100 - x)$ g.

Combining the above we get

$$x \text{ g} \times \frac{1 \text{ mol. X}}{410 \text{ g}} + (0.100 \text{ g} - x \text{ g}) \times \frac{1 \text{ mol sugar}}{342 \text{ g}} = 2.69 \times 10^{-4} \text{ mol.}$$

$$\frac{x \text{ mol.}}{410} + \frac{0.100}{342} - \frac{x \text{ mol.}}{342} = 2.69 \times 10^{-4} \text{ mol.} \Rightarrow \frac{x \text{ mol.}}{410} - \frac{x \text{ mol.}}{342} = -2.34 \times 10^{-5} \text{ mol.}$$

$$\frac{(410)(x \text{ mol.})}{410} - \frac{(410)(x \text{ mol.})}{342} = (410)(-2.34 \times 10^{-5} \text{ mol.})$$

x mol. $- 1.20x$ mol. $= -9.59 \times 10^{-3}$ mol.

$-0.20x$ mol. $= -9.59 \times 10^{-3}$ mol.

$x = 0.048$ and

mass percent of $X = \dfrac{0.048\,g}{0.100\,g} \times 100\% = 48\%$

---

76. *Refer to Chapters 1 and 3.*

Calculate the amount of alcohol that enters the blood (recall that to convert a percent to a decimal, one divides by 100). Then calculate the concentration of alcohol in the blood.

142 g martini x 2 martinis x 0.30 (30% alcohol) x 0.15 (15% enter bloodstream) = 13 g alcohol.

$7.0\,L\,blood \times \dfrac{1000\,mL}{1\,L} = 7000\,mL$

Conc. of alcohol in blood $= \dfrac{13\,g\,alcohol}{7000\,mL\,blood} = 0.0019\,g/cm^{3}$

Thus the person is legally intoxicated.

---

77. *Refer to Section 10.1 and Chapter 5.*

a. $49.92\,g\,NaOH \times \dfrac{1\,mol.}{40.00\,g\,NaOH} = 1.248\,mol.\,NaOH$

$\dfrac{1.248\,mol.\,NaOH}{0.600\,L} = 2.08\,M$

b. $1.248\,mol.\,NaOH \times \dfrac{1\,mol.\,OH^-}{1\,mol.\,NaOH} \times \dfrac{3\,mol.\,H_2}{2\,mol.\,OH^-} = 1.872\,mol.\,H_2$

c.   Determine the limiting reactant first.

$41.28\,g\,Al \times \dfrac{1\,mol.\,Al}{26.98\,g\,Al} \times \dfrac{3\,mol.\,H_2}{2\,mol.\,Al} = 2.30\,mol.\,H_2$

The amount of H_2 that could be produced by the Al is less than the actual amount produced, thus the limiting reactant is NaOH.

$P = (758.6\,mm\,Hg - 23.8\,mm\,Hg) \times \dfrac{1\,atm}{760\,mm\,Hg} = 0.967\,atm$

$$V = \frac{nRT}{P} = \frac{(1.872 \text{ mol.})(0.0821 \text{ L} \cdot \text{atm/mol.} \cdot \text{K})(298 \text{ K})}{0.967 \text{ atm}} = 47.4 \text{ L}$$

78. Refer to Section 10.3.

The ideal gas law is: $V = \dfrac{nRT}{P}$

Henry's law is: $C_g = kP_g$, where C_g is the concentration of the gas in solution.

If we hold the volume of solution constant, then $n_g = kP_g$, where n_g is the moles of gas.

Substituting gives: $V = \dfrac{(kP) RT}{P} = kRT$

Thus, the volume of gas is directly proportional only to the temperature; pressure is not a factor.

Chapter 11: Rate of Reaction

2. *Refer to Section 11.1.*

a. $\text{rate} = \left(\dfrac{1}{2}\right)\dfrac{\Delta[\text{HOCl}]}{\Delta t} = \dfrac{\Delta[\text{HOCl}]}{2\Delta t}$

b. $\text{rate} = \dfrac{-\Delta[\text{O}_2]}{\Delta t}$

4. *Refer to Section 11.1.*

$\text{rate} = \dfrac{-\Delta[\text{C}_2\text{H}_6]}{\Delta t} = \dfrac{-\Delta[\text{O}_2]}{7\Delta t} = \dfrac{\Delta[\text{CO}_2]}{4\Delta t} = \dfrac{\Delta[\text{H}_2\text{O}]}{6\Delta t}$

$\dfrac{-\Delta[\text{C}_2\text{H}_6]}{\Delta t} = 0.20\,\text{mol./L}\cdot\text{s}$

$\dfrac{0.20\,\text{mol.}}{\text{L}\cdot\text{s}} = \dfrac{\Delta[\text{CO}_2]}{4\Delta t} \quad \Rightarrow \quad \dfrac{\Delta[\text{CO}_2]}{\Delta t} = 4\left(\dfrac{0.20\,\text{mol.}}{\text{L}\cdot\text{s}}\right) = 0.80\,\text{mol./L}\cdot\text{s}$

$\dfrac{\Delta[\text{H}_2\text{O}]}{6\Delta t} = \dfrac{0.20\,\text{mol.}}{\text{L}\cdot\text{s}} \quad \Rightarrow \quad \dfrac{\Delta[\text{H}_2\text{O}]}{\Delta t} = 6\left(\dfrac{0.20\,\text{mol.}}{\text{L}\cdot\text{s}}\right) = 1.2\,\text{mol./L}\cdot\text{s}$

6. *Refer to Section 11.1.*

a. $\text{N}_2(g) + 3\text{H}_2(g) \rightarrow 2\text{NH}_3(g)$

b. $\text{rate} = \dfrac{\Delta[\text{NH}_3]}{2\Delta t}$

c. $\text{rate} = \dfrac{0.815\,M - 0.257\,M}{2(15 - 0)\text{min}} = 1.86 \times 10^{-2}\,M\,/\text{min}$

8. *Refer to Section 11.1 and Figure 11.2.*

Time (min)	0	2	4	6	8	10
[B] (*M*)	0	0.100	0.130	0.150	0.165	0.175

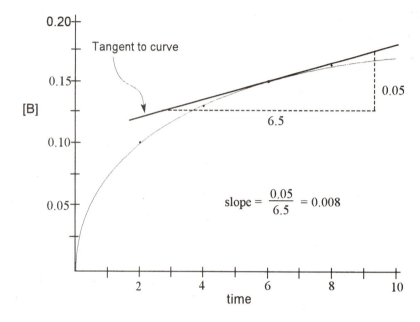

a. The graph shows the change in concentration of B over time.

b. The slope of the tangent gives the instantaneous rate (0.008 *M*/min) at that point in time.

c. The average rate over the 2 to 8 minute time frame is the change in concentration over that given time.

$$\text{rate}_{avg} = \frac{\Delta[B]}{\Delta t} = \frac{0.165\,M - 0.100\,M}{8\,\text{min} - 2\,\text{min}} = 0.011\,M\,/\text{min}$$

d. The instantaneous rate (0.008 min) is less than the average rate (0.011 min) over that six minute interval.

---

10. *Refer to Section 11.2.*

a. rate = $k_1[X]^2[Y]$
   $m = 2$; so the reaction is **2nd order in X**.
   $n = 1$; so the reaction is **1st order in Y**.
   $m + n = 3$; so the reaction is **3rd order overall**.

b. rate = $k_2[X]$
   $m = 1$; so the reaction is **1st order in X**.
   $m = 1$; so the reaction is **1st order overall**.

c. rate = $k_3[X]^2[Y]^2$
    $m = 2$; so the reaction is **2nd order in X**.
    $n = 2$; so the reaction is **2nd order in Y**.
    $m + n = 4$; so the reaction is **4th order overall**.

d. rate = k_4
    $m = 0$; so the reaction is **0th order in X**.
    $n = 0$; so the reaction is **0th order in Y**.
    $m + n = 0$; so the reaction is **0th order overall**.

---

12. Refer to Section 11.2.

---

a. rate = $k_1[X]^2[Y]$

$$k_1 = \frac{\text{rate}}{[M]^2[M]} = \frac{\text{mol./L} \cdot \text{s}}{(\text{mol./L})^2(\text{mol./L})} = \frac{1/\text{s}}{(\text{mol./L})^2} = L^2/\text{mol.}^2 \cdot \text{s}$$

b. rate = $k_2[X]$

$$k_1 = \frac{\text{rate}}{[M]} = \frac{\text{mol./L} \cdot \text{s}}{\text{mol./L}} = \frac{1/\text{s}}{1} = 1/\text{s} = \text{s}^{-1}$$

c. rate = $k_3[X]^2[Y]^2$

$$k_1 = \frac{\text{rate}}{[M]^2[M]^2} = \frac{\text{mol./L} \cdot \text{s}}{(\text{mol./L})^2(\text{mol./L})^2} = \frac{1/\text{s}}{(\text{mol./L})^3} = L^3/\text{mol.}^3 \cdot \text{s}$$

d. rate = k_4

k_4 = rate = mol./L·s

---

14. Refer to Section 11.2.

	[A]	k (h^{-1})	Rate (mol./L·h)
a.	0.115	0.417	**4.80×10^{-2}**
b.	0.025	**67**	1.67
c.	**20**	2.0×10^{-3}	0.0392

a. rate = $k[A]$
    rate = $(0.417 \text{ h}^{-1})(0.115 \, M)$ = **4.80×10^{-2} M/h**

b. rate = $k[A]$
    $1.67 \, M/\text{h} = (k)(0.025 \, M)$
    k = **67 h^{-1}**

c. rate = $k[A]$
    $0.0392 \, M/\text{h} = (2.0 \times 10^{-3} \text{ h}^{-1})[A]$
    $[A]$ = **20 M**

a. rate = $k[NO_2]^2$

b. $1.17 \, \text{mol./L} \cdot \text{min} = k(0.250 \, \text{mol./L})^2$

$$k = \frac{1.17 \, \text{mol./L} \cdot \text{min}}{0.0625 \, \text{mol.}^2/\text{L}^2} = 18.7 \, \text{L/mol.} \cdot \text{min}$$

c. rate = $k[NO_2]^2$

rate = $(18.7 \, \text{L/mol.} \cdot \text{min})(0.800 \, \text{mol./L})^2 = 12.0 \, \text{mol./L} \cdot \text{min}$

---

18. Refer to Section 11.2.

a. rate = $k[ICl][H_2]$
$4.89 \times 10^{-5} \, \text{mol./L} \cdot \text{s} = k(0.100 \, M)(0.030 \, M)$
$k = 0.016 \, \text{L/mol.} \cdot \text{s}$

b. rate = $k[ICl][H_2]$
$5.00 \times 10^{-4} \, \text{mol./L} \cdot \text{s} = (0.016 \, \text{L/mol.} \cdot \text{s})(0.233 \, M)[H_2]$
$[H_2] = 0.13 \, M$

c. rate = $k[ICl][H_2]$
$3[ICl] = [H_2]$
substituting gives: rate = $k[ICl]3[ICl]$
$0.0934 \, \text{mol./L} \cdot \text{s} = (0.016 \, \text{L/mol.} \cdot \text{s}) \, [ICl]3[ICl]$
$5.8 = [ICl]3[ICl] = 3[ICl]^2$
$1.9 = [ICl]^2$
$[ICl] = 1.4 \, M$

---

20. Refer to Section 11.2 and Example 11.1.

A quick look at the data indicates that the reaction is zero order. This is due to the fact that the rate is largely invariant with changing concentration of A. This can be confirmed mathematically.

a. $rate_1 = k[A]^m$

$rate_2 = k[A]^m$

$$\frac{rate_1}{rate_2} = \left(\frac{[A]_1}{[A]_2}\right)^m \quad \Rightarrow \quad \frac{0.020 \, \text{mol./L} \cdot \text{s}}{0.020 \, \text{mol./L} \cdot \text{s}} = \left(\frac{0.100 \, \text{mol./L}}{0.090 \, \text{mol./L}}\right)^m$$

$1 = (1.1)^m \quad \Rightarrow \quad \log(1) = \log(1.1)^m = m \cdot \log(1.1)$

$m = 0$

b. rate $= k[A]^0$
    rate $= k$

c. To get k, take an average of the four values given in the table.

$$\text{rate} = \frac{0.020 + 0.020 + 0.019 + 0.021}{4} = 0.020 \text{ mol./L} \cdot \text{min.}$$

---

22. Refer to Section 11.1 and Example 11.2.

---

a. To determine the reaction order in $[I^-]$, select two experiments in which $[S_2O_8^{2-}]$ is constant (such as experiments 3 and 4, as used below).

$$\frac{\text{rate}_3}{\text{rate}_4} = \frac{k[I^-]_3^m[S_2O_8^{2-}]_3^n}{k[I^-]_4^m[S_2O_8^{2-}]_4^n} \Rightarrow \frac{\text{rate}_1}{\text{rate}_2} = \left(\frac{[I^-]}{[I^-]}\right)^m \left(\frac{[S_2O_8^{2-}]}{[S_2O_8^{2-}]}\right)^n$$

$$\frac{2.22 \times 10^{-4} \text{ mol./L} \cdot \text{min}}{3.06 \times 10^{-4} \text{ mol./L} \cdot \text{min}} = \left(\frac{0.0200 \text{ mol./L}}{0.0275 \text{ mol./L}}\right)^m \left(\frac{0.0300 \text{ mol./L}}{0.0300 \text{ mol./L}}\right)^n$$

$0.725 = (0.727)^m(1)^n$
$0.725 = (0.727)^m$
$m = 1$ (The reaction is 1st order in I^-.)

To determine the reaction order in $[S_2O_8^{2-}]$, select two experiments in which $[I^-]$ is constant (such as experiments 2 and 3) and repeat the process above.

$$\frac{1.85 \times 10^{-4} \text{ mol./L} \cdot \text{min}}{2.22 \times 10^{-4} \text{ mol./L} \cdot \text{min}} = \left(\frac{0.0200 \text{ mol./L}}{0.0200 \text{ mol./L}}\right)^m \left(\frac{0.0250 \text{ mol./L}}{0.0300 \text{ mol./L}}\right)^n$$

$0.833 = (1)^m(0.833)^n$
$0.833 = (0.833)^n$
$n = 1$ (The reaction is 1st order in $S_2O_8^{2-}$.)

Overall reaction order $= m + n = 1 + 1 = 2$

b. Rate $= k[S_2O_8^{2-}][I^-]$

c. Substitute values from any of the experiments into the rate equation and solve for k.
    $1.15 \times 10^{-4} \text{ } M/\text{min} = k(0.0200 \text{ } M)(0.0155 \text{ } M)$
    $1.15 \times 10^{-4} \text{ } M/\text{min} = k(3.10 \times 10^{-4} \text{ } M^2)$
    $k = 0.371 \text{ min}^{-1} \cdot M^{-1} = 0.371 \text{ L/mol} \cdot \text{min}$

d. Rate $= k[S_2O_8^{2-}][I^-]$
    Rate $= (0.371 \text{ L/mol} \cdot \text{min})(0.105 \text{ } M)(0.0875 \text{ } M)$
    Rate $= 3.41 \times 10^{-3} \text{ } M/\text{min}$

a. To determine the reaction order in [OH⁻], select two experiments in which [ClO₂] is constant (such as experiments 1 and 2, as used below).

$$\frac{rate_1}{rate_2} = \frac{k[ClO_2]_1^m[OH^-]_1^n}{k[ClO_2]_2^m[OH^-]_2^n} \Rightarrow \frac{rate_1}{rate_2} = \left(\frac{[ClO_2]}{[ClO_2]}\right)^m\left(\frac{[OH^-]}{[OH^-]}\right)^n$$

$$\frac{6.00 \times 10^{-4}\ mol./L \cdot min}{1.50 \times 10^{-3}\ mol./L \cdot min} = \left(\frac{0.010\ mol./L}{0.010\ mol./L}\right)^m\left(\frac{0.030\ mol./L}{0.075\ mol./L}\right)^n$$

$0.400 = (1)^m(0.40)^n$

$0.400 = (0.40)^n$

$n = 1$ (The reaction is 1st order in OH⁻.)

To determine the reaction order in [ClO₂], select two experiments in which [OH⁻] is constant (such as experiments 1 and 3, as used below) and repeat the process above.

$$\frac{6.00 \times 10^{-4}\ mol./L \cdot min}{1.82 \times 10^{-2}\ mol./L \cdot min} = \left(\frac{0.010\ mol./L}{0.055\ mol./L}\right)^m\left(\frac{0.030\ mol./L}{0.030\ mol./L}\right)^n$$

$0.0330 = (0.182)^m(1)^n$

$0.0330 = (0.182)^m$

$\log(0.0330) = \log(0.182)^m = m \cdot \log(0.182)$

$m = 2$ (The reaction is 2nd order in ClO₂.)

Overall reaction order = $m + n = 2 + 1 = 3$

b. Rate = $k[ClO_2]^2[I^-]$

c. Substitute values from any of the experiments into the rate equation and solve for k.

$6.00 \times 10^{-4}\ M/s = k(0.010\ M)^2(0.030\ M)$

$6.00 \times 10^{-4}\ M/s = k(3.0 \times 10^{-6}\ M^3)$

$k = 2.0 \times 10^2\ s^{-1} \cdot M^{-2} = 2.0 \times 10^2\ L^2/mol.^2 \cdot s$

d. Rate = $k[ClO_2]^2[I^-]$

Rate = $(2.0 \times 10^2\ s^{-1} \cdot M^{-2})(0.25\ M)^2(0.036\ M)$

Rate = $4.5 \times 10^{-1}\ M/s = 4.5 \times 10^{-1}\ mol./L \cdot s$

26. Refer to Section 11.2, Example 11.2 and Problem 24 above.

a. To determine the reaction order in [I⁻], select two experiments in which [BrO₃⁻] and [H⁺] are constant (such as experiments 1 and 2, as used below).

$$\frac{rate_1}{rate_2} = \frac{k[I^-]_1^m[BrO_3^-]_1^n[H^+]_1^p}{k[I^-]_2^m[BrO_3^-]_2^n[H^+]_2^p} \Rightarrow \frac{rate_1}{rate_2} = \left(\frac{[I^-]}{[I^-]}\right)^m\left(\frac{[BrO_3^-]}{[BrO_3^-]}\right)^n\left(\frac{[H^+]}{[H^+]}\right)^p$$

$$\frac{8.89 \times 10^{-5} \text{ mol./L} \cdot \text{min}}{1.78 \times 10^{-4} \text{ mol./L} \cdot \text{min}} = \left(\frac{0.0020 \text{ mol./L}}{0.0040 \text{ mol./L}}\right)^m \left(\frac{0.0080 \text{ mol./L}}{0.0080 \text{ mol./L}}\right)^n \left(\frac{0.0020 \text{ mol./L}}{0.0020 \text{ mol./L}}\right)^p$$

$0.499 = (0.50)^m (1)^n (1)^p$
$0.499 = (0.50)^m$
$m = 1$ (The reaction is 1st order in I^-.)

To determine the reaction order in $[BrO_3^-]$, select two experiments in which $[I^-]$ and $[H^+]$ are constant (such as experiments 1 and 3, as used below) and repeat the process above.

$$\frac{\text{rate}_1}{\text{rate}_2} = \frac{k[I^-]_1^m [BrO_3^-]_1^n [H^+]_1^p}{k[I^-]_2^m [BrO_3^-]_2^n [H^+]_2^p} \quad \Rightarrow \quad \frac{\text{rate}_1}{\text{rate}_2} = \left(\frac{[I^-]}{[I^-]}\right)^m \left(\frac{[BrO_3^-]}{[BrO_3^-]}\right)^n \left(\frac{[H^+]}{[H^+]}\right)^p$$

$$\frac{8.89 \times 10^{-5} \text{ mol./L} \cdot \text{min}}{1.78 \times 10^{-4} \text{ mol./L} \cdot \text{min}} = \left(\frac{0.0020 \text{ mol./L}}{0.0020 \text{ mol./L}}\right)^m \left(\frac{0.0080 \text{ mol./L}}{0.0160 \text{ mol./L}}\right)^n \left(\frac{0.0020 \text{ mol./L}}{0.0020 \text{ mol./L}}\right)^p$$

$0.499 = (1)^m (0.50)^n (1)^p$
$0.499 = (0.50)^n$
$n = 1$ (The reaction is 1st order in BrO_3^-.)

To determine the reaction order in $[H^+]$, select two experiments in which $[I^-]$ and $[BrO_3^-]$ are constant (such as experiments 1 and 4, as used below) and repeat the process above.

$$\frac{\text{rate}_1}{\text{rate}_2} = \frac{k[I^-]_1^m [BrO_3^-]_1^n [H^+]_1^p}{k[I^-]_2^m [BrO_3^-]_2^n [H^+]_2^p} \quad \Rightarrow \quad \frac{\text{rate}_1}{\text{rate}_2} = \left(\frac{[I^-]}{[I^-]}\right)^m \left(\frac{[BrO_3^-]}{[BrO_3^-]}\right)^n \left(\frac{[H^+]}{[H^+]}\right)^p$$

$$\frac{8.89 \times 10^{-5} \text{ mol./L} \cdot \text{min}}{3.56 \times 10^{-4} \text{ mol./L} \cdot \text{min}} = \left(\frac{0.0020 \text{ mol./L}}{0.0020 \text{ mol./L}}\right)^m \left(\frac{0.0080 \text{ mol./L}}{0.0080 \text{ mol./L}}\right)^n \left(\frac{0.0020 \text{ mol./L}}{0.0040 \text{ mol./L}}\right)^p$$

$0.250 = (1)^m (1)^n (0.50)^p$
$0.250 = (0.50)^p$
$p = 2$ (The reaction is 2nd order in H^+.)

Overall reaction order $= m + n + p = 1 + 1 + 2 = 4$

b. Rate $= k[I^-][BrO_3^-][H^+]^2$

c. Substitute values from any of the experiments into the rate equation and solve for k.
$8.89 \times 10^{-5} \text{ M/s} = k(0.0020 \text{ M})(0.0080 \text{ M})(0.020 \text{ M})^2$
$8.89 \times 10^{-5} \text{ M/s} = k(6.4 \times 10^{-9} \text{ M}^4)$
$k = 1.4 \times 10^4 \text{ s}^{-1} \cdot \text{M}^{-3} = 1.4 \times 10^4 \text{ L}^3/\text{mol.}^3 \cdot \text{s}$

d. Rate $= k[I^-][BrO_3^-][H^+]^2$
$[I^-] = \frac{1}{2}[BrO_3^-] = 0.0075 \text{ M} \Rightarrow [BrO_3^-] = 0.015 \text{ M}$
$5.00 \times 10^{-4} \text{ M/s} = (1.4 \times 10^4 \text{ L}^3/\text{mol.}^3 \cdot \text{s})(0.0075 \text{ M})(0.015 \text{ M})[H^+]^2$
$5.00 \times 10^{-4} \text{ M/s} = (1.6 \text{ L/mol.} \cdot \text{s})[H^+]^2$
$[H^+]^2 = 3.2 \times 10^{-4} \text{ M}^2$
$[H^+] = 1.8 \times 10^{-2} \text{ M}$

a. To determine the reaction order in [SCN⁻], select two experiments in which $[Cr(H_2O)_6^{3+}]$ is constant (such as experiments 1 and 2, as used below).

$$\frac{rate_1}{rate_2} = \frac{k[Cr(H_2O)_6^{3+}]_1^m[SCN^-]_1^n}{k[Cr(H_2O)_6^{3+}]_2^m[SCN^-]_2^n} \Rightarrow \frac{rate_1}{rate_2} = \left(\frac{[Cr(H_2O)_6^{3+}]}{[Cr(H_2O)_6^{3+}]}\right)^m \left(\frac{[SCN^-]}{[SCN^-]}\right)^n$$

$$\frac{8.4 \times 10^{-4}\ mol./L \cdot min}{6.5 \times 10^{-4}\ mol./L \cdot min} = \left(\frac{0.025\ mol./L}{0.025\ mol./L}\right)^m \left(\frac{0.077\ mol./L}{0.060\ mol./L}\right)^n$$

$1.29 = (1)^m(1.28)^n$

$1.29 = (1.28)^n$

$n = 1$ (The reaction is 1st order in SCN⁻.)

To determine the reaction order in $[Cr(H_2O)_6^{3+}]$, select two experiments in which [SCN⁻] is constant (such as experiments 2 and 3, as used below) and repeat the process above.

$$\frac{8.4 \times 10^{-4}\ mol./L \cdot min}{1.4 \times 10^{-2}\ mol./L \cdot min} = \left(\frac{0.025\ mol./L}{0.042\ mol./L}\right)^m \left(\frac{0.077\ mol./L}{0.077\ mol./L}\right)^n$$

$0.60 = (0.60)^m(1)^n$

$0.60 = (0.60)^m$

$m = 1$ (The reaction is 1st order in $Cr(H_2O)_6^{3+}$.)

Overall reaction order $= m + n = 1 + 1 = 2$

Rate $= k[Cr(H_2O)_6^{3+}][SCN^-]$

b. Substitute values from any of the experiments into the rate equation and solve for k.

$6.5 \times 10^{-4}\ M/s = k(0.025\ M)(0.060\ M)$

$k = 0.43\ min^{-1} \cdot M = 0.43\ L/mol. \cdot min$

c. Calculate $[Cr(H_2O)_6^{3+}]$ and [SCN⁻]. Then use the rate law and k determined above to calculate the rate.

$$15\ mg\ KSCN \times \frac{1\ g\ KSCN}{1000\ mg\ KSCN} \times \frac{1\ mol.\ KSN}{97.19\ g\ KSCN} = 1.5 \times 10^{-4}\ mol.\ KSCN$$

$$1.5 \times 10^{-4}\ mol.\ KSCN \times \frac{1\ mol.\ SCN^-}{1\ mol.\ KSCN} = 1.5 \times 10^{-4}\ mol.\ SCN^-$$

$$[SCN^-] = \frac{1.5 \times 10^{-4}\ mol.\ SCN^-}{1.50\ L} = 1.0 \times 10^{-4}\ M\ SCN^-$$

Rate $= k[Cr(H_2O)_6^{3+}][SCN^-]$

Rate $= (0.43\ L/mol. \cdot min)(0.0500\ M)(1.0 \times 10^{-4}\ M)$

Rate $= 2.2 \times 10^{-6}\ M/min = 2.2 \times 10^{-6}\ mol./L \cdot min$

Generate 2 graphs, ln $[C_{12}H_{22}O_{11}]$ vs. time and $1/[C_{12}H_{22}O_{11}]$ vs. time. The one that yields a straight line is the one that indicates the order of the reaction (see Table 11.2).

time (min)	$[C_{12}H_{22}O_{11}]$	$\ln[C_{12}H_{22}O_{11}]$	$1/[C_{12}H_{22}O_{11}]$
0	0.368	-1.000	2.72
20	0.333	-1.100	3.00
60	0.287	-1.248	3.48
120	0.235	-1.448	4.26
160	0.208	-1.570	4.81

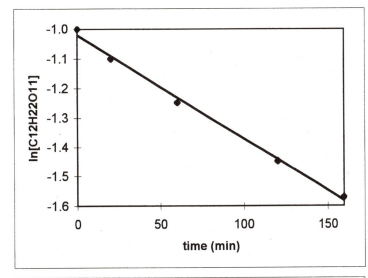

Note how the data points curve around the straight line plotted through the points. This graph is not linear, so the reaction is not 1st order.

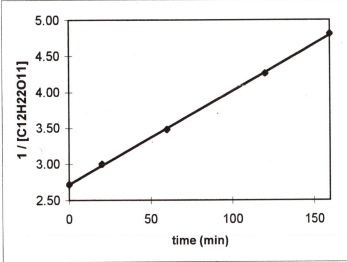

These data points are well represented by the straight line drawn through them. This graph is linear and the reaction is 2nd order.

rate $= k[C_{12}H_{22}O_{11}]^2$

a. A 1st order reaction will yield a straight line when time is plotted against ln[C_2H_6]. Note that the plot yields a straight line.

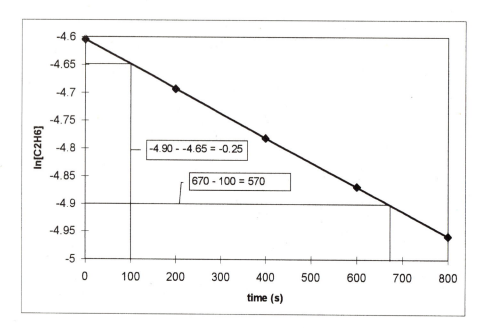

b. $k = \text{- slope} = -\dfrac{\Delta y}{\Delta x} = -\dfrac{-0.25}{570} = 4.4 \times 10^{-4}$

c. $k \cdot t = \ln\dfrac{[C_2H_6]_o}{[C_2H_6]} \implies (4.4 \times 10^{-4}\ s^{-1})(t) = \ln\dfrac{0.01000\,M}{0.00500\,M}$

$t = \dfrac{0.693}{4.4 \times 10^{-4}\ s^{-1}} = 1.6 \times 10^3\ s$

Note that $0.00500\,M$ is half the initial concentration, thus we just calculated the half-life.

d. rate = k[C_2H_6]
rate = $(4.4 \times 10^{-4}\ s^{-1})(0.00400\,M) = 1.8 \times 10^{-6}\ M/s$

a. We will assume 31 day months, thus 2 months = 62 days.

$k \cdot t = \ln\dfrac{[A]_o}{[A]} \implies (3.42 \times 10^{-4}\ d^{-1})(62\ d) = \ln\dfrac{0.0200\,M}{[A]}$

$2.12 \times 10^{-2} = \ln(0.0200\,M) - \ln[A]$
$3.93 = -\ln[A]$
$[A] = e^{-3.93} = 1.96 \times 10^{-2}\,M$

b. $k \cdot t = \ln \frac{[A]_o}{[A]} \implies (3.42 \times 10^{-4} \text{ d}^{-1})(t) = \ln \frac{0.0200 \, M}{0.00350 \, M}$

$\ln(5.71) = (3.42 \times 10^{-4} \text{ d}^{-1})(t)$
$t = 5.10 \times 10^3 \text{ d}$

$5.10 \times 10^3 \text{ d} \times \frac{1 \text{ yr}}{365 \text{ d}} = 14.0 \text{ yr}$

c. $t_{1/2} = \frac{0.693}{k} = \frac{0.693}{3.42 \times 10^{-4} \text{ d}} = 2.03 \times 10^3 \text{ d}$

$2.03 \times 10^3 \text{ d} \times \frac{1 \text{ yr}}{365 \text{ d}} = 5.56 \text{ yr}$

---

36. Refer to Section 11.3 and Problem 34 (above).

---

a. Since the volume of solution and molar mass of the drug are constant, we can substitute mass of the drug for concentration (volume and molar mass cancel).

$k \cdot t = \ln \frac{[A]_o}{[A]} \implies (0.215 \text{ mon.}^{-1})(12 \text{ mon.}) = \ln \frac{10.0 \text{ g}}{m_{drug}}$

$2.58 = \ln(10.0) - \ln(m_{drug})$

$0.28 = -\ln(m_{drug})$

$m_{drug} = e^{-0.28} = 0.758 \text{ g}$

b. $t_{1/2} = \frac{0.693}{k} = \frac{0.693}{0.215 \text{ mon.}^{-1}} = 3.22 \text{ mon.}$

c. 65% of 10.0 g is 6.5 g. Thus, if 65% decomposed, 3.5 g (35%) are left.

$k \cdot t = \ln \frac{[A]_o}{[A]} \implies (0.215 \text{ mon.}^{-1})(t) = \ln \frac{10.0 \text{ g}}{3.5 \text{ g}}$

$t = 4.9 \text{ mon.}$

---

38. Refer to Section 11.3 and Problem 36 above.

---

a. $k \cdot t = \ln \frac{[SO_2Cl_2]_o}{[SO_2Cl_2]} \implies (k)(98 \text{ min}) = \ln \frac{0.300 M}{0.274 M}$

$\ln(1.09) = k(98 \text{ min})$
$k = 9.3 \times 10^{-4} \text{ min}^{-1}$

b. $t_{1/2} = \dfrac{0.693}{k} = \dfrac{0.693}{9.3 \times 10^{-4}\ min^{-1}} = 7.5 \times 10^2\ min$

c. rate $= k[SO_2Cl_2] = (9.3 \times 10^{-4}\ min^{-1})(0.750\ M) = 7.0 \times 10^{-4}\ mol./L \cdot min$

d. 33% of 0.300 M is $(0.33)(0.300\ M) = 0.099\ M$

$k \cdot t = \ln \dfrac{[SO_2Cl_2]_o}{[SO_2Cl_2]} \implies (9.3 \times 10^{-4}\ min^{-1})(t) = \ln \dfrac{0.300\ M}{0.099\ M}$

$\ln(3.03) = (9.3 \times 10^{-4}\ min^{-1})(t)$

$t = 1.2 \times 10^3\ min \times \dfrac{1\ hr}{60\ min} = 20\ hr$

---

40. Refer to Section 11.3 and Example 11.4.

$k \cdot t = \ln \dfrac{[Cu]_o}{[Cu]} \implies (0.0546\ hr^{-1})(8\ hr) = \ln \dfrac{5.00\ mg}{x}$

$0.437 = \ln(5.00\ mg) - \ln[x]$

$-1.17 = -\ln[x]$

$x = e^{1.17} = 3.23\ mg$

---

42. Refer to Section 11.3 and Problem 36 (above).

$k \cdot t = \ln \dfrac{[Na\text{-}24]_o}{[Na\text{-}24]} \implies (k)(24.9\ hr) = \ln \dfrac{0.050\ mg}{0.016\ mg}$

$\ln(3.1) = k(24.9\ hr)$

$k = 0.046\ hr^{-1}$

$t_{1/2} = \dfrac{0.693}{k} = \dfrac{0.693}{0.046\ hr^{-1}} = 15\ hr$

---

44. Refer to Section 11.3.

a. For a zero order reaction, the rate is independent of the concentration.

Rate $= k = 2.08 \times 10^{-4}\ mol./L \cdot s$

$t_{1/2} = \dfrac{[A]_o}{2k} = \dfrac{0.250\ M}{2(2.08 \times 10^{-4}\ mol./L \cdot s)} = 6.01 \times 10^2\ s$

$6.01 \times 10^2\ s \times \dfrac{1\ min}{60\ s} = 10.0\ min$

b. $[A]_o - [A] = kt$

$1.25\,M - 0.388\,M = (2.08 \times 10^{-4}\ \text{mol./L·s})(t)$

$t = 4.1 \times 10^3\ \text{s}$

$4.1 \times 10^3\ \text{s} \times \dfrac{1\,\text{min}}{60\,\text{s}} \times \dfrac{1\,\text{hr}}{60\,\text{min}} = 1.2\,\text{hr}$

---

46. Refer to Section 11.3.

a. $\dfrac{1}{[NO_2]} - \dfrac{1}{[NO_2]_o} = kt$

$\dfrac{1}{0.0104\,M} - \dfrac{1}{0.800\,M} = (k)(125\,\text{s})$

$94.9\,M^{-1} = (k)(125\ \text{s})$

$k = 0.759\ \text{L/mol.·s}$

b. $t_{1/2} = \dfrac{1}{k[NO_2]_o} = \dfrac{1}{(0.759\ \text{L/mol.·s})(0.500\ \text{mol./L})} = 2.64\,\text{s}$

---

48. Refer to Section 11.3 and Problem 36 (above).

If 90.0% has decomposed, then 10.0% is left. 10.0% of 0.0200 is $(0.100)(0.0200\,M) = 0.00200\,M$.

$\dfrac{1}{[NOBr]} - \dfrac{1}{[NOBr]_o} = kt$

$\dfrac{1}{0.00200\,M} - \dfrac{1}{0.0200\,M} = (48\ \text{L/mol.·min})(t)$

$450\,M^{-1} = (48\ \text{L/mol.·min})(t)$

$t = 9.4\,\text{min}$

---

50. Refer to Section 11.5 and Figure 11.10.

Convert temperatures to kelvins, calculate $\ln(k)$, plot the data and calculate E_a from the slope.

k (L/mol.·s)	0.048	2.3	49	590
$\ln(k)$	-3.04	0.833	3.89	6.38
T (K)	773	873	973	1073
$1/T$	1.29×10^{-3}	1.15×10^{-3}	1.03×10^{-3}	9.32×10^{-4}

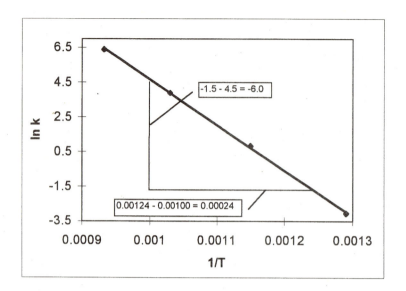

$$\text{slope} = \frac{-6.0}{0.00024} = -2.5 \times 10^4$$

$$\text{slope} = -\frac{E_a}{R}$$

$E_a = -(R)(\text{slope}) = -(8.31 \text{ J/mol.·K})(-2.5 \times 10^4 \text{ K}^{-1}) = 2.1 \times 10^5 \text{ J/mol.}$

$E_a = 2.1 \times 10^2 \text{ kJ/mol.}$

---

52. *Refer to Sections 11.3 and 11.5 and Examples 11.13 and 11.6.*

a. $t_{1/2} = \dfrac{0.693}{k} = \dfrac{0.693}{4.90 \times 10^{-4} \text{ min}^{-1}} = 1.41 \times 10^3 \text{ min}$

b. $\ln \dfrac{k_2}{k_1} = \dfrac{E_a}{R}\left(\dfrac{1}{T_1} - \dfrac{1}{T_2} \right)$

$\ln \dfrac{k_2}{4.90 \times 10^{-4} \text{ min}^{-1}} = \dfrac{4.2 \times 10^5 \text{ J/mol.}}{8.31 \text{ J/mol.·K}} \left(\dfrac{1}{310} - \dfrac{1}{313} \right)$

$\ln(k_2) - (-7.62) = 1.6$

$\ln(k_2) = -6.0$

$k_2 = e^{-6.0} = 2.5 \times 10^{-3} \text{ min}^{-1}$

$t_{1/2} = \dfrac{0.693}{k} = \dfrac{0.693}{2.5 \times 10^{-3} \text{ min}^{-1}} = 2.8 \times 10^2 \text{ min}$

c. The rate increased by a factor of k_2/k_1 for 3°C.

$$\frac{2.5 \times 10^{-3} \text{ min}^{-1}}{4.9 \times 10^{-4} \text{ min}^{-1}} = 5.1 \quad (\text{per } 3°C)$$

$$5.1/3°C = 1.7/°C$$

---

54. Refer to Section 11.5.

a. $X_{25°C} = 7.2(25) - 32 = 148 \text{ chirps/min}$
    $X_{35°C} = 7.2(35) - 32 = 220 \text{ chirps/min}$

b. Use the two point form of the Arrhenius equation and solve for E_a.

$$\ln \frac{k_{25°C}}{k_{35°C}} = \ln \frac{X_{35°C}}{X_{25°C}} = \frac{E_a}{R}\left(\frac{1}{T_1} - \frac{1}{T_2}\right)$$

$$\ln \frac{220 \text{ chirps/min}}{148 \text{ chirps/min}} = \frac{E_a}{8.31 \text{ J/mol.} \cdot \text{K}}\left(\frac{1}{298 \text{ K}} - \frac{1}{308 \text{ K}}\right)$$

$$0.396 = E_a(1.3 \times 10^{-5} \text{ J/mol.})$$

$$E_a = 3.0 \times 10^4 \text{ J/mol.} = 30 \text{ kJ/mol.}$$

c. Percent increase for a 10°C rise in temperature is:

$$148 + 148x = 220$$

$$x = 0.49 = 49\% \text{ increase}$$

---

56. Refer to Section 11.5.

a. Use the two point form of the Arrhenius equation, setting $k_2 = \frac{1}{3}k_1$ (where k_1 corresponds to $T_1 = 500°C = 773$ K).

$$\ln \frac{k_2}{k_1} = \ln \frac{\frac{1}{3}k_1}{k_1} = \frac{E_a}{R}\left(\frac{1}{T_1} - \frac{1}{T_2}\right)$$

$$\ln \frac{1}{3} = \frac{1.81 \times 10^5 \text{ J/mol.}}{8.31 \text{ J/mol.} \cdot \text{K}}\left(\frac{1}{773 \text{ K}} - \frac{1}{T_2}\right)$$

$$-1.099 = (2.18 \times 10^4 \text{ K}^{-1})\left(\frac{1}{773 \text{ K}} - \frac{1}{T_2}\right)$$

$$-5.05 \times 10^{-5} \text{ K} = 1.29 \times 10^{-3} - \frac{1}{T_2}$$

$$T_2 = 746 \text{ K} = 473°C$$

b. $\ln \dfrac{k_2}{k_1} = \dfrac{E_a}{R}\left(\dfrac{1}{T_1} - \dfrac{1}{T_2}\right)$

$\ln \dfrac{k_2}{0.025\ \text{L/mol.}\cdot\text{s}} = \dfrac{1.81\times10^5\ \text{J/mol.}}{8.31\ \text{J/mol.}\cdot\text{K}}\left(\dfrac{1}{773\ \text{K}} - \dfrac{1}{1098\ \text{K}}\right)$

$\ln(k_2) - (-3.7) = 8.34$

$\ln(k_2) = 4.6$

$k_2 = e^{4.6} = 99\ \text{L/mol.}\cdot\text{s}$

---

58. Refer to Sections 11.3 and 11.5.

$t_{1/2} = \dfrac{0.693}{k} \;\Rightarrow\; k = \dfrac{0.693}{t_{1/2}}$

$\ln \dfrac{k_2}{k_1} = \ln \dfrac{\frac{0.693}{t_{1/2}(2)}}{\frac{0.693}{t_{1/2}(1)}} = \ln \dfrac{t_{1/2}(1)}{t_{1/2}(2)} = \dfrac{E_a}{R}\left(\dfrac{1}{T_1} - \dfrac{1}{T_2}\right)$

$\ln \dfrac{40.8\ \text{min}}{t_{1/2}(2)} = \dfrac{2.48\times10^5\ \text{J/mol.}}{8.31\ \text{J/mol.}\cdot\text{K}}\left(\dfrac{1}{743\ \text{K}} - \dfrac{1}{698\ \text{K}}\right) = -2.6$

$3.71 - \ln(t_{1/2}) = -2.6$

$-\ln(t_{1/2}) = -6.3$

$t_{1/2} = e^{6.3} = 5.4\times10^2\ \text{min}$

---

60. Refer to Section 11.7.

a. rate = $k[NO_3][CO]$

b. rate = $k[I_2]$

c. rate = $k[NO][O_2]$

---

62. Refer to Section 11.7 and Example 11.7.

The slow step is the rate determining step (RDS) and determines the rate equation. Thus:

rate$_2$ = $k_2[N_2O_2][H_2]$

However, N_2O_2 is an intermediate, and rate equations should not have intermediates. Use the fast equilibrium in the first step to develop an equation for $[N_2O_2]$. Bear in mind that the forward rate must equal the reverse rate for this to be an equilibrium. Consequently:

214

$k_1[NO]^2 = \text{rate}_1 = \text{rate}_{-1} = k_{-1}[N_2O_2]$

$[N_2O_2] = k_1/k_{-1}[NO]^2$

Substituting this into the first equation gives:

$\text{rate}_2 = k_2k_1/k_{-1}[NO]^2[H_2]$
$\text{rate}_2 = k[NO]^2[H_2]$ (*where k is the combined rate constant*)

Thus the proposed mechanism is consistent with the rate expression.

---

64. *Refer to Section 11.7 and Example 11.7.*

Write a rate law for each of the proposed mechanisms as compare them to the experimental rate law.

a. $\text{rate} = k[CO][NO_2]$

b.   The slow step is the rate determining step (RDS) and determines the rate equation. Thus:

$\text{rate}_2 = k_2[N_2O_4][CO]^2$

However, N_2O_4 is an intermediate, and rate equations should not have intermediates. Use the fast equilibrium in the first step to develop an equation for $[N_2O_4]$. Bear in mind that the forward rate must equal the reverse rate for this to be an equilibrium. Consequently:

$k_1[NO_2]^2 = \text{rate}_1 = \text{rate}_{-1} = k_{-1}[N_2O_4]$

$[N_2O_4] = k_1/k_{-1}[NO_2]^2$

Substituting this into the first equation gives:

$\text{rate}_2 = k_2k_1/k_{-1}[NO_2]^2[CO]^2$
$\text{rate}_2 = k[NO]^2[CO]^2$ (*where k is the combined rate constant*)

c. $\text{rate} = k[NO_2]^2$

d. $\text{rate} = k[NO_2]^2$

Thus proposed mechanisms c and d are both consistent with the experimental results.

---

66. *Refer to Section 11.3 and Chapter 8.*

Calculate $\Delta H°$. Then calculate the amount of NOBr consumed in one minute. Use those values to calculate heat consumed/evolved.

$\Delta H° = \Sigma \Delta H_f°_{\text{(products)}} - \Sigma \Delta H_f°_{\text{(reactants)}}$
$\Delta H° = [\Delta H_f° \, NO(g) + \frac{1}{2}\Delta H_f° \, Br_2(g)] - [\Delta H_f° \, NOBr(g)]$
$\Delta H° = [(1 \text{ mol.})(90.2 \text{ kJ/mol.}) + (\frac{1}{2} \text{ mol.})(30.9 \text{ kJ/mol.})] - [(1 \text{ mol.})(82.2 \text{ kJ/mol.})]$
$\Delta H° = 23.5 \text{ kJ}$

$$\frac{1}{[NOBr]} - \frac{1}{[NOBr]_o} = kt$$

$$\frac{1}{[NOBr]} - \frac{1}{0.250\,M} = (0.80\,\text{L/mol}\cdot\text{s})(60\,\text{s})$$

$$\frac{1}{[NOBr]} = (48\,\text{L/mol.}) - (4\,\text{L/mol.})$$

$[NOBr] = 0.0227\,M$ (this is the final concentration of NOBr)

$[NOBr]_{consumed} = [NOBr]_o - [NOBr] = 0.250 - 0.0227 = 0.227\,M$

$$0.227\,\text{mol. NOBr} \times \frac{23.5\,\text{kJ}}{1\,\text{mol. NOBr}} = 5.3\,\text{kJ}$$

Thus 5.3 kJ of heat would be consumed per liter in one minute.

---

68. Refer to Section 11.5 and Problem 52 (above).

$$\ln\frac{k_2}{k_1} = \frac{E_a}{R}\left(\frac{1}{T_1} - \frac{1}{T_2}\right)$$

$$\ln\frac{k_2}{k_1} = \frac{1.33\times10^5\,\text{J/mol.}}{8.31\,\text{J/mol.}\cdot\text{K}}\left(\frac{1}{300} - \frac{1}{318}\right)$$

$$\ln\frac{k_2}{k_1} = 3.02$$

$$\frac{k_2}{k_1} = 20.5$$

Thus the reaction would proceed approximately 20 times faster.

---

70. Refer to Section 11.3.

$-\Delta[A] = k\Delta t$

Now expand the terms $\Delta[A]$ and Δt and substitute the expanded equations into the one above.

    $\Delta[A] = [A] - [A]_o$

    $\Delta t = t - t_o = t - 0 = t$ ($t_o = 0$ since the reaction starts at time zero)

$-([A] - [A]_o) = kt$

$-[A] + [A]_o = kt$

$[A]_o - [A] = kt$ (see Table 11.2)

    or

$[A] = [A]_o - kt$

In both the trials depicted, the amount of reactant is reduced to half, and thus the time is the half-life. Calculate k for each trial assuming a 0 order reaction, a 1st order reaction and a 2nd order reaction. The equation that gives a consistent k for both trials indicates the order of the reaction.

0 order: half-life = $[A]_o/2k$

    trial 1: 1 min = $8/2k$ $\Rightarrow$ $k = 4$

    trial 2: 2 min = $4/2k$ $\Rightarrow$ $k = 1$

        The reaction is not 0 order.

1st order: half-life = $0.693/k$

    trial 1: 1 min = $0.693/k$ $\Rightarrow$ $k = 0.693$

    trial 2: 2 min = $0.693/k$ $\Rightarrow$ $k = 0.347$

        The reaction is not 1st order.

2nd order: half-life = $1/k[A]_o$

    trial 1: 1 min = $1/k(8)$ $\Rightarrow$ $k = 0.125$

    trial 2: 2 min = $1/k(4)$ $\Rightarrow$ $k = 0.125$

        The reaction is 2nd order.

a. The fastest reaction will be the one with the smallest activation energy barrier: **A**.

b. The reaction with the largest half-life will be the slowest reaction, which will be the one with the largest activation energy barrier: **C**.

c. The reaction with the largest rate constant will be the fastest reaction, which will be the one with the smallest activation energy barrier: **A**.

a. rate = $k[A]^2[B]$

b. rate = $k(3\,M)^2(4\,M)$
    rate = $k \cdot 36\,M^3$

c. Combine the equations from a and b above, substitute for A (squares) and solve for B (circles).

$$k[A]^2[B] = \text{rate} = k(36\ M^3)$$
$$k[2]^2[B] = k(36\ M^3)$$
$$[B] = 9\ M$$

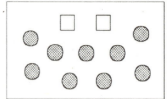

---

78. Refer to Section 11.2 and 11.5.

a. (1) rate = $k(0.10\ M)(0.10\ M) = 0.010k$
    (2) rate = $k(0.15\ M)(0.15\ M) = 0.023k$
    (3) rate = $k(0.06\ M)(1.0\ M) = 0.06k$

Test tube (1) will have the smallest rate.

b. When the temperature increases, rate increases.
Since rate and k are directly proportional, k also increases.
E_a is constant for a given reaction, thus E_a remains the same. The increase in rate is due to an increase in the kinetic motion of the molecules.

---

80. Refer to Sections 11.2, 11.4 and 11.6 and Chapter 8.

a. $\Delta H° = \Sigma\ \Delta H°_{f(prod)} - \Sigma\ \Delta H°_{f(react.)}$

$\Delta H° = (2\ \text{mol.})(26.48\ \text{kJ/mol.}) - [(1\ \text{mol.})(62.44\ \text{kJ/mol.}) + (1\ \text{mol.})(0.0\ \text{kJ/mol.})]$

$\Delta H° = -9.48\ \text{kJ}$

$\Delta H° = E_{a(forward)} - E_{a(reverse)}$

$E_{a(reverse)} = 165\ \text{kJ/mol.} - (-9.48\ \text{kJ/mol.}) = 174\ \text{kJ/mol.}$

b. $k_{forward} = Ae^{-Ea/RT}$

$138\ \text{L/mol.·s} = Ae^{(-165000\ \text{J})/(8.31\ \text{J/mol.·K})(973\ \text{K})}$

$138\ \text{L/mol.·s} = Ae^{-20.4}$

$138\ \text{L/mol.·s} = A(1.38 \times 10^{-9})$

$A = 1.00 \times 10^{11}$

$k_{reverse} = Ae^{-Ea/RT}$

$k_{reverse} = (1.00 \times 10^{11})\ e^{(-174000\ \text{J})/(8.31\ \text{J/mol.·K})(973\ \text{K})}$

$k_{reverse} = 46\ \text{L/mol.·s}$

c. $\text{rate}_{\text{reverse}} = k_{\text{reverse}} [\text{HI}]^2$
    $\text{rate}_{\text{reverse}} = (46 \text{ L/mol.·s})(0.200 \text{ } M)^2$
    $\text{rate}_{\text{reverse}} = 1.8 \text{ mol./L·s}$

---

81. Refer to Section 11.3 and Footnote on page 318.

For first order reactions $\text{rate} = \dfrac{-d[\text{A}]}{\text{a} \cdot dt} = k[\text{A}]$

$$-d[\text{A}] = ak[\text{A}]dt \quad \Rightarrow \quad \int \dfrac{-d[\text{A}]}{[\text{A}]} = \int akdt$$

$$\ln \dfrac{[\text{A}]_0}{[\text{A}]} = \text{a } kt$$

Since both a and k are constants, they can be replaced with a new constant k.

$$\ln \dfrac{[\text{A}]_0}{[\text{A}]} = kt$$

---

82. Refer to Section 11.2 and Example 11.2.

To determine the order with respect to A, use two trials in which [A] changes and [B] and [C] are constant (such as experiments 1 and 4, as used below), and solve for m.

$$\dfrac{\text{rate}_1}{\text{rate}_2} = \dfrac{k[\text{A}]_1^m[\text{B}]_1^n[\text{C}]_1^p}{k[\text{A}]_2^m[\text{B}]_2^n[\text{C}]_2^p}$$

$$\dfrac{4x}{x} = \left(\dfrac{0.40}{0.20}\right)^m \left(\dfrac{0.40}{0.40}\right)^n \left(\dfrac{0.10}{0.10}\right)^p$$

$$4 = (2)^m$$

$m = 2$ Thus the reaction is 2nd order in A

Determining the order with respect to B involves a small twist of the usual procedure. Use two trials in which [B] changes and [C] is constant (such as experiments 2 and 3, as used below), and substitute $m = 2$ into the equation.

$$\dfrac{8x}{x} = \left(\dfrac{0.40}{0.20}\right)^2 \left(\dfrac{0.40}{0.20}\right)^n \left(\dfrac{0.20 \text{ mol./L}}{0.20 \text{ mol./L}}\right)^p$$

$$2 = (2)^n$$

$n = 1$ Thus the reaction is 1st order in B

To determine the order with respect to C, use two trials in which [C] changes (such as experiments 1 and 2, as used below) and substitute $m = 2$ and $n = 1$ into the equation.

$$\frac{8x}{x} = \left(\frac{0.40}{0.20}\right)^2 \left(\frac{0.40}{0.40}\right)^1 \left(\frac{0.20}{0.10}\right)^p$$

$$2 = (2)^p$$

$p = 1$ Thus the reaction is 1st order in C

rate $= k[A]^2[B][C]$

---

83. Refer to Section 11.3, Table 11.2 and Problem 81 (above).

---

a. rate $= \dfrac{-d[A]}{a \cdot dt} = k[A]^2$

$$\int \frac{-d[A]}{[A]^2} = \int akdt$$

$$\frac{1}{[A]} - \frac{1}{[A]_o} = a\,kt$$

Since both a and k are constants, they can be replaced with a new constant k.

$$\frac{1}{[A]} - \frac{1}{[A]_o} = kt$$

b. rate $= \dfrac{-d[A]}{a \cdot dt} = k[A]^3$

$$\int \frac{-d[A]}{[A]^3} = \int akdt$$

$$\frac{1}{2[A]^2} - \frac{1}{2[A]_o^{\,2}} = a\,kt$$

Since both a and k are constants, they can be replaced with a new constant k.

$$\frac{1}{[A]^2} - \frac{1}{[A]_o^{\,2}} = 2\,kt$$

At 3 times per day, $\quad \Delta t = \dfrac{24\,\text{hr}}{3} = 8\,\text{hr}$

$t_{1/2} = 2.0\ \text{days} = 48\ \text{hr}$

$$\text{saturation value} = \frac{X}{1-10^{-0.30\frac{\Delta t}{t_{1/2}}}} = \frac{0.100\,\text{g}}{1-10^{-0.30\frac{8}{48}}}$$

$$\text{saturation value} = \frac{0.100\,\text{g}}{1-10^{-0.05}} = \frac{0.100\,\text{g}}{1-0.89} = 0.91\,\text{g}$$

In the second part of the problem, the goal is to find a mass (X) of drug that gives a saturation value of 0.500g

$$\text{saturation value} = 0.500\,\text{g} = \frac{X}{1-0.89} = \frac{X}{0.11}$$

$X = 0.055\ \text{g} = 55\ \text{mg}$

Thus a dose of 55 mg, three times a day would produce an accumulation of 0.500 g drug in the body, and thus the patient would experience side effects. To prevent the side effects, a pharmacist would assign a dosage of 54 mg, three times a day.

Chapter 12: Gaseous Chemical Equilibrium

---

2. *Refer to Section 12.1.*

a. 80 s. After 80 s, P_A and P_B remain constant.

b. *After 30 seconds*, [A] is still decreasing, so the rate of the forward reaction must be **greater** than the rate of the reverse reaction.
    After 90 s, the [A] is no longer changing, so the rates of the forward and reverse reactions must be **equal**.

---

4. *Refer to Section 12.1.*

Time (min)	0	1	2	3	4	5	6
P_A (atm)	1.000	0.778	**0.580**	**0.415**	**0.355**	0.325	**0.325**
P_B (atm)	0.400	**0.326**	0.260	**0.205**	0.185	**0.175**	0.175
P_C (atm)	0.000	**0.148**	**0.280**	0.390	**0.430**	**0.450**	**0.450**

1 min: 1.000 atm A – 0.788 atm A = 0.222 atm A

$$0.222 \, \text{atm A} \times \frac{1B}{3A} = 0.0740 \, \text{atm B}$$

0.400 atm B – 0.0740 atm B = 0.326 atm B

$$0.222 \, \text{atm A} \times \frac{2C}{3A} = 0.148 \, \text{atm C}$$

2 min: 0.400 atm B – 0.260 atm B = 0.140 atm B

$$0.140 \, \text{atm B} \times \frac{3A}{1B} = 0.420 \, \text{atm A}$$

1.000 atm A – 0.420 atm A = 0.580 atm A

$$0.140 \, \text{atm B} \times \frac{2C}{1B} = 0.280 \, \text{atm C}$$

3 min: $0.390 \, \text{atm C} \times \dfrac{1B}{2C} = 0.195 \, \text{atm B}$

0.400 atm B – 0.195 atm B = 0.205 atm B

$$0.390 \, \text{atm C} \times \frac{3A}{2C} = 0.585 \, \text{atm A}$$

1.000 atm A − 0.585 atm A = 0.415 atm A

4 min: 0.400 atm B − 0.185 atm B = 0.215 atm B

$$0.215\,\text{atm B} \times \frac{3A}{1B} = 0.645\,\text{atm A}$$

1.000 atm A − 0.645 atm A = 0.355 atm A

$$0.215\,\text{atm B} \times \frac{2C}{1B} = 0.430\,\text{atm C}$$

5 min: 1.000 atm A − 0.325 atm A = 0.675 atm A

$$0.675\,\text{atm A} \times \frac{1B}{3A} = 0.225\,\text{atm B}$$

0.400 atm B − 0.225 atm B = 0.175 atm B

$$0.675\,\text{atm A} \times \frac{2C}{3A} = 0.450\,\text{atm C}$$

6 min: Since P_B has not changed from 5 min, P_A and P_C will also equal the values from 5 min.

---

6. Refer to Section 12.2 and Example 12.1.

a. $K = \dfrac{(P_{CH_4})(P_{H_2S})^2}{(P_{CS_2})(P_{H_2})^4}$

b. $K = \dfrac{(P_{H_2O})^2(P_{O_2})}{(P_{H_2O_2})^2}$

c. $K = \dfrac{(P_{H_2O})^4(P_{CO_2})^3}{(P_{C_3H_8})(P_{O_2})^5}$

---

8. Refer to Section 12.2 and Example 12.2.

Remember, only gases and dissolved species enter the equilibrium constant expression.

a. $K = \dfrac{(P_{NO})^2[Cu^{2+}]^3}{[NO_3^-]^2[H^+]^8}$

b. $K = \dfrac{(P_{SO_2})^2}{(P_{O_2})^3}$

c. $K = \dfrac{1}{[CO_3^{2-}][Ca^{2+}]}$

224

10. Refer to Section 12.2 and Example 12.2.

a. $C_3H_6O(l) \rightleftharpoons C_3H_6O(g)$ $\qquad\qquad K = P_{C_3H_6O}$

Note that the reactant is a liquid and therefore does not appear in the equation.

b. $7H_2(g) + 2NO_2(g) \rightleftharpoons 2NH_3(g) + 4H_2O(g)$ $\qquad K = \dfrac{(P_{NH_3})^2 (P_{H_2O})^4}{(P_{H_2})^7 (P_{NO_2})^2}$

c. $H_2S(g) + Pb^{2+}(aq) \rightleftharpoons PbS(s) + 2H^+(aq)$ $\qquad K = \dfrac{[H^+]^2}{(P_{H_2S})[Pb^{2+}]}$

12. Refer to Section 12.2.

Those species not in the K_{eq} expression, but needed to balance the reactions, must be present in the reaction as either a solid or a liquid.

a. $3CO_2(g) + 4H_2O(g) \rightleftharpoons C_3H_8(g) + 5O_2(g)$

b. $2Fe^{3+}(aq) + 2Cl^-(g) \rightleftharpoons 2Fe^{2+}(aq) + Cl_2(g)$

c. $2NH_3(g) \rightleftharpoons N_2(g) + 3H_2(g)$

d. $4HCl(g) + O_2(g) \rightleftharpoons 2H_2O(g) + 2Cl_2(g)$

e. $2H_2O(g) + O_2(g) \rightleftharpoons 2H_2O_2(g)$

14. Refer to Section 12.2.

Write the balanced reaction for the equilibrium and relate that to the one given.

a. $2ICl(g) \rightleftharpoons I_2(g) + Cl_2(g)$

This equation is 2 times the given equation, so the given K must be raised to the 2nd power.

$K = (2.2 \times 10^{-3})^2 = 4.8 \times 10^{-6}$

b. $I_2(g) + Cl_2(g) \rightleftharpoons 2ICl(g)$

This equation is the reverse of the equation in part a, so take the reciprocal of that K.

$K = 1/(4.8 \times 10^{-6}) = 2.1 \times 10^{+5}$

16. Refer to Section 12.2.

Multiply the second equation through by two, then add the two equations to get the desired equation. Remember that multiplying through by two means K must be squared.

$$SnO_2(s) + 2H_2(g) \rightleftharpoons Sn(s) + 2H_2O(g) \qquad\qquad K_1 = 21$$
$$\underline{2CO(g) + 2H_2O(g) \rightleftharpoons 2CO_2(g) + 2H_2(g) \qquad\qquad K_2 = (0.034)^2}$$
$$SnO_2(s) + 2H_2(g) + 2CO(g) + 2H_2O(g)$$
$$\rightleftharpoons 2CO_2(g) + 2H_2(g) + Sn(s) + 2H_2O(g)$$
$$SnO_2(s) + 2CO(g) \rightleftharpoons 2CO_2(g) + Sn(s) \qquad\qquad K = K_1K_2 = 0.024$$

18. Refer to Section 12.2 and Chapter 8.

Reversing the first equation and adding it to the second gives the equation for the formation of 2 moles NOBr. Divide that equation by two and take the square root of K. Remember that reversing a reaction inverts the equilibrium constant.

$$N_2(g) + O_2(g) \rightleftharpoons 2NO(g) \qquad\qquad K_1 = (1 \times 10^{-30})^{-1} = 1 \times 10^{30}$$
$$\underline{2NO(g) + Br_2(g) \rightleftharpoons 2NOBr(g) \qquad\qquad K_2 = 8 \times 10}$$
$$N_2(g) + O_2(g) + 2NO(g) + Br_2(g) \rightleftharpoons 2NOBr(g) + 2NO(g) \qquad K = K_1K_2 = 8 \times 10^{31}$$
$$\tfrac{1}{2}N_2(g) + \tfrac{1}{2}O_2(g) + \tfrac{1}{2}Br_2(g) \rightleftharpoons NOBr(g) \qquad K = (8 \times 10^{31})^{\frac{1}{2}} = 9 \times 10^{15}$$

20. Refer to Section 12.3 and Example 12.3.

$$K = \frac{(P_{CH_4})(P_{H_2O})}{(P_{CO})(P_{H_2})^3} = \frac{(0.0391\,atm)(0.0124\,atm)}{(0.921\,atm)(1.21\,atm)^3} = 2.97 \times 10^{-4}$$

Note that K is usually expressed without units.

22. Refer to Section 12.2 and Chapter 5.

a. $CH_4(g) + 2H_2S(g) \rightleftharpoons 2CS_2(g) + 4H_2(g)$

b. Use the ideal gas law equation to calculate the pressure of each gas, then substitute those values into the equation for the equilibrium constant.

$$P_{CH_4} = \frac{n}{V}RT = \frac{0.00142\,mol.}{1\,L}(0.0821\,L \cdot atm/mol. \cdot K)(1123\,K) = 0.131\,atm$$

$$P_{H_2S} = \frac{n}{V}RT = \frac{6.14 \times 10^{-4}\,mol.}{1\,L}(0.0821\,L \cdot atm/mol. \cdot K)(1123\,K) = 0.0566\,atm$$

$$P_{CS_2} = \frac{n}{V}RT = \frac{0.00266 \, \text{mol.}}{1 \, \text{L}}(0.0821 \, \text{L} \cdot \text{atm/mol.} \cdot \text{K})(1123 \, \text{K}) = 0.245 \, \text{atm}$$

$$P_{H_2} = \frac{n}{V}RT = \frac{0.00943 \, \text{mol.}}{1 \, \text{L}}(0.0821 \, \text{L} \cdot \text{atm/mol.} \cdot \text{K})(1123 \, \text{K}) = 0.869 \, \text{atm}$$

$$K = \frac{(P_{H_2})^4(P_{CS_2})}{(P_{CH_4})(P_{H_2S})^2} = \frac{(0.869 \, \text{atm})^4(0.245 \, \text{atm})}{(0.131 \, \text{atm})(0.0566 \, \text{atm})^3} = 334$$

---

24. Refer to Section 12.3 and Example 12.4.

	2NO(g)	+	Br$_2$(g)	⇌	2NOBr(g)
P_o	1.577 atm		0.427 atm		0 atm
ΔP	-0.624 atm		-0.312 atm		0.624 atm
P_{eq}	0.953 atm		0.115 atm		0.624 atm

If $P_{eq \, (NOBr)} = 0.624$, then $\Delta P_{(NOBr)} = 0.624$ atm

$\Delta P_{(NOBr)} = -\Delta P_{(NO)} = -0.624$ atm

$\Delta P_{(NOBr)} = -\frac{1}{2}\Delta P_{(Br2)} = -0.312$ atm

$$K = \frac{(P_{NOBr})^2}{(P_{NO})^2(P_{Br_2})} = \frac{(0.624 \, \text{atm})^2}{(0.953 \, \text{atm})^2(0.115 \, \text{atm})} = 3.73$$

---

26. Refer to Section 12.4 and Example 12.5.

a. $K = \dfrac{(P_{SO_2})(P_{Cl_2})}{(P_{SO_2Cl_2})}$

$Q = \dfrac{(P_{SO_2})(P_{Cl_2})}{(P_{SO_2Cl_2})} = \dfrac{(0.16)(0.30)}{0.50} = 0.096$

Since $Q \neq K$, the system is **not** at equilibrium.

b. $Q > K$, so the reaction will proceed to the **left** (to more reactants).

28. *Refer to Section 12.4 and Example 12.5.*

a. $K = \dfrac{(P_{NO})^2 (P_{O_2})}{(P_{NO_2})^2}$

$Q = \dfrac{(P_{NO})^2 (P_{O_2})}{(P_{NO_2})^2} = \dfrac{(0.10)^2 (0.10)}{(0.10)^2} = 0.10$

$Q < K$, therefore the reaction proceeds to the **right** (more products)

b. $Q = \dfrac{(P_{NO})^2 (P_{O_2})}{(P_{NO_2})^2} = \dfrac{(0.0116)^2 (0)}{(0.0848)^2} = 0$

$Q < K$, therefore the reaction proceeds to the right (more products). If the partial pressure of one of the products is zero, then the reaction can only proceed to the **right**, and calculations were actually unnecessary.

c. $Q = \dfrac{(P_{NO})^2 (P_{O_2})}{(P_{NO_2})^2} = \dfrac{(0.040)^2 (0.010)}{(0.20)^2} = 4.0 \times 10^{-4}$

$Q < K$, therefore the reaction proceeds to the **right** (more products)

30. *Refer to Section 12.4.*

$K = \dfrac{(P_{NO})(P_{O_2})^{1/2}}{P_{NO_2}} \Rightarrow 0.0132 = \dfrac{(0.281)(P_{O_2})^{1/2}}{0.817}$

$(P_{O_2})^{1/2} = 0.0384 \text{ atm}$

$(P_{O_2}) = 0.00147 \text{ atm}$

32. *Refer to Section 12.4*

$K = \dfrac{(P_{H_2S})^2}{(P_{H_2})^2 (P_{S_2})} \Rightarrow 1.3 \times 10^5 = \dfrac{(P_{H_2S})^2}{(0.103)^2 (0.417)}$

$(P_{H_2S})^2 = 5.8 \times 10^2 \text{ atm}$

$P_{H_2S} = 24 \text{ atm}$

First, use the equilibrium equation to establish a ratio of the partial pressures of NO_2 to NO. Then substitute that ratio into Dalton's law of partial pressures and solve for the partial pressure of NO_2.

$$K = \frac{(P_{NO})^2(P_{O_2})}{(P_{NO_2})^2} \implies 0.87 = \frac{(P_{NO})^2(0.515)}{(P_{NO_2})^2}$$

$$\sqrt{1.69} = \sqrt{\frac{(P_{NO})^2}{(P_{NO_2})^2}}$$

$$1.3 = \frac{P_{NO}}{P_{NO_2}} \implies 1.3 P_{NO_2} = P_{NO} \qquad\qquad \text{Equation 1}$$

$$P_{total} = P_{NO} + P_{NO_2} + P_{O_2} \implies 1.25 = P_{NO} + P_{NO_2} + 0.515$$

$$0.735 = P_{NO} + P_{NO_2} \qquad\qquad \text{Equation 2}$$

Now substitute equation 1 into equation two, and solve for pressure of NO_2.

$$0.735 = 1.3 P_{NO_2} + P_{NO_2} \implies 0.735 = 2.3 P_{NO_2}$$

$$P_{NO_2} = 0.32$$

$$P_{NO} = 1.3 P_{NO_2} = 0.42$$

Set ΔP to x, calculate P_{eq} and then apply the resulting expressions to the equilibrium equation. Then solve for x.

	$C_2N_2(g)$	+	$H_2(g)$	$\rightleftharpoons$	$2HCN(g)$
P_o	0.500 atm		0.500 atm		0 atm
ΔP	-x atm		-x atm		+2x atm
P_{eq}	0.500 - x atm		0.500 - x atm		2x atm

$$K = \frac{(P_{HCN})^2}{(P_{C_2N_2})(P_{H_2})} \implies 47 = \frac{(2x)^2}{(0.500 - x)(0.500 - x)}$$

$$\sqrt{47} = \sqrt{\frac{(2x)^2}{(0.500 - x)^2}} \implies 6.9 = \frac{2x}{0.500 - x}$$

$x = 0.39$

P_{eq} (C_2N_2) = P_{eq} (H_2) = 0.500 – 0.39 = 0.11 atm
P_{eq} (HCN) = 2(0.39) = 0.77 atm

---

38. Refer to Section 12.4, Example 12.6 and Problem 36 (above).

	CO(g)	+	H$_2$O(g)	$\rightleftharpoons$	H$_2$(g)	+	CO$_2$(g)
P_o	0.485 atm		0.485 atm		0.159 atm		0.159 atm
ΔP	-x atm		-x atm		+x atm		+x atm
P_{eq}	0.485 - x atm		0.485 - x atm		0.159 + x atm		0.159 + x atm

a. $K = \dfrac{(P_{H_2})(P_{CO_2})}{(P_{CO})(P_{H_2O})} \Rightarrow 1.30 = \dfrac{(0.159+x)(0.159+x)}{(0.485-x)(0.485-x)}$

$\sqrt{1.30} = \sqrt{\dfrac{(0.159+x)^2}{(0.485-x)^2}} \Rightarrow 1.14 = \dfrac{0.159+x}{0.485-x}$

$0.553 - 1.14x = 0.159 - x$

$x = 0.184$

P_{eq} (CO) = P_{eq} (H$_2$O) = 0.485 – 0.184 = 0.301 atm
P_{eq} (H$_2$) = P_{eq} (CO$_2$) = 0.159 + 0.184 = 0.343 atm

b. $P_{init} = \Sigma\, P_o$ = 0.485 + 0.485 + 0.159 + 0.159 = 1.288 atm
$P_{final} = \Sigma\, P_{eq}$ = 0.301 + 0.301 + 0.343 + 0.343 = 1.288 atm
$P_{init} = P_{final}$; This is always true when $\Delta n_g = 0$.

---

40. Refer to Section 12.4 and Chapter 5.

a. Write the equilibrium expression, bearing in mind that NH$_4$I is a solid, and therefore does not take part in the equation.

$K = \dfrac{(P_{NH_3})(P_{HI})}{1} = (P_{NH_3})(P_{HI})$

Since equimolar amounts of NH$_3$ and HI are produced the pressure of each will be equal:

$K = (P_{NH_3})(P_{HI}) = x^2 \Rightarrow 0.215 = x^2$

$P_{NH_3} = P_{HI} = x = 0.464\,\text{atm}$

$P_{total} = P_{NH_3} + P_{HI} = 0.464 + 0.464 = 0.928\,\text{atm}$

b. Use the ideal gas equation to calculate the amount of HI (or NH_3) that was produced, then calculate the amount of NH_4I this represents.

$PV = nRT$
$(0.464 \text{ atm})(5.0 \text{ L}) = (n)(0.0821 \text{ L·atm/mol.·K})(673 \text{ K})$
$n = 0.042 \text{ mol.}$

$$0.042 \text{ mol. NH}_3 \times \frac{1 \text{ mol. NH}_4\text{I}}{1 \text{ mol. NH}_3} \times \frac{145 \text{ g NH}_4\text{I}}{1 \text{ mol. NH}_4\text{I}} = 6.1 \text{ g NH}_4\text{I}$$

---

42. Refer to Section 12.4 and Appendix 3.

	$N_2O_4(g)$	$\rightleftharpoons$	$2NO_2(g)$
P_o	0.863 atm		0 atm
ΔP	$-x$ atm		$+2x$ atm
P_{eq}	$0.863 - x$ atm		$2x$ atm

$$K = 0.144 = \frac{(P_{NO_2})^2}{P_{N_2O_4}} = \frac{(2x)^2}{(0.863 - x)}$$

To solve for x, one must put the equation in the form of $ax^2 + bx + c = 0$ and employ the quadratic equation.

$$0.124 - 0.144x = 4x^2 \quad \Rightarrow \quad 4x^2 + 0.144x - 0.124 = 0$$

$$x = \frac{-b \pm \sqrt{b^2 - 4ac}}{2a} = \frac{-0.144 \pm \sqrt{(0.144)^2 - 4(4)(-0.124)}}{2(4)} = 0.159$$

There is actually a second value for x, -0.195, but a negative pressure is impossible, so this value is rejected.

$P_{eq} (N_2O_4) = 0.863 - 0.159 = 0.704 \text{ atm}$
$P_{eq} (NO_2) = 2(0.159) = 0.318 \text{ atm}$

---

44. Refer to Section 12.5.

a. (1) Adding O_2 means one is adding product, thus the equilibrium will shift towards the reactant side, causing a **reverse** reaction.

(2) Compressing the system will result in higher pressure. This stress can be offset by the equilibrium shifting to the side with the least moles of gas, the reactant side, which would **reverse** the reaction.

(3) Adding an inert gas will not change the partial pressures of the gases involved in the reaction, thus adding Ar gas has **no effect** on the equilibrium.

(4) Removing SO_2 means one is removing product, thus the equilibrium will shift towards the product side, causing a **forward** reaction to occur.

(5) Decreasing the temperature in an endothermic reaction will cause the **reverse** reaction to occur, which then generates heat to offset the cooling.

b. K is a constant for any given temperature. If the changes occur at a constant temperature, then K will not change. Thus only **(5)** would change (decrease) K.

---

46. Refer to Section 12.5 and Example 12.9.

---

a. 4 mol. (g) $\rightleftharpoons$ 1 mol. (g)
A pressure decrease would shift the equilibrium to more moles of gas, to the **left**.

b. 1 mol. (g) $\rightleftharpoons$ 2 mol. (g)
A pressure decrease would shift the equilibrium to more moles of gas, to the **right**.

c. 1 mol. (g) $\rightleftharpoons$ 1 mol. (g)
Since there is an equal number of moles, no change would **occur**.

---

48. Refer to Section 12.5 and Example 12.8.

---

a. $K = \dfrac{(P_{CO})(P_{H_2})}{P_{H_2O}} = \dfrac{(0.22)(0.63)}{(0.31)} = 0.45$

b. Set ΔP to x, calculate P_{eq} and then apply the resulting expressions to the equilibrium equation. Then solve for x.

	C(s)	+	$H_2O(g)$	$\rightleftharpoons$	$H_2(g)$	+	CO(g)
P_o	---		0.50 atm		0.22 atm		0.63 atm
ΔP	---		-x atm		+x atm		+x atm
P_{eq}	---		0.50 - x atm		0.22 + x atm		0.63 + x atm

$K = \dfrac{(P_{CO})(P_{H_2})}{P_{H_2O}} \Rightarrow 0.45 = \dfrac{(0.22 + x)(0.63 + x)}{(0.50 - x)}$

$0.23 - 0.45x = 0.14 + 0.85x + x^2 \Rightarrow x^2 + 1.3x - 0.09 = 0$

$$x = \frac{-b \pm \sqrt{b^2 - 4ac}}{2a} = \frac{-1.3 \pm \sqrt{(1.3)^2 - 4(1)(-0.09)}}{2(1)} = 0.066$$

There is actually a second value for x, -1.36, but a negative pressure is impossible, so this value is rejected.

P_{eq} (H_2O) = 0.50 – 0.066 = 0.43 atm
P_{eq} (CO) = 0.22 + 0.066 = 0.29 atm
P_{eq} (H_2) = 0.63 + 0.066 = 0.70 atm

---

50. Refer to Section 12.5.

Use the van't Hoff equation, and solve for K_2.

$$\ln\frac{K_2}{K_1} = \frac{\Delta H^\circ}{R}\left[\frac{1}{T_1} - \frac{1}{T_2}\right] \Rightarrow \ln\frac{K_2}{62.5} = \frac{-9.4 \times 10^3 \text{ J/mol.}}{8.31 \text{ J/mol.}\cdot K}\left[\frac{1}{800 \text{ K}} - \frac{1}{606 \text{ K}}\right] = 0.453$$

$\ln(K_2) - \ln(62.5) = 0.453$

$\ln(K_2) = 0.453 + 4.14 = 4.59$

$K_2 = e^{4.59} = 98$

---

52. Refer to Section 12.5 and Problem 50 (above).

Set $K_1 = 0.40K_2$ and substitute into the van't Hoff equation.

$$\ln\frac{K_2}{K_1} = \frac{\Delta H^\circ}{R}\left[\frac{1}{T_1} - \frac{1}{T_2}\right] \Rightarrow \ln\frac{K_2}{0.40K_2} = \frac{\Delta H^\circ}{8.31 \text{ J/mol.}\cdot K}\left[\frac{1}{323 \text{ K}} - \frac{1}{310 \text{ K}}\right]$$

$0.92 = \Delta H^\circ(-1.56 \times 10^{-5} \text{ mol./J})$

$\Delta H^\circ = -5.9 \times 10^4 \text{ J/mol.}$

$\Delta H^\circ = -59 \text{ kJ/mol.}$

---

54. Refer to Section 12.5.

Calculate $K_{700^\circ C}$ using the ideal gas law equation to determine the partial pressures. Then repeat the calculation for $K_{600^\circ C}$.

$$K_{700°C} = \frac{(P_{CO})^2}{P_{CO_2}} = \frac{\left(\dfrac{(0.10\,\text{mol.})(0.0821\,\text{L}\cdot\text{atm/mol}\cdot\text{K})(973\,\text{K})}{2.0\,\text{L}}\right)^2}{\dfrac{(0.20\,\text{mol.})(0.0821\,\text{L}\cdot\text{atm/mol}\cdot\text{K})(973\,\text{K})}{2.0\,\text{L}}} = 2.0$$

If 0.040 moles additional C(s) are formed, then an additional 0.040 mol. $CO_2(g)$ are also formed (giving 0.24 mol.), and the amount of CO(g) is decreased by 2(0.040) to 0.020 mol.

$$K_{600°C} = \frac{(P_{CO})^2}{P_{CO_2}} = \frac{\left(\dfrac{(0.020\,\text{mol.})(0.0821\,\text{L}\cdot\text{atm/mol}\cdot\text{K})(873\,\text{K})}{2.0\,\text{L}}\right)^2}{\dfrac{(0.24\,\text{mol.})(0.0821\,\text{L}\cdot\text{atm/mol}\cdot\text{K})(873\,\text{K})}{2.0\,\text{L}}} = 0.06$$

One should also be able to calculate $K_{600°C}$ from $K_{700°C}$ by calculating $\Delta H°$ and applying the van't Hoff equation. Due to an error in the data, however, this gives $K_{600°C} = 0.17$.

56. *Refer to Section 12.1 and Figure 12.2.*

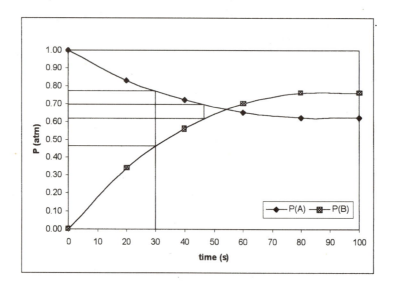

a. $P_A = 0.78$, $P_B = 0.47$

b. 0.62 atm; equilibrium was established after 80 seconds.

c. 0.61 atm; see the graph above.

Given that K is quite high, we can estimate that at equilibrium, the mixture is mostly product (CO). This can be verified by calculation.

	C(s)	+	CO$_2$(g)	⇌	2CO(g)
P_o	---		0.3 atm		0 atm
ΔP	---		-x atm		+2x atm
P_{eq}	---		0.3 - x atm		2x atm

$$K = \frac{(P_{CO})^2}{P_{CO_2}} \Rightarrow 168 = \frac{(2x)^2}{0.3 - x}$$

$$50.4 - 168x = 4x^2 \Rightarrow 4x^2 + 168x - 50.4 = 0$$

$$x = \frac{-b \pm \sqrt{b^2 - 4ac}}{2a} = \frac{-168 \pm \sqrt{(168)^2 - 4(4)(-50.4)}}{2(4)} = 0.25$$

There is actually a second value for x, -189, but a negative pressure is impossible, so this value is rejected.

P_{eq} (CO$_2$) = 0.30 − 0.25 = 0.05 atm
P_{eq} (CO) = 0 + 2(0.25) = 0.50 atm

As predicted, the mixture is mostly product (CO), with 10 times as much CO as CO$_2$. Thus option b is correct.

At equilibrium, there is much more NO$_2$ (product) than N$_2$O$_4$ (reactant), thus $K > 1$ and the equilibrium equation is:

$$N_2O_4 \rightleftharpoons 2NO_2$$

If the pressure of N$_2$O$_4$ is suddenly increased to 1.0 atm, the system will no longer be at equilibrium. Since reactant had been added, the reaction will proceed to the right, increasing the amount of products (and reducing the reactants) until equilibrium is reestablished.

Thus the graph will show a sudden spike at 100 s (where the partial pressure was increased) and then, over time, a decrease, followed by an eventual leveling off as equilibrium is reestablished. The NO$_2$, on the other hand will initially increase, then level off.

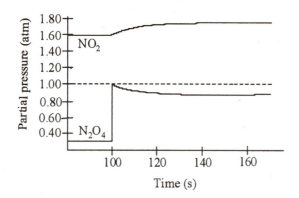

For the statement to be useful, one would also need a balanced equation and a temperature.

	$2NH_3(g)$	$\rightleftharpoons$	$N_2(g)$	$+$	$3H_2(g)$
P_o	1.00 atm		0 atm		0 atm
ΔP	$-2x$ atm		x atm		$3x$ atm
P_{eq}	$1.00 - 2x$ atm		x atm		$3x$ atm

$$K = 2.5 = \frac{(P_{N_2})(P_{H_2})^3}{(P_{NH_3})^2} = \frac{(x)(3x)^3}{(1-2x)^2} = \frac{27x^4}{1.00 - 4x + 4x^2}$$

$$2.5 - 10x + 10x^2 = 27x^4 \quad \Rightarrow \quad 2.5 = 10x - 10x^2 + 27x^4$$

This is best solved by successive approximations. As a first approximation $x \sim 2.5/10$. Successive approximations gives: $x = 0.33$.

$$P_{NH_3} = 1 - 2(0.33) = 0.34 \text{ atm}$$

$$P_{H_2} = 3(0.33) = 0.99 \text{ atm}$$

$$P_{N_2} = 0.33 \text{ atm}$$

The symbol $\prod$ is the product symbol, where : $\prod x_i = x_1 \times x_2 \times x_3 \times \ldots x_n$

$[products] = n_p/V$ and $[reactants] = n_r/V$

$$K_c = \frac{\prod [\text{products}]^{\text{coef}}}{\prod [\text{reactants}]^{\text{coef}}} = \frac{\prod (n_p/V)^{n_p}}{\prod (n_r/V)^{n_r}}$$

$$n_p = P_p V/RT \quad \text{and} \quad n_r = P_r V/RT$$

$$K_c = \frac{\prod (P_p/RT)^{n_p}}{\prod (P_r/RT)^{n_r}} = \frac{\left(\dfrac{1}{RT}\right)^{n_p} \prod (P_p)^{n_p}}{\left(\dfrac{1}{RT}\right)^{n_r} \prod (P_r)^{n_r}}$$

Since, $K = \dfrac{\prod (P_p)^{n_p}}{\prod (P_r)^{n_r}}$, substituting gives

$$K_c = K \times \frac{\left(\dfrac{1}{RT}\right)^{n_p}}{\left(\dfrac{1}{RT}\right)^{n_r}} = K \times \left(\frac{1}{RT}\right)^{n_p - n_r} \qquad \text{but:} \quad n_p - n_r = \Delta n_g$$

$$K_c = K \times \left(\frac{1}{RT}\right)^{\Delta n_g} \quad \Rightarrow \quad K = K_c \times (RT)^{\Delta n_g}$$

Also see the solution given in the answers in Appendix 6.

---

66. *Refer to Section 12.4.*

Use the titration data to calculate the amount of I_2 that was present at equilibrium. Then calculate the moles of each species and calculate K using moles instead of pressures (this is acceptable since the R, T and V terms would cancel in the equilibrium constant expression).

$$0.0370 \, \text{L} \times \frac{0.200 \, \text{mol. } S_2O_3^{2-}}{1 \, \text{L}} \times \frac{1 \, \text{mol. } I_2}{2 \, \text{mol. } S_2O_3^{2-}} = 3.70 \times 10^{-3} \, \text{mol. } I_2$$

$$3.20 \, \text{g HI} \times \frac{1 \, \text{mol.}}{127.9 \, \text{g}} = 2.50 \times 10^{-2} \, \text{mol. HI}$$

At equilibrium: $\text{HI} = 2.50 \times 10^{-2} - 2(3.70 \times 10^{-3}) = 0.0176 \, \text{mol.}$
$I_2 = 3.70 \times 10^{-3} \, \text{mol.}$
$H_2 = 3.70 \times 10^{-3} \, \text{mol.}$

$$K = \frac{(P_{I_2})(P_{H_2})}{(P_{HI})^2} = \frac{(3.7 \times 10^{-3})(3.7 \times 10^{-3})}{(0.0176)^2} = 0.0442$$

67. Refer to Section 12.4.

	$SO_3(g)$	$\rightleftharpoons$	$SO_2(g)$	$+$	$\frac{1}{2}O_2(g)$
P_o	1.00 atm		0 atm		0 atm
ΔP	-2x atm		2x atm		x atm
P_{eq}	1.00 - 2x atm		2x atm		x atm

$$K = 0.45 = \frac{(P_{SO_2})(P_{O_2})^{\frac{1}{2}}}{(P_{SO_3})} = \frac{(2x)(x)^{\frac{1}{2}}}{1.00-2x} = \frac{2x^{\frac{3}{2}}}{1.00-2x}$$

$$0.45 - 0.90x = 2x^{\frac{3}{2}}$$

This is best solved by successive approximations.

$x = 0.24$

$P_{SO_3} = 1.00 - 2(0.24) = 0.52$

68. Refer to Section 12.4.

A 50% yield indicates that half the reactants were converted to product, and the partial pressure changes accordingly

$$K = \frac{(P_{XeF_4})}{(P_{Xe})(P_{F_2})^2} = \frac{(0.10)}{(0.10)(0.20)^2} = 25$$

A 75% yield would mean that the pressure of Xe would decrease by 75% (to 0.05 atm) and the amount of XeF_4 would increase by that amount (to 0.15 atm). The F_2 would decrease by twice the amount of Xe (to x - 2(0.15)).

$$K = 25 = \frac{(P_{XeF_4})}{(P_{Xe})(P_{F_2})^2} = \frac{(0.15)}{(0.050)(x-0.30)^2} \quad \Rightarrow \quad 0 = x^2 - 0.60x - 0.03$$

$$x = \frac{-b \pm \sqrt{b^2 - 4ac}}{2a} = \frac{0.60 \pm \sqrt{(0.60)^2 - 4(1)(-0.03)}}{2(1)} = 0.65$$

69. Refer to Section 12.4.

	$I_2(g)$	$\rightleftharpoons$	$2I(g)$
P_o	1.00 atm		0 atm
ΔP	-x atm		2x atm
P_{eq}	1.00 - x atm		2x atm

Total pressure at equilibrium is 40% greater than initial pressure, so total pressure equals 1.4 and $1 - x + 2x = 1.4$, and $x = 0.4$

$$K = \frac{(P_I)^2}{P_{I_2}} = \frac{(0.8)^2}{0.6} = 1.1$$

70. Refer to Section 12.4.

a. Use the ideal gas law to determine the pressure of the benzyl alcohol, then use an equilibrium table and the equilibrium expression to solve for x.

$$1.50\,g \times \frac{1\,mol.}{108.1\,g} = 0.0139\,mol.$$

$$P = \frac{nRT}{V} = \frac{(0.0139\,mol.)(0.0821\,L \cdot atm/mol. \cdot K)(523\,K)}{2.0\,L} = 0.30\,atm$$

	$C_6H_5CH_2OH(g)$	$\rightleftharpoons$	$C_6H_5CHO(g)$	+	$H_2(g)$
P_o	0.30 atm		0 atm		0 atm
ΔP	-x atm		x atm		x atm
P_{eq}	0.30 - x atm		x atm		x atm

$$K = 0.56 = \frac{x^2}{0.30 - x} \quad \Rightarrow \quad x^2 + 0.56x - 0.168 = 0$$

$$x = \frac{-b \pm \sqrt{b^2 - 4ac}}{2a} = \frac{-0.56 \pm \sqrt{(0.56)^2 - 4(1)(-0.168)}}{2(1)} = 0.22\,atm$$

b. $$n = \frac{PV}{RT} = \frac{(0.30 - 0.22\,atm)(2.0\,L)}{(0.0821\,L \cdot atm/mol. \cdot K)(523\,K)} = 4 \times 10^{-3}\,mol.$$

$$4 \times 10^{-3}\,mol. \times \frac{108.1\,g}{1\,mol.} = 0.4\,g$$

Chapter 13: Acids and Bases

---

2. *Refer to Section 13.1 and Example 13.1.*

The acid is the reactant that donates an H^+, while the base is the reactant that accepts the H^+. The conjugate acid/base is the product that received/donated an H^+.

	Acid	Base	acid/conj. base pair	conj. acid/base pair
a.	H_3O^+	CN^-	H_3O^+/H_2O	HCN/CN^-
b.	HNO_2	OH^-	HNO_2/NO_2^-	H_2O/OH^-
c.	$HCHO_2$	H_2O	$HCHO_2/CHO_2^-$	H_3O^+/H_2O

---

4. *Refer to Section 13.1.*

a. CHO_2^- **Base**. As an anion, it is likely to act as an H^+ acceptor.

b. NH_4^+ **Acid**. Since the molecule has a positive charge, it is unlikely to accept an H^+ with another positive charge.

c. HSO_3^- **Acid or base** (amphoteric). As an anion, it can act as an H^+ acceptor. With an H^+ attached to an oxygen, it can also act as an H^+ donor.

---

6. *Refer to Section 13.1.*

Write the equation for the dissociation of the species (HB) into H^+ and B^-. The resulting B^- will be the conjugate base of the acid HB.

a. $HC_2H_3O_2 \rightarrow H^+ + \mathbf{C_2H_3O_2^-}$

b. $[Zn(H_2O)_4]^{2+} \rightarrow H^+ + \mathbf{[Zn(H_2O)_3(OH)]^+}$

c. $HBrO_2 \rightarrow H^+ + \mathbf{BrO_2^-}$

d. $CH_3NH_3^+ \rightarrow H^+ + \mathbf{CH_3NH_2}$

e. $H_2S \rightarrow H^+ + \mathbf{HS^-}$

8. _Refer to Section 13.3 and Example 13.3._

a. pH = -log[H^+] = -log(6.0) = -0.78   pH < 7, thus the solution is acidic.
    pOH = 14.00 – pH = 14.00 + 0.78 = 14.78

b. pH = -log[H^+] = -log(0.33) = 0.48   pH < 7, thus the solution is acidic.
    pOH = 14.00 – pH = 14.00 – 0.48 = 13.52

c. pH = -log[H^+] = -log(4.6 x 10^{-8}) = 7.34   pH > 7, thus the solution is basic.
    pOH = 14.00 – pH = 14.00 – 7.34 = 6.66

d. pH = -log[H^+] = -log(7.2 x 10^{-14}) = 13.14   pH > 7, thus the solution is basic.
    pOH = 14.00 – pH = 14.00 – 13.14 = 0.86

10. _Refer to Section 13.3 and Example 13.3._

a. $[H^+] = 10^{-pH} = 10^{-9.0} = 1.0 \times 10^{-9} \, M$

$$[OH^-] = \frac{K_w}{[H^+]} = \frac{1.0 \times 10^{-14}}{1.0 \times 10^{-9}} = 1.0 \times 10^{-5} \, M$$

b. $[H^+] = 10^{-pH} = 10^{-3.2} = 6.3 \times 10^{-4} \, M$

$$[OH^-] = \frac{K_w}{[H^+]} = \frac{1.0 \times 10^{-14}}{6.3 \times 10^{-4}} = 1.6 \times 10^{-11} \, M$$

c. $[H^+] = 10^{-pH} = 10^{+1.05} = 11.2 \, M$

$$[OH^-] = \frac{K_w}{[H^+]} = \frac{1.0 \times 10^{-14}}{11.2} = 8.9 \times 10^{-16} \, M$$

d. $[H^+] = 10^{-pH} = 10^{-7.46} = 3.5 \times 10^{-8} \, M$

$$[OH^-] = \frac{K_w}{[H^+]} = \frac{1.0 \times 10^{-14}}{3.5 \times 10^{-8}} = 2.9 \times 10^{-7} \, M$$

12. _Refer to Sections 13.2 and 13.3._

Calculate the pH and pOH of both solutions and compare.

Sol. 1:   pH = 4.3
            pOH = 14.00 – pH = 14.00 – 4.3 = 9.7

242

Sol. 2: $pOH = -\log[OH^-] = -\log(3.4 \times 10^{-7}) = 6.47$
            $pH = 14.00 - pOH = 14.00 - 6.47 = 7.53$

Solution 1 is more acidic (lower pH and thus higher $[H^+]$)
Solution 1 also has a higher pOH (again, indicating it is more acidic).

---

14. Refer to Sections 13.2 and 13.3.

Calculate $[H^+]$ for solutions X, Y and Z. Then use the results to answer parts a, b and c.

$$[H^+]_X = 10^{-pH} = 10^{-4.35} = 4.5 \times 10^{-5} \, M$$

$$[OH^-]_X = \frac{K_w}{[H^+]_X} = \frac{1.0 \times 10^{-14}}{4.5 \times 10^{-5}} = 2.2 \times 10^{-10} \, M$$

$$[OH^-]_Y = 10[OH^-]_X = 10(2.2 \times 10^{-10}) = 2.2 \times 10^{-9} \, M$$

$$[H^+]_Y = \frac{K_w}{[OH^-]_Y} = \frac{1.0 \times 10^{-14}}{2.2 \times 10^{-9}} = 4.5 \times 10^{-6} \, M$$

$$pH_Z = pH_X + 4 = 4.35 + 4 = 8.35$$

$$[H^+]_Z = 10^{-pH} = 10^{-8.35} = 4.5 \times 10^{-9} \, M$$

a. $$\frac{[H^+]_X}{[H^+]_Y} = \frac{4.5 \times 10^{-5}}{4.5 \times 10^{-6}} = 10$$

$$\frac{[H^+]_X}{[H^+]_Z} = \frac{4.5 \times 10^{-5}}{4.5 \times 10^{-9}} = 1.0 \times 10^4 = 10,000$$

b. $pH_Y = -\log[H^+] = -\log(4.5 \times 10^{-6}) = 5.35$
    $pH_Z = 8.35$ (see preliminary calculations above)

c. $pH_X = 4.35$ pH is less than 7, therefore solution X is acidic
    $pH_Y = 5.35$ pH is less than 7, therefore solution Y is acidic
    $pH_Z = 8.35$ pH is greater than 7, therefore solution Z is basic

---

16. Refer to Section 13.3 and Example 13.3.

a. $[H^+]_{\text{milk of magnesia}} = 10^{-pH} = 10^{-10.5} = 3.2 \times 10^{-11} \, M$

b. $[H^+]_{gastric\ juice} = 10^{-pH} = 10^{-1.5} = 3.2 \times 10^{-2}\ M$

$$\frac{[H^+]_{gastric\ juice}}{[H^+]_{milk\ of\ magnesia}} = \frac{3.2 \times 10^{-2}}{3.2 \times 10^{-11}} = 1.0 \times 10^9$$

---

18. Refer to Section 13.3 and Example 13.4.

Calculate the concentration of the perchloric acid. $[H^+]$ will equal $[HClO_4]$ since this is a strong monoprotic acid.

a. $\dfrac{25.0\ g\ HClO_4}{100\ g\ solution} \times \dfrac{1.00\ g}{1.00\ mL} \times \dfrac{1000\ mL}{1\ L} \times \dfrac{1\ mol.\ HClO_4}{100.5\ g\ HClO_4} = 2.49\ M$

pH = -log[H$^+$] = -log(2.49) = -0.396

One liter would have the same concentration as two liters, and thus the same pH.

b. $\dfrac{12.0\ g\ HClO_4}{2.00\ L} \times \dfrac{1\ mol.\ HClO_4}{100.5\ g\ HClO_4} = 0.0600\ M$

pH = -log[H$^+$] = -log(0.0600) = 1.22

For 1.00 L: $\dfrac{12.0\ g\ HClO_4}{1.00\ L} \times \dfrac{1\ mol.\ HClO_4}{100.5\ g\ HClO_4} = 0.119\ M$

pH = -log[H$^+$] = -log(0.119) = 0.923

---

20. Refer to Section 13.3.

Calculate the moles of H$^+$ in each solution. Add those values together and divide by the total volume (L) to get concentration. Remember that both acids are strong acids and ionize completely. Consequently, $[H^+] = [HX]$

HI: $5.00\ g\ HI \times \dfrac{1\ mol.\ HI}{128\ g\ HI} \times \dfrac{1\ mol.\ H^+}{1\ mol.\ HI} = 0.0391\ mol.\ H^+$

HNO$_3$: $0.295\ L\ HNO_3 \times \dfrac{0.786\ mol.\ HNO_3}{1\ L} \times \dfrac{1\ mol.\ H^+}{1\ mol.\ HNO_3} = 0.232\ mol.\ H^+$

moles H$^+$ = 0.0391 mol. + 0.232 mol. = 0.271 mol.

$[H^+] = \dfrac{0.0391\ mol. + 0.232\ mol.}{0.295\ L} = 0.919\ M$

pH = -log[H$^+$] = -log(0.919) = 0.0367

22. *Refer to Section 13.3 and Problem 20 (above).*

     Calculate the concentration of OH⁻ in the final solution, and from that the pOH, then pH and finally [H⁺].

a. $0.0450\,L \times \dfrac{0.0921\,mol.\ Ba(OH)_2}{1\,L} \times \dfrac{2\,mol.\ OH^-}{1\,mol.\ Ba(OH)_2} = 8.29 \times 10^{-3}\,mol.\ OH^-$

    $[OH^-] = \dfrac{8.29 \times 10^{-3}\,mol.\ OH^-}{0.3500\,L} = 0.0237\,M$

    $pOH = -\log(0.0237) = 1.63$

    $pH = 14.00 - pOH = 14.00 - 1.63 = 12.37$

    $[H^+] = 10^{-12.37} = 4.3 \times 10^{-13}\,M$

b. $4.68\,g\ NaOH \times \dfrac{1\,mol.\ NaOH}{40.0\,g\ NaOH} \times \dfrac{1\,mol.\ OH^-}{1\,mol.\ NaOH} = 0.117\,mol.\ OH^-$

    $[OH^-] = \dfrac{0.117\,mol.\ OH^-}{0.635\,L} = 0.184\,M$

    $pOH = -\log(0.184) = 0.74$

    $pH = 14.00 - pOH = 14.00 - 0.74 = 13.26$

    $[H^+] = 10^{-13.26} = 5.5 \times 10^{-14}\,M$

24. *Refer to Section 13.3.*

     Calculate the moles of OH⁻ in each solution. Then calculate [OH⁻]$_{final}$ by dividing the total moles of OH⁻ by the total volume of solution. Then calculate pOH and convert to pH.

CsOH: As a strong base, [CsOH] = [OH⁻]

        $0.0750\,L \times \dfrac{0.366\,mol.\ CsOH}{1\,L} = 0.0275\,mol.\ OH^-$

$Ba(OH)_2$: $pOH = 14.00 - pH = 14.00 - 11.65 = 2.35$

        $[OH^-] = 10^{-pOH} = 10^{-2.35} = 4.47 \times 10^{-3}\,M$

        $0.2500\,L \times \dfrac{4.47 \times 10^{-3}\,mol.\ OH^-}{1\,L} = 0.00112\,mol.\ OH^-$

$[OH^-]_{final} = \dfrac{0.0275\,mol.\ OH^- + 0.00112\,mol.\ OH^-}{0.0750\,L + 0.2500\,L} = 0.0881\,M$

$pOH = -log[OH^-] = -log(0.0881) = 1.06$

$pH = 14.00 - pOH = 14.00 - 1.06 = 12.94$

---

26. Refer to Section 13.4 and Example 13.5.

a. $[Ni(H_2O)_5OH]^+(aq) + H_2O(l) \rightleftharpoons [Ni(H_2O)_4(OH)_2](aq) + H_3O^+(aq)$

b. $[Al(H_2O)_6]^{3+}(aq) + H_2O(l) \rightleftharpoons [Al(H_2O)_5OH]^{2+}(aq) + H_3O^+(aq)$

c. $H_2S(aq) + H_2O(l) \rightleftharpoons HS^-(aq) + H_3O^+(aq)$

d. $HPO_4^{2-}(aq) + H_2O(l) \rightleftharpoons PO_4^{3-}(aq) + H_3O^+(aq)$

e. $HClO_2(aq) + H_2O(l) \rightleftharpoons ClO_2^-(aq) + H_3O^+(aq)$

f. $[Cr(H_2O)_5OH]^+(aq) + H_2O(l) \rightleftharpoons [Cr(H_2O)_4(OH)_2](aq) + H_3O^+(aq)$

---

28. Refer to Section 13.4.

Since each of the species is identified as an acid, the ionization equation will follow the general format: $HB(aq) \rightleftharpoons H^+(aq) + B^-(aq)$, and $K = K_a$.

a. $HSO_3^-(aq) \rightleftharpoons H^+(aq) + SO_3^{2-}(aq)$

$$K_a = \frac{[H^+][SO_3^{2-}]}{[HSO_3^-]}$$

b. $HPO_4^{2-}(aq) \rightleftharpoons H^+(aq) + PO_4^{3-}(aq)$

$$K_a = \frac{[H^+][PO_4^{3-}]}{[HPO_4^{2-}]}$$

c. $HNO_2(aq) \rightleftharpoons H^+(aq) + NO_2^-(aq)$

$$K_a = \frac{[H^+][NO_2^-]}{[HNO_2]}$$

---

30. Refer to Section 13.4.

a. $pK_a = -logK_a = -log(6.3 \times 10^{-4}) = 3.20$

b. $pK_a = -logK_a = -log(1.7 \times 10^{-11}) = 10.77$

c. $pK_a = -logK_a = -log(1.9 \times 10^{-8}) = 7.72$

a. The stronger the acid, the greater the degree of ionization and thus the larger the K_a. This correlates with a smaller pK_a. Thus the smallest pK_a correlates with the strongest acid, and the largest pK_a with the weakest acid.

    D > A > C > B

b. The largest K_a corresponds to the smallest pK_a.

    D has the largest K_a.

---

34. *Refer to Section 13.4 and Example 13.6.*

Calculate $[HC_7H_6NO_2]_0$, and then the equilibrium concentrations for all three species. Use those values to calculate K_a.

$$[HC_7H_6NO_2]_0 = \frac{0.263\,\text{mol.}}{0.7500\,\text{L}} = 0.351\,M$$

	$HC_7H_6NO_2(aq)$	$\rightleftharpoons$	$C_7H_6NO_2^-(aq)$	+	$H^+(aq)$
[ ]$_0$	0.351		0		0
Δ[ ]	-2.6 x 10^{-3}		+2.6 x 10^{-3}		+2.6 x 10^{-3}
[ ]$_{eq}$	0.348		2.6 x 10^{-3}		2.6 x 10^{-3}

$$K_a = \frac{[H^+][C_7H_6NO_2^-]}{[HC_7H_6NO_2]} = \frac{(2.60 \times 10^{-3})(2.6 \times 10^{-3})}{0.348} = 1.9 \times 10^{-5}$$

---

36. *Refer to Section 13.4 and Example 13.6.*

Calculate $[HC_7H_5O_2]_0$ from the mass and $[H^+]_0$ from the pH. Then the equilibrium concentrations for all three species. Use those values to calculate K_a.

$$13.7\,\text{g}\,HC_7H_5O_2 \times \frac{1\,\text{mol.}}{122.1\,\text{g}} = 0.112\,\text{mol.}$$

$$[HC_7H_5O_2]_0 = \frac{0.112\,\text{mol.}}{0.2500\,\text{L}} = 0.448\,M$$

$$[H^+]_{eq} = 10^{-pH} = 10^{-2.27} = 5.37 \times 10^{-3}$$

	$HC_7H_5O_2(aq)$	$\rightleftharpoons$	$C_7H_5O_2^-(aq)$	+	$H^+(aq)$
[ ]$_0$	0.448		0		0
Δ[ ]	-5.37×10^{-3}		$+5.37 \times 10^{-3}$		$+5.37 \times 10^{-3}$
[ ]$_{eq}$	0.443		5.37×10^{-3}		5.37×10^{-3}

$$K_a = \frac{[H^+][C_7H_5O_2^-]}{[HC_7H_5O_2]} = \frac{(5.37 \times 10^{-3})(5.37 \times 10^{-3})}{0.443} = 6.51 \times 10^{-5}$$

---

38. Refer to Section 13.4 and Example 13.7.

Consider the 5% rule: if $\dfrac{x}{a} \leq 0.05$, then $x \leq 0.05a$. Substituting this relationship into the

equation for K_a gives: $K_a = \dfrac{x^2}{a} = \dfrac{(0.05a)^2}{a} = (2.5 \times 10^{-3})a$ and finally: $K_a/a < 2.5 \times 10^{-3}$.

Thus one can apply the 5% rule when $K_a/a < 2.5 \times 10^{-3}$. This provides a method for determining *a priori* if the 5% rule applies.

a. $[HC_4H_7O_2]_0 = \dfrac{0.279 \text{ mol.}}{1.30 \text{ L}} = 0.215 M$

	$HC_4H_7O_2(aq)$	$\rightleftharpoons$	$C_4H_7O_2^-(aq)$	+	$H^+(aq)$
[ ]$_0$	0.215		0		0
Δ[ ]	$-x$		$+x$		$+x$
[ ]$_{eq}$	$0.215 - x$		x		x

$$K_a = \frac{[H^+][C_4H_7O_2^-]}{[HC_4H_7O_2]} \Rightarrow 1.51 \times 10^{-5} = \frac{(x)(x)}{0.215 - x} = \frac{x^2}{0.215 - x}$$

$K_a/a = 1.51 \times 10^{-5} / 0.215 = 7.02 \times 10^{-5}$
$K_a/a = 7.02 \times 10^{-5} < 2.5 \times 10^{-3}$ (thus the 5% rule applies)

$$K_a = 1.51 \times 10^{-5} = \frac{x^2}{0.215} \Rightarrow x = 1.80 \times 10^{-3}$$

$[H^+] = x = 1.80 \times 10^{-3} M$

b. $[HC_4H_7O_2]_0 = \dfrac{13.5\,g}{1.30\,L} \times \dfrac{1\,mol.}{88.10\,g} = 0.118\,M$

	$HC_4H_7O_2(aq)$	$\rightleftharpoons$	$C_4H_7O_2^-(aq)$	$+$	$H^+(aq)$
$[\ ]_0$	0.118		0		0
$\Delta[\ ]$	$-x$		$+x$		$+x$
$[\ ]_{eq}$	$0.118 - x$		x		x

$K_a = \dfrac{[H^+][C_4H_7O_2^-]}{[HC_4H_7O_2]} \Rightarrow 1.51 \times 10^{-5} = \dfrac{x^2}{0.118 - x}$

$K_a/a = 1.51 \times 10^{-5} / 0.118 = 1.28 \times 10^{-4}$
$K_a/a = 1.28 \times 10^{-4} < 2.5 \times 10^{-3}$ (thus the 5% rule applies)

$K_a = 1.51 \times 10^{-5} = \dfrac{x^2}{0.118} \Rightarrow x = 1.33 \times 10^{-3}$

$[H^+] = x = 1.33 \times 10^{-3}\,M$

---

40. Refer to Section 13.4 and Problem 38 (above).

$HBarb(aq) \rightarrow Barb^-(aq) + H^+(aq)$

a. $K_a = \dfrac{[H^+][Barb^-]}{[HBarb]} \Rightarrow 1.1 \times 10^{-4} = \dfrac{x^2}{0.673 - x}$

$K_a/a = 1.1 \times 10^{-4} / 0.673 = 1.63 \times 10^{-4}$
$K_a/a = 1.63 \times 10^{-4} < 2.5 \times 10^{-3}$ (The 5% rule applies. See the explanation accompanying problem 38.)

$K_a = 1.1 \times 10^{-4} = \dfrac{x^2}{0.673} \Rightarrow x = 8.6 \times 10^{-3}$

$[H^+] = x = 8.6 \times 10^{-3}\,M$

b. $[OH^-] = \dfrac{K_w}{[H^+]} = \dfrac{1.00 \times 10^{-14}}{8.6 \times 10^{-3}} = 1.2 \times 10^{-12}\,M$

c. $pH = -\log[H^+] = -\log(8.6 \times 10^{-3}) = 2.07$

d. $\%\ ionization = \dfrac{[H^+]_{eq}}{[HBarb]_0} \times 100\% = \dfrac{8.6 \times 10^{-3}}{0.673} \times 100\% = 1.3\%$

42. *Refer to Section 13.4 and Example 13.8.*

Let HTCA represent trichloroacetic acid, then $HTCA(aq) \rightleftharpoons H^+(aq) + TCA^-(aq)$

$$K_a = 2.0 \times 10^{-1} = \frac{[H^+][TCA^-]}{[HTCA]} = \frac{x^2}{4.0}$$

$x = 0.89$

$x/a = 0.89/4.0 = 0.22$ (thus the 5% rule does not apply, see page 395)

$$K_a = 2.0 \times 10^{-1} = \frac{[H^+][TCA^-]}{[HTCA]} = \frac{x^2}{4.0 - x}$$

$x^2 + 0.20x - 0.80 = 0$

$$x = \frac{-b \pm \sqrt{b^2 - 4ac}}{2a} = \frac{-0.20 \pm \sqrt{(0.20)^2 - 4(1)(-0.80)}}{2(1)} = 0.80$$

$[H^+] = 0.80 \ M$

$pH = -\log[H^+] = -\log(0.80) = 0.097$

$$\% \text{ ionization} = \frac{[H^+]_{eq}}{[HTCA]_o} \times 100\% = \frac{0.80}{4.0} \times 100\% = 20\%$$

---

44. *Refer to Section 13.4.*

$$[Al(H_2O)_6]^{3+}(aq) + H_2O(l) \rightleftharpoons [Al(H_2O)_5OH]^{2+}(aq) + H_3O^+(aq)$$

$$K_a = \frac{[H_3O^+][Al(H_2O)_5OH^{2+}]}{[Al(H_2O)_6^{3+}]} \Rightarrow 1.2 \times 10^{-5} = \frac{x^2}{1.75 - x}$$

$K_a/a = 1.2 \times 10^{-5} / 1.75 = 6.9 \times 10^{-6}$
$K_a/a = 6.9 \times 10^{-6} < 2.5 \times 10^{-3}$ (The 5% rule applies. See the explanation accompanying problem 38.)

$$K_a = 1.2 \times 10^{-5} = \frac{x^2}{1.75} \Rightarrow x = 4.6 \times 10^{-3}$$

$[H^+] = x = 4.6 \times 10^{-3} \ M$

$pH = -\log[H^+] = -\log(4.6 \times 10^{-3}) = 2.34$

46. Refer to Sections 12.2 and 13.4 and Table 13.3.

$$H_3PO_4(aq) \rightleftharpoons H^+(aq) + H_2PO_4^-(aq) \qquad K_1 = 7.1 \times 10^{-3}$$
$$H_2PO_4^-(aq) \rightleftharpoons H^+(aq) + HPO_4^{2-}(aq) \qquad K_2 = 6.2 \times 10^{-8}$$
$$\underline{HPO_4^{2-}(aq) \rightleftharpoons H^+(aq) + PO_4^{3-}(aq) \qquad K_3 = 4.5 \times 10^{-13}}$$
$$H_3PO_4(aq) \rightleftharpoons 3H^+(aq) + PO_4^{3-}(aq) \qquad K = K_1 \times K_2 \times K_3 = 2.0 \times 10^{-22}$$

48. Refer to Section 13.4 and Example 13.9.

Assume that all the H^+ comes from the dissociation of H_2CO_3 and calculate $[H^+]$ and $[HCO_3^-]$

$$H_2CO_3(aq) \rightleftharpoons H^+(aq) + HCO_3^-(aq)$$

$$K_{a_1} = \frac{[H^+][HCO_3^-]}{[H_2CO_3]} \quad \Rightarrow \quad 4.4 \times 10^{-7} = \frac{x^2}{0.63 - x}$$

$K_a/a = 4.4 \times 10^{-7} / 0.63 = 7.0 \times 10^{-7}$
$K_a/a = 7.0 \times 10^{-7} < 2.5 \times 10^{-3}$ (The 5% rule applies. See the explanation accompanying problem 38.)

$$K_a = 4.4 \times 10^{-7} = \frac{x^2}{0.63} \quad \Rightarrow \quad x = 5.3 \times 10^{-4}$$

$[H^+] = [HCO_3^-] = x = 5.3 \times 10^{-4} M$

pH = $-\log[H^+] = -\log(5.3 \times 10^{-4}) = 3.28$

Now use the second dissociation constant to calculate $[CO_3^{2-}]$.

$$K_{a_2} = \frac{[H^+][CO_3^{2-}]}{[HCO_3^-]} = [CO_3^{2-}] \qquad \text{Recall that } [H^+] = [HCO_3^-], \text{ thus these terms cancel.}$$

$[CO_3^{2-}] = 4.7 \times 10^{-11} M$

50. Refer to Section 13.5 and Example 13.10.

a. $NH_3(aq) + H_2O(l) \rightleftharpoons NH_4^+(aq) + OH^-(aq)$

b. $NO_2^-(aq) + H_2O(l) \rightleftharpoons HNO_2(aq) + OH^-(aq)$

c. $C_6H_5NH_2(aq) + H_2O(l) \rightleftharpoons C_6H_5NH_3^+(aq) + OH^-(aq)$

d. $CO_3^{2-}(aq) + H_2O(l) \rightleftharpoons HCO_3^-(aq) + OH^-(aq)$

e. $F^-(aq) + H_2O(l) \rightleftharpoons HF(aq) + OH^-(aq)$

f. $HCO_3^-(aq) + H_2O(l) \rightleftharpoons H_2CO_3(aq) + OH^-(aq)$

---

52. Refer to Section 13.5 and Table 13.4.

The stronger the base, the greater the K_b. Also, stronger bases have higher pH's.

a. KOH is a strong base, thus K_b will be large and pH will be high.

b. NaCN $K_b = 1.7 \times 10^{-5}$

c. HCO_3^- $K_b = 2.3 \times 10^{-8}$

d. $Ba(OH)_2$ is a strong base and will have a large K_b and high pH. Upon dissociating, however, this base produces $2OH^-$. Therefore, $0.1\ M\ Ba(OH)_2$ will have a higher pH than a $0.1\ M\ NaOH$ solution.

   $Ba(OH)_2 > KOH > NaCN > NaHCO_3$

---

54. Refer to Section 13.5 and Example 13.12.

At 25°C, $K_a \times K_b = K_w = 1.0 \times 10^{-14}$. Substitute the given value for K_b and solve for K_a.

a. $K_a = \dfrac{K_w}{K_b} = \dfrac{1.0 \times 10^{-14}}{7.4 \times 10^{-7}} = 1.4 \times 10^{-8}$

b. $K_a = \dfrac{K_w}{K_b} = \dfrac{1.0 \times 10^{-14}}{1.4 \times 10^{-4}} = 7.1 \times 10^{-11}$

---

56. Refer to Section 13.5.

a. Since codeine is a weak base (like ammonia), it will react with water to produce OH^-.
    $Cod(aq) + H_2O(l) \rightleftharpoons HCod^+(aq) + OH^-(aq)$

b. $K_b = \dfrac{K_w}{K_a} = \dfrac{1.0 \times 10^{-14}}{1.2 \times 10^{-8}} = 8.3 \times 10^{-7}$

c. Use K_b to calculate $[OH^-]$. Then calculate pOH and convert that to pH.

   $K_b = \dfrac{[HCod^+][OH^-]}{[Cod]} \Rightarrow 8.3 \times 10^{-7} = \dfrac{x^2}{0.0020 - x}$

$K_b/b = 8.3 \times 10^{7} / 0.0020 = 4.2 \times 10^{-4}$

$K_b/b = 4.2 \times 10^{-4} < 2.5 \times 10^{-3}$ (The 5% rule applies. See the explanation accompanying problem 38.)

$$K_b = 8.3 \times 10^{-7} = \frac{x^2}{0.0020} \Rightarrow x = 4.1 \times 10^{-5}$$

$[OH^-] = x = 4.1 \times 10^{-5} M$

$pOH = -\log[OH^-] = -\log(4.1 \times 10^{-5}) = 4.39$

$pH = 14.00 - pOH = 14.00 - 4.39 = 9.61$

---

58. Refer to Section 13.5.

Calculate pOH from pH, then calculate $[OH^-]$. [HCN] will equal $[OH^-]$. Use K_b to calculate [NaCN]. Finally, calculate the mass of NaCN necessary to give the calculated concentration.

$NaCN(aq) + H_2O(l) \rightleftharpoons HCN(aq) + NaOH(aq)$

$pOH = 14.00 - pH = 14.00 - 12.10 = 1.90$

$[OH^-] = 10^{-pOH} = 10^{-1.90} = 0.013 M$

$$K_b = \frac{[HCN][NaOH]}{[NaCN]} \Rightarrow 1.7 \times 10^{-5} = \frac{(0.013)^2}{x}$$

$x = [NaCN] = 9.9 M$

$$0.425 \, L \times \frac{9.9 \, mol. \, NaCN}{1 \, L} \times \frac{49.01 \, g \, NaCN}{1 \, mol. \, NaCN} = 2.1 \times 10^2 \, g$$

---

60. Refer to Sections 13.5 and 13.6, Example 13.13 and Table 13.5.

Consider the ions that form when the salt dissolves. Label the ions as acidic, basic or spectator, and add the effects.

	Salt	Cation	Anion	Solution
a.	$Sr(NO_3)_2$	Sr^{2+} (spectator)	NO_3^- (spectator)	neutral
b.	Li_2SO_3	Li^+ (spectator)	SO_3^{2-} (basic)	basic
c.	$NaHCO_3$	Na^+ (spectator)	HCO_3^- (acidic or basic)*	basic
d.	NH_4NO_2	NH_4^+ (acidic)	NO_2^- (basic)	acidic*
e.	Na_3PO_4	Na^+ (spectator)	PO_4^{3-} (basic)	basic

* For these two problems, one must compare K_a to K_b to determine if the acid or the base dominates (see problem 62 below).

253

Write the equations showing the reactions of the acidic and basic ions (from problem 60 above) with water. The spectator ions can be ignored.

a. $H^+(aq) + OH^-(aq) \rightleftharpoons H_2O(l)$

b. $SO_3^{2-}(aq) + H_2O(l) \rightleftharpoons HSO_3^-(aq) + OH^-(aq)$

c. $HCO_3^-(aq) \rightleftharpoons CO_3^{2-}(aq) + H^+(aq)$ $K_a = 4.7 \times 10^{-11}$
   $HCO_3^-(aq) + H_2O(l) \rightleftharpoons H_2CO_3(aq) + OH^-(aq)$ $K_b = 2.3 \times 10^{-8}$
      $K_b > K_a$, so the solution is basic.

d. $NH_4^+(aq) \rightleftharpoons NH_3(aq) + H^+(aq)$ $K_a = 5.6 \times 10^{-10}$
   $NO_2^-(aq) + H_2O(l) \rightleftharpoons HNO_2(aq) + OH^-(aq)$ $K_b = 1.7 \times 10^{-11}$
      $K_a > K_b$, so the solution is acidic.

e. $PO_4^{3-}(aq) + H_2O(l) \rightleftharpoons HPO_4^{2-}(aq) + OH^-(aq)$

The first three salts contain the same spectator cation (K^+), thus one can consider the K_b's of the anions. The last three salts contain the same spectator anion (Cl^-), thus one can consider the K_a's of the cations.

OH^-	strong base	K_b = very large	pH ≈ 14
F^-	weak base	$K_b = 1.4 \times 10^{-11}$	pH = 8.1
Cl^-	spectator	neutral	pH = 7
K^+	spectator	neutral	pH = 7
Zn^{2+}	weak acid	$K_a = 3.3 \times 10^{-10}$	pH = 5.2
H^+	strong acid	K_a = very large	pH ≈ 0

$HCl < ZnCl_2 < KCl < KF < KOH$

a. Since K^+ is a spectator ion, salts with basic anions will give basic salts.
      $KC_2H_3O_2$, K_2CO_3, KF, K_3PO_4

b. Since K^+ is a spectator ion, salts with neutral anions will give neutral salts.
      KCl, KBr, KNO_3, $KClO_4$

c. Since ClO_4^- is a spectator ion, salts with neutral cations will give neutral salts.

   $LiClO_4$, $NaClO_4$, $Ca(ClO_4)_2$, $KClO_4$

d. Since ClO_4^- is a spectator ion, salts with acidic cations will give acidic salts.

   $NH_4ClO_4^-$, $Al(ClO_4)_3$, $Zn(ClO_4)_2$, $Mg(ClO_4)_2$

---

68. Refer to Section 13.4.

Convert the volume of solution to liters and calculate $[\ ]_0$ of aspirin. Set up a table and calculate $[H^+]$, and from that, the pH.

$$\frac{1}{16}\,qt \times \frac{1\,L}{1.057\,qt} = 0.059\,L$$

$$648\,mg \times \frac{1\,g}{1000\,mg} \times \frac{1\,mol.}{180.15\,g} = 3.60 \times 10^{-3}\,mol.$$

$$\frac{3.60 \times 10^{-3}\,mol.}{0.059\,L} = 0.061 M$$

	HB(aq)	$\rightleftharpoons$	B⁻(aq)	+	H⁺(aq)
$[\ ]_0$	0.061		0		0
$\Delta[\ ]$	-x		+x		+x
$[\ ]_{eq}$	0.061 - x		x		x

$$K_a = 3.6 \times 10^{-4} = \frac{[H^+][B^-]}{[HB]} = \frac{x^2}{0.061 - x}$$

$K_a/a = 3.6 \times 10^{-3} / 0.061 = 5.9 \times 10^{-2}$
$K_a/a = 5.9 \times 10^{-2} > 2.5 \times 10^{-3}$ (The 5% rule does **not** apply. See the explanation accompanying problem 38.)

$x^2 + 3.6 \times 10^{-4}x + -2.2 \times 10^{-5} = 0$

$$x = \frac{-b \pm \sqrt{b^2 - 4ac}}{2a} = \frac{-3.6 \times 10^{-4} \pm \sqrt{(3.6 \times 10^{-4})^2 - 4(1)(-2.2 \times 10^{-5})}}{2(1)} = 4.5 \times 10^{-3}$$

$pH = -\log[H^+] = -\log(4.5 \times 10^{-3}) = 2.35$

Calculate [NH_3] from the pH, then use the ideal gas equation to calculate the volume.

$$NH_3(aq) + H_2O(l) \rightleftharpoons NH_4^+(aq) + OH^-(aq)$$

$$pOH = 14.00 - pH = 14.00 - 11.55 = 2.45$$
$$[OH^-] = [NH_4^+] = 10^{-pOH} = 10^{-2.45} = 3.55 \times 10^{-3}\, M$$

$$K_b = \frac{[NH_4^+][OH^-]}{[NH_3]} \Rightarrow 1.8 \times 10^{-5} = \frac{(3.55 \times 10^{-3})^2}{x}$$

$$x = [NH_3] = 0.70\, M$$

$$4.00\,L \times \frac{0.70\,mol.\,NH_3}{1\,L} = 2.8\,mol.\,NH_3$$

$$V = \frac{nRT}{P} = \frac{(2.8\,mol.)(0.0821\,L \cdot atm/mol. \cdot K)(298\,K)}{1\,atm} = 69\,L$$

Na^+ is the cation of a strong base, thus it creates a neutral solution.
Cl^- is the anion of a strong acid, thus it creates a neutral solution.
Therefore, a solution of these two ions will be **neutral**.

a. **True**. Strong acids dissociate completely, thus [Y^-] = [H^+], which equals the concentration of the acid.

b. **False**. Strong acids dissociate completely, thus the concentration of the undissociated acid is approximately zero ([HY] $\approx$ 0)

c. **True**. See (a) above.

d. **True**. pH = -log[H^+] = -log(0.1) = 1.

e. **True**. See (a) above.

Box one shows $2H^+$, $2B^-$ and $5HB$. This is an acid that is partially ionized in solution, and thus must be a **weak acid**.

Box two shows $5H^+$, $5B^-$ and no HB. This is an acid that is completely ionized in solution, and thus must be a **strong acid**.

Figure (a) shows ten molecules in the box. If 10% dissociates, then one of those molecules will dissociate into H^+ and A^-, leaving the remaining nine as HA.

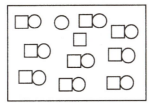

Write the net ionic equation for the reaction, then calculate $\Delta H°$ for the reaction. ($\Delta H° = \Sigma \Delta H_f°\,_{(products)} - \Sigma \Delta H_f°\,_{(reactants)}$). Calculate $\Delta H°$ for the 0.10 mol. given in the problem.

a. $Na^+(aq) + OH^-(aq) + H^+(aq) + Cl^-(aq) \rightleftharpoons Na^+(aq) + Cl^-(aq) + H_2O(l)$

$OH^-(aq) + H^+(aq) \rightleftharpoons H_2O(l)$

$\Delta H° = [(1 \text{ mol.})(-285.8 \text{ kJ/mol.})] - [(1 \text{ mol.})(-230.0 \text{ kJ/mol.}) + (1\text{mol.})(0.0 \text{ kJ/mol.})]$

$\Delta H° = -55.8 \text{ kJ}$ (This is for 1 mole, calculate for 0.10 moles.)

$\Delta H°\,_{(0.10 \text{ mol.})} = -55.8 \text{ kJ/mol.} \times 0.10 \text{ mol.} = -5.58 \text{ kJ}$

b. $HF(aq) + Na^+(aq) + OH^-(aq) \rightleftharpoons Na^+(aq) + F^-(aq) + H_2O(l)$

$HF(aq) + OH^-(aq) \rightleftharpoons F^-(aq) + H_2O(l)$

$\Delta H° = [(1 \text{ mol.})(-332.6 \text{ kJ/mol.}) - (1 \text{ mol.})(285.5 \text{ kJ/mol.})]$
$\qquad - [(1 \text{ mol.})(-320.1 \text{ kJ/mol.}) - (1 \text{ mol.})(230.0 \text{ kJ/mol.})]$

$\Delta H° = -68.3 \text{ kJ}$ (This is for 1 mole, calculate for 0.10 moles.)

$\Delta H°\,_{(0.10 \text{ mol.})} = -68.3 \text{ kJ/mol.} \times 0.10 \text{ mol.} = -6.83 \text{ kJ}$

81. Refer to Section 13.4.

Write K_a expressions for the concentrated acid $[HA]_c$ and the diluted acid $[HA]_d$, assuming that $[HA]_0 = [HA]_{eq}$ (which is true for a weak acid). Recalling that K_a is constant for a given acid, equate the two expressions and solve for $[H^+]$ and substitute the results into the percent ionization expression. Note the we are using the equation:

$$HA(aq) \rightleftharpoons H^+(aq) + A^-(aq).$$

$$K_a = \frac{[H^+]_c[A^-]_c}{[HA]_c} = \frac{x^2}{[HA]_c}, \text{ where } [H^+]_c = [A^-]_c = x$$

$$K_a = \frac{[H^+]_d[A^-]_d}{[HA]_d} = \frac{y^2}{[HA]_d}, \text{ where } [H^+]_d = [A^-]_d = y$$

If $[HA]_d$ is one tenth the concentration of $[HA]_c$, then $10[HA]_d = [HA]_c$ and

$$K_a = \frac{x^2}{[HA]_c} = \frac{y^2}{[HA]_d} \Rightarrow \frac{x^2}{10[HA]_d} = \frac{y^2}{[HA]_d} \Rightarrow x^2 = 10y^2$$

$$x = y\sqrt{10}$$

Substituting into the percent ionization expression:

$$\% \text{ ion.}_{(conc.)} = \frac{[H^+]_c}{[HA]_c} \times 100\% = \frac{100 \cdot x}{[HA]_c} = \frac{100(y\sqrt{10})}{10[HA]_d} = \frac{y \cdot 10\sqrt{10}}{[HA]_d}$$

$$\% \text{ ion.}_{(dil.)} = \frac{[H^+]_d}{[HA]_d} \times 100\% = \frac{y}{[HA]_d} \times 100\% = \frac{100 \cdot y}{[HA]_d}$$

Obviously the two terms differ by a factor of $\sqrt{10}$,

$$\frac{y \cdot 100}{[HA]_d} = \frac{y \cdot 10\sqrt{10}}{[HA]_d} \cdot \sqrt{10} \quad \text{thus: } \% \text{ ion.}_{(dil.)} = \left(\% \text{ ion.}_{(conc.)}\right) \times \sqrt{10}$$

Recall that freezing point lowering is a colligative property and thus depends on the total moles of solute (ions and molecules) present in solution. Determine the moles of H^+, B^- and HB in solution, calculate the molality of the solution and the freezing point depression. Assume one liter of solution.

$$1\,L \times \frac{1000\,mL}{1\,L} \times \frac{1.006\,g}{1\,mL} = 1006\,g\ solution$$

$$1006\,g\ solution \times \frac{5.00\,g\ HC_2H_3O_2}{100\,g\ solution} = 50.3\,g\ HC_2H_3O_2$$

$$50.3\,g\ HC_2H_3O_2 \times \frac{1\,mol.}{60.05\,g} = 0.838\,mol.\ HC_2H_3O_2$$

	$HC_2H_3O_2(aq)$	$\rightleftharpoons$	$C_2H_3O_2^-(aq)$	+	$H^+(aq)$
[ ]$_0$	0.838		0		0
Δ[ ]	-x		+x		+x
[ ]$_{eq}$	0.838 - x		x		x

$$K_a = 1.8 \times 10^{-5} = \frac{[H^+][C_2H_3O_2^-]}{[HC_2H_3O_2]} = \frac{x^2}{0.838 - x}$$

$K_a/a = 1.8 \times 10^{-5} / 0.838 = 2.1 \times 10^{-5}$
$K_a/a = 2.1 \times 10^{-5} < 2.5 \times 10^{-3}$ (The 5% rule applies. See the explanation accompanying problem 38.)

$$1.8 \times 10^{-5} = \frac{x^2}{0.838} \implies 1.5 \times 10^{-5} = x^2$$

$$x = 3.9 \times 10^{-3}\,M$$

moles solute = $[HC_2H_3O_2] + [C_2H_3O_2^-] + [H^+]$ = (0.838 - x) + x + x = 0.838 + x
moles solute = 0.838 + 0.0039 = 0.842

$$molality = \frac{mol.\ solute}{kg\ solvent} = \frac{0.842\,mol.}{1.006\,kg\ solution - 0.0503\,kg\ HC_2H_3O_2} = 0.881\,m$$

$\Delta T_f = k_f\,(m) = 1.86\,°C/m\,(0.877\,m) = 1.64°C$

$T_f = 0°C - 1.64°C = -1.64°C$

Assume 1 L of solution and complete dissociation of $Ca(OH)_2$ solution. Then calculate $[OH^-]$ for a saturated $Ca(OH)_2$ solution. Then pOH and pH.

$$\frac{0.153\,g\,Ca(OH)_2}{100\,mL} \times \frac{1000\,mL}{1\,L} \times \frac{1\,mol.\,Ca(OH)_2}{74.096\,g} \times \frac{2\,mol.\,OH^-}{1\,mol.\,Ca(OH)_2} = 0.0413\,M\,OH^-$$

pOH = -log$[OH^-]$ = -log(0.0413) = 1.38
pH = 14.00 - pOH = 14.00 - 1.38 = 12.62

Chapter 14: Equilibria in Acid-Base Solutions

---

2. *Refer to Section 14.2 and Chapter 13.*

Write the reactions, eliminating the spectator ions. Bear in mind that soluble salts ionize, thus they exist in solution as ions. Weak acids and bases (Table 13.2) do not ionize significantly, thus they exist in solution as the undissociated acid or base.

a. $NH_3(aq) + HF(aq) \rightleftharpoons NH_4^+(aq) + F^-(aq)$
Hydrofluoric acid and ammonium ion are weak acids and thus do not dissociate.

b. $HClO_4(aq) + RbOH(aq) \rightleftharpoons RbClO_4(aq) + H_2O(l)$
$HClO_4$ and $RbOH$ are strong acids and bases (respectively) and thus dissociate completely in solution.
$H^+(aq) + ClO_4^-(aq) + Rb^+(aq) + OH^-(aq) \rightleftharpoons Rb^+(aq) + ClO_4^-(aq) + H_2O(l)$
$H^+(aq) + OH^-(aq) \rightleftharpoons H_2O(l)$

c. $Na_2SO_3(aq) + 2HI(aq) \rightleftharpoons H_2SO_3(aq) + 2NaI(aq)$
Na_2SO_3 and NaI are salts and HI is a strong acid, thus all three dissociate in solution, while H_2SO_3 is a weak acid and thus remains undissociated.
$2Na^+(aq) + SO_3^{2-}(aq) + 2H^+(aq) + 2I^-(aq) \rightleftharpoons H_2SO_3(aq) + 2Na^+(aq) + 2I^-(aq)$
$SO_3^{2-}(aq) + 2H^+(aq) \rightleftharpoons H_2SO_3(aq)$

(Note: the answer given in the first printing of Appendix 6 assumes that only one mole of HI is added per mole of Na_2SO_3.)

d. $2HNO_3(aq) + Ca(OH)_2(aq) \rightleftharpoons Ca(NO_3)_2(aq) + 2H_2O(l)$
HNO_3 is a strong acid, $Ca(OH)_2$ is a strong acid and $Ca(NO_3)_2$ is a salt, thus all three dissociate in solution.
$2H^+(aq) + 2NO_3^-(aq) + Ca^{2+}(aq) + 2OH^-(aq) \rightleftharpoons Ca^{2+}(aq) + 2NO_3^-(aq) + 2H_2O(l)$
$2H^+(aq) + 2OH^-(aq) \rightleftharpoons 2H_2O(l)$
$H^+(aq) + OH^-(aq) \rightleftharpoons H_2O(l)$

---

4. *Refer to Chapters 4 and 13 and Problem 2 above.*

Write the reactions, eliminating the spectator ions. Bear in mind that weak acids and bases (Table 13.2) do not ionize significantly, thus they exist in solution as the undissociated acid or base.

a. $H^+(aq) + NaC_2H_3O_2(aq) \rightleftharpoons Na^+(aq) + HC_2H_3O_2(aq)$

 $NaC_2H_3O_2$ is a salt and thus dissociates in solution, while $HC_2H_3O_2$ is a weak acid and does not dissociate.

$$H^+(aq) + Na^+(aq) + C_2H_3O_2^-(aq) \rightleftharpoons Na^+(aq) + HC_2H_3O_2(aq)$$
$$H^+(aq) + C_2H_3O_2^-(aq) \rightleftharpoons HC_2H_3O_2(aq)$$

b. $H^+(aq) + CsOH(aq) \rightleftharpoons Cs^+(aq) + H_2O(l)$

CsOH is a stong base and thus dissociates in solution.
$$H^+(aq) + Cs^+(aq) + OH^-(aq) \rightleftharpoons Cs^+(aq) + H_2O(l)$$
$$H^+(aq) + OH^-(aq) \rightleftharpoons H_2O(l)$$

c. $H^+(aq) + NaHCO_3(aq) \rightleftharpoons Na^+(aq) + H_2CO_3(aq)$

$NaHCO_3$ is a salt and thus dissociates in solution, while H_2CO_3 is a weak acid and does not dissociate.
$$H^+(aq) + Na^+(aq) + HCO_3^-(aq) \rightleftharpoons Na^+(aq) + H_2CO_3(aq)$$
$$H^+(aq) + HCO_3^-(aq) \rightleftharpoons H_2CO_3(aq)$$

---

6. Refer to Chapter 13.

Recall that K for the reverse reaction equals the inverse of K for the forward reaction, and that when adding reactions, the K's are multiplied.

a.
$NH_3(aq) + H_2O(aq) \rightleftharpoons NH_4^+(aq) + OH^-(aq)$	$K_b = 1.8 \times 10^{-5}$
$HF(aq) \rightleftharpoons H^+(aq) + F^-(aq)$	$K_a = 6.9 \times 10^4$
$H^+(aq) + OH^-(aq) \rightleftharpoons H_2O(l)$	$1/K_w = 1.0 \times 10^{14}$

$$NH_3(aq) + H_2O(l) + HF(aq) + H^+(aq) + OH^-(aq)$$
$$\rightleftharpoons H_2O(l) + H^+(aq) + F^-(aq) + NH_4^+(aq) + OH^-(aq) \quad K = K_a \times K_b \times 1/K_w$$
$$NH_3(aq) + HF(aq) \rightleftharpoons NH_4^+(aq) + F^-(aq) \quad K = 1.2 \times 10^6$$

b. $H^+(aq) + OH^-(aq) \rightleftharpoons H_2O(l)$ $K = 1/K_w = 1.0 \times 10^{14}$

c. $SO_3^{2-}(aq) + 2H^+(aq) \rightleftharpoons H_2SO_3(aq)$ $K = 1/K_{a1} \times 1/K_{a2} = 9.8 \times 10^8$

d. $H^+(aq) + OH^-(aq) \rightleftharpoons H_2O(l)$ $K = 1/K_w = 1.0 \times 10^{14}$

---

8. Refer to Chapter 13 and Problem 6 above.

---

a. $H^+(aq) + C_2H_3O_2^-(aq) \rightleftharpoons HC_2H_3O_2(aq)$ $K = 1/K_a = 5.6 \times 10^4$

b. $H^+(aq) + OH^-(aq) \rightleftharpoons H_2O(l)$ $K = 1/K_w = 1.0 \times 10^{14}$

c. $H^+(aq) + HCO_3^-(aq) \rightleftharpoons H_2CO_3(aq)$ $K = 1/K_a = 2.3 \times 10^6$

10. Refer to Section 14.1.

Calculate $[H^+]$, and from that, pH and $[OH^-]$.

a. $[H^+] = K_a \times \dfrac{[HB]}{[B^-]} = 2.8 \times 10^{-8} \times \dfrac{0.25}{0.50} = 1.4 \times 10^{-8} \, M$

      pH $= -\log[H^+] = -\log(1.4 \times 10^{-8}) = 7.85$

      $[OH^-] = \dfrac{K_w}{[H^+]} = \dfrac{1.0 \times 10^{-14}}{1.4 \times 10^{-8}} = 7.1 \times 10^{-7} \, M$

b. $[H^+] = K_a \times \dfrac{[HB]}{[B^-]} = 2.8 \times 10^{-8} \times \dfrac{0.25}{0.30} = 2.3 \times 10^{-8} \, M$

      pH $= -\log[H^+] = -\log(2.3 \times 10^{-8}) = 7.63$

      $[OH^-] = \dfrac{K_w}{[H^+]} = \dfrac{1.0 \times 10^{-14}}{2.3 \times 10^{-8}} = 4.3 \times 10^{-7} \, M$

c. $[H^+] = K_a \times \dfrac{[HB]}{[B^-]} = 2.8 \times 10^{-8} \times \dfrac{0.25}{0.15} = 4.7 \times 10^{-8} \, M$

      pH $= -\log[H^+] = -\log(4.7 \times 10^{-8}) = 7.33$

      $[OH^-] = \dfrac{K_w}{[H^+]} = \dfrac{1.0 \times 10^{-14}}{4.7 \times 10^{-8}} = 2.1 \times 10^{-7} \, M$

d. $[H^+] = K_a \times \dfrac{[HB]}{[B^-]} = 2.8 \times 10^{-8} \times \dfrac{0.25}{0.075} = 9.3 \times 10^{-8} \, M$

      pH $= -\log[H^+] = -\log(9.3 \times 10^{-8}) = 7.03$

      $[OH^-] = \dfrac{K_w}{[H^+]} = \dfrac{1.0 \times 10^{-14}}{9.3 \times 10^{-8}} = 1.1 \times 10^{-7} \, M$

12. Refer to Section 14.1.

    A buffer should work until all the acid (HClO) or all the base (ClO$^-$) is consumed. The buffer capacity will equal to the moles of the acid or base present in the buffer. If we assume one liter of buffer, then the capacity will equal the concentration of the acid or base.

    The concentration of acid is the same in each case (0.25 M), thus one liter of solution will be able to buffer 0.25 mol. of base.

a. $[ClO^-] = 0.50\ M$, thus 1 L will be able to buffer 0.50 mol. acid.

b. $[ClO^-] = 0.30\ M$, thus 1 L will be able to buffer 0.30 mol. acid.

c. $[ClO^-] = 0.15\ M$, thus 1 L will be able to buffer 0.15 mol. acid.

d. $[ClO^-] = 0.075\ M$, thus 1 L will be able to buffer 0.075 mol. acid.

14. Refer to Section 14.1.

$[HF] = 0.0237\ M$

$$[KF] = \frac{0.037\ mol.}{0.135\ L} = 0.274\ M$$

$$[H^+] = K_a \times \frac{[HB]}{[B^-]} = 6.9 \times 10^{-4} \times \frac{0.0237}{0.274} = 6.0 \times 10^{-5}\ M$$

$pH = -\log[H^+] = -\log(6.0 \times 10^{-5}) = 4.22$

16. Refer to Section 14.1, Example 14.2 and Tables 13.2 and 14.1.

For the ideal buffer, $pH \approx pK_a$. Look for an acid with a pK_a approximately equal to the desired pH.

a. pK_a of HSO_4^- is 2, thus use HSO_4^-/SO_4^{2-}.

b. pK_a of lactic acid is 3.85 and that of benzoic acid is 4.18, thus either HLac/Lac⁻ or HBzO/BzO⁻ should work.

c. pK_a of NH_4^+ is 9.25 and that of HCN is 9.23, thus either NH_4^+/NH_3 or HCN/CN⁻ should work, although the HCN/CN⁻ system is a recipe for disaster, given the toxicity of HCN.

18. Refer to Section 14.1 and Example 14.2.

Calculate the moles of the acid and base, then calculate the $[H^+]$ and pH.

a. Calculate $[H^+]$ from pH, then calculate the acid/base ratio of the buffer.

$[H^+] = 10^{-pH} = 10^{-9.40} = 4.0 \times 10^{-10}\ M$

$$[H^+] = K_a \times \frac{[HB]}{[B^-]} \Rightarrow 4.0 \times 10^{-10} = 4.7 \times 10^{-11} \times \frac{[HCO_3^-]}{[CO_3^{2-}]}$$

$$\frac{[HCO_3^-]}{[CO_3^{2-}]} = 8.5$$

b. The Na_2CO_3 will dissociate to Na^+ and CO_3^{2-}. Use this fact and the equation from part (a) to calculate $[HCO_3^-]$.

$$\frac{[HCO_3^-]}{[CO_3^{2-}]} = 8.5 \implies \frac{[HCO_3^-]}{0.225\,M} = 8.5$$

$[HCO_3^-] = 1.9\,M$, thus 1.9 mol. would need to be added to the liter of solution.

c. $\frac{[HCO_3^-]}{[CO_3^{2-}]} = 8.5 \implies \frac{0.336}{[CO_3^{2-}]} = 8.5$

$[CO_3^{2-}] = 0.040\,M$

$$0.475\,L \times \frac{0.040\,\text{mol } CO_3^{2-}}{1\,L} \times \frac{1\,\text{mol. } Na_2CO_3}{1\,\text{mol. } CO_3^{2-}} \times \frac{106\,g\ Na_2CO_3}{1\,\text{mol. } Na_2CO_3} = 2.0\,g\ Na_2CO_3$$

d. Note that $\dfrac{[HB]}{[B^-]} = \dfrac{n_{HB}}{n_{B^-}} = 8.5$. It is, therefore, the mole ratio of HB to B^- (or HCO_3^- to CO_3^{2-}) that is important. Calculate the moles of CO_3^{2-} present in the solution. Then calculate the moles of HCO_3^- required, and finally, the volume of solution necessary to deliver that number of moles.

$$0.735\,L \times \frac{0.139\,\text{mol. } Na_2CO_3}{1\,L} = 0.102\,\text{mol. } Na_2CO_3 = 0.102\,\text{mol. } CO_3^{2-}$$

$$\frac{n_{HB}}{n_{B^-}} = \frac{n_{HCO_3^-}}{n_{HCO_3^{2-}}} = \frac{n_{HCO_3^-}}{0.102\,\text{mol. } CO_3^{2-}} = 8.5$$

$$n_{HCO_3^-} = 0.87\,\text{mol. } HCO_3^-$$

$$0.87\,\text{mol. } HCO_3^- \times \frac{1\,L}{0.200\,\text{mol. } HCO_3^-} = 4.3\,L$$

20. Refer to Section 14.1.

a. $5.50 \, g \times \dfrac{1 \, mol. \, NH_4Cl}{53.49 \, g \, NH_4Cl} = 0.103 \, mol. \, NH_4Cl$

$[H^+] = K_a \times \dfrac{n_{NH_4^+}}{n_{NH_3}} = 5.6 \times 10^{-10} \times \dfrac{0.103 \, mol. \, NH_4^+}{0.0188 \, mol. \, NH_3} = 3.1 \times 10^{-9} \, M$

$pH = -\log[H^+] = -\log(3.1 \times 10^{-9}) = 8.51$

b. Note that volume was not used in the calculations above. Thus we would expect the pH to remain the same, independent of the volume of solution.

22. Refer to Section 14.1 and Chapter 13.

Calculate $[H^+]$ and moles of acetic acid in 300 mL of vinegar. Then calculate moles of acetate ion in solution, and from that, mass of $KC_2H_3O_2$.

$[H^+] = 10^{-pH} = 10^{-4.10} = 7.9 \times 10^{-5} \, M$

$300.0 \, mL \, vinegar \times \dfrac{1.006 \, g \, vinegar}{1 \, mL \, vinegar} \times \dfrac{5.00 \, g \, HC_2H_3O_2}{100 \, g \, vinegar} = 15.1 \, g \, HC_2H_3O_2$

$15.1 \, g \, HC_2H_3O_2 \times \dfrac{1 \, mol. \, HC_2H_3O_2}{60.05 \, g \, HC_2H_3O_2} = 0.251 \, mol. \, HC_2H_3O_2$

$[H^+] = K_a \times \dfrac{n_{HC_2H_3O_2}}{n_{C_2H_3O_2^-}} \Rightarrow 7.9 \times 10^{-5} = 1.8 \times 10^{-5} \times \dfrac{0.251 \, mol.}{n_{C_2H_3O_2^-}}$

$n_{C_2H_3O_2^-} = 5.7 \times 10^{-2} \, mol. \, C_2H_3O_2^-$

$5.7 \times 10^{-2} \, mol. \, C_2H_3O_2^- \times \dfrac{1 \, mol. \, KC_2H_3O_2}{1 \, mol. \, C_2H_3O_2^-} \times \dfrac{98.1 \, g \, KC_2H_3O_2}{1 \, mol. \, KC_2H_3O_2} = 5.6 \, g \, KC_2H_3O_2$

24. Refer to Section 14.1 and Chapter 13.

The equation for the reaction is: $X^- + H_2O \rightleftharpoons HX + OH^-$. Calculate pOH, $[OH^-]$ and K_b. Convert K_b to K_a and then calculate $[H^+]$ and pH for the solution after addition of HX.

$[X^-]_o = \dfrac{0.614 \, mol.}{2.50 \, L} = 0.246 \, M$

$pOH = 14.00 - pH = 14.00 - 8.73 = 5.27$

$$[OH^-] = 10^{-pOH} = 10^{-5.27} = 5.37 \times 10^{-6} \, M$$

	$X^-(aq)$	+ $H_2O(l)$	$\rightleftharpoons$	$HX(aq)$	+	$OH^-(aq)$
$[\ ]_0$	0.246	---		0		0
$\Delta[\ ]$	-5.37×10^{-6}	---		$+5.37 \times 10^{-6}$		$+5.37 \times 10^{-6}$
$[\ ]_{eq}$	0.246	---		5.37×10^{-6}		5.37×10^{-6}

$$K_b = \frac{[HX][OH^-]}{[X^-]} = \frac{(5.37 \times 10^{-6})^2}{0.246} = 1.17 \times 10^{-10}$$

$$K_a = \frac{K_w}{K_b} = \frac{1.0 \times 10^{-14}}{1.17 \times 10^{-10}} = 8.5 \times 10^{-5}$$

$$[H^+] = K_a \times \frac{n_{HX}}{n_{X^-}} = 8.5 \times 10^{-5} \times \frac{0.219}{0.614} = 3.03 \times 10^{-5} \, M$$

$$pH = -\log(3.03 \times 10^{-5}) = 4.52$$

---

26. Refer to Section 14.1 and Example 14.3.

a. Calculate the moles of HCO_3^- and CO_3^{2-} and then calculate $[H^+]$ and pH.

$$0.355 \, L \times \frac{0.200 \, mol.}{1 \, L} = 0.0710 \, mol. \, HCO_3^-$$

$$0.355 \, L \times \frac{0.134 \, mol.}{1 \, L} = 0.0476 \, mol. \, CO_3^{2-}$$

$$[H^+] = K_a \times \frac{n_{HCO_3^-}}{n_{CO_3^{2-}}} = 4.7 \times 10^{-11} \times \frac{0.0710}{0.0476} = 7.0 \times 10^{-11} \, M$$

$$pH = -\log(7.0 \times 10^{-11}) = 10.15$$

b. The HCl will convert CO_3^{2-} to HCO_3^-. Calculate the moles of each that are present after the HCl addition. Then calculate $[H^+]$ and pH.

$$n_{HCO_3^-} = 0.0710 \, mol. + 0.0300 \, mol. = 0.101 \, mol.$$

$$n_{CO_3^{2-}} = 0.0476 \, mol. - 0.0300 \, mol. = 0.0176 \, mol.$$

$$[H^+] = K_a \times \frac{n_{HCO_3^-}}{n_{CO_3^{2-}}} = 4.7 \times 10^{-11} \times \frac{0.101}{0.0176} = 2.7 \times 10^{-10} \, M$$

$$pH = -\log(2.7 \times 10^{-10}) = 9.57$$

c. The NaOH will convert HCO_3^- to CO_3^{2-}. Calculate the moles of each that are present after the NaOH addition. Then calculate $[H^+]$ and pH.

$$n_{HCO_3^-} = 0.0710 \text{ mol.} - 0.0300 \text{ mol.} = 0.0410 \text{ mol.}$$

$$n_{CO_3^{2-}} = 0.0476 \text{ mol.} + 0.0300 \text{ mol.} = 0.0776 \text{ mol.}$$

$$[H^+] = K_a \times \frac{n_{HCO_3^-}}{n_{CO_3^{2-}}} = 4.7 \times 10^{-11} \times \frac{0.410}{0.0776} = 2.5 \times 10^{-11} \, M$$

$$pH = -\log(2.5 \times 10^{-11}) = 10.60$$

---

28. Refer to Section 14.1 and Problem 26 (above).

---

a. $$[H^+] = K_a \times \frac{n_{HCO_3^-}}{n_{CO_3^{2-}}} = 4.7 \times 10^{-11} \times \frac{0.0710}{0.0476} = 7.0 \times 10^{-11} \, M$$

$$pH = -\log(7.0 \times 10^{-11}) = 10.15$$

Note that the pH hasn't changed. This should be expected since the *moles* of HCO_3^- and CO_3^{2-} did not change upon dilution.

b. Begin by calculating the moles of HCO_3^- and CO_3^{2-} in 0.710 L of the diluted buffer.

$$0.710 \text{ L} \times \frac{0.0710 \text{ mol.}}{10.0 \text{ L}} = 0.00504 \text{ mol. } HCO_3^-$$

$$0.710 \text{ L} \times \frac{0.0476 \text{ mol.}}{10.0 \text{ L}} = 0.00338 \text{ mol. } CO_3^{2-}$$

The HCl will convert CO_3^{2-} to HCO_3^-. However, the moles of HCl are greater than the moles of CO_3^{2-}, thus all the CO_3^{2-} will be converted to HCO_3^- and excess HCl will remain. Calculate the moles of HCO_3^- and HCl that are present.

$$n_{HCl \text{ (remaining)}} = n_{HCl} - n_{CO_3^{2-}} = 0.0300 \text{ mol.} - 0.00338 \text{ mol.} = 0.0266 \text{ mol. } HCl$$

$$n_{HCO_3^- \text{ (final)}} = n_{HCO_3^- \text{ (initial)}} + n_{CO_3^{2-}} = 0.00504 \text{ mol.} + 0.00338 \text{ mol.} = 0.00842 \text{ mol. } HCO_3^-$$

The excess HCl will then react with the HCO_3^-, converting it to H_2CO_3. But again, an excess of acid will remain. Given the low concentration of the H_2CO_3 (relative to the HCl) and the fact that H_2CO_3 is a weak acid, it can be ignored, and the pH calculations can be based entirely on the remaining HCl.

$$n_{HCl \text{ (remaining)}} = n_{HCl} - n_{HCO_3^-} = 0.0266 \text{ mol.} - 0.00842 \text{ mol.} = 0.0182 \text{ mol. } HCl$$

$$[H^+] = \frac{n_{HCl}}{L \text{ solution}} = \frac{0.0182 \text{ mol.}}{0.710 \text{ L}} - 0.0256\, M$$

$$pH = -\log(0.0256) = 1.59$$

Note: If one neglects $HCO_3^- + H^+ \rightarrow H_2CO_3$, then one would get:

$$[H^+] = \frac{n_{HCl}}{L \text{ solution}} = \frac{0.0266 \text{ mol.}}{0.710 \text{ L}} = 0.0375\, M$$

$$pH = -\log(0.0375) = 1.43$$

c. The addition of 0.0030 mol. NaOH will convert HCO_3^- to CO_3^{2-}. Calculate the moles of each that are present after the NaOH addition. Then calculate $[H^+]$ and pH.

$$n_{HCO_3^-} = 0.00504 \text{ mol.} - 0.00300 \text{ mol.} = 0.0020 \text{ mol.}$$

$$n_{CO_3^{2-}} = 0.00338 \text{ mol.} + 0.00300 \text{ mol.} = 0.0064 \text{ mol.}$$

$$[H^+] = K_a \times \frac{n_{HCO_3^-}}{n_{CO_3^{2-}}} = 4.7 \times 10^{-11} \times \frac{0.0020}{0.0064} = 1.5 \times 10^{-11}\, M$$

$$pH = -\log(1.5 \times 10^{-11}) = 10.82$$

d. The pH of the diluted buffer is unchanged, but the pH of 28b is dramatically different from that of 26b, due to the decreased capacity of the diluted buffer. The pH's after base addition are little different since the buffer capacity was not exceeded.

e. Diluting the buffer does not affect the pH, but a diluted buffer has less capacity than an equal volume of a more concentrated buffer.

---

30. Refer to Section 14.1 and Table 13.2.

a. $$[H^+] = K_a \times \frac{[HNO_2]}{[NO_2^-]} = 6.0 \times 10^{-4} \times 2.50 = 1.5 \times 10^{-3}\, M$$

$$pH = -\log(1.5 \times 10^{-3}) = 2.82$$

b. Set $[HNO_2]_i = 2.50$ and $[NO_2^-]_i = 1.00$. If 33% of the HNO_2 is converted to NO_2^-, we get:

$$[HNO_2]_f = [HNO_2]_i - 0.33[HNO_2]_i = 2.50 - 0.33(2.50) = 1.68\, M$$

$$[NO_2^-]_f = [NO_2^-]_i + 0.33[HNO_2]_i = 1.00 + 0.33(2.50) = 1.83\, M$$

$$[H^+] = K_a \times \frac{[HNO_2]}{[NO_2^-]} = 6.0 \times 10^{-4} \times \frac{1.68}{1.83} = 5.5 \times 10^{-4} M$$

pH = -log(5.5 x 10^{-43}) = 3.26

c. $[H^+] = 10^{-pH} = 10^{-3.13} = 7.4 \times 10^{-4} M$

$$[H^+] = K_a \times \frac{[HNO_2]}{[NO_2^-]} \Rightarrow 7.4 \times 10^{-4} = 6.0 \times 10^{-4} \times \frac{[HNO_2]}{[NO_2^-]}$$

$$\frac{[HNO_2]}{[NO_2^-]} = 1.2$$

---

32. Refer to Section 14.1 and Example 14.1.

A buffer is made of a weak acid and its conjugate base, or a weak base and its conjugate acid. SnF_2 is a salt, therefore look for a weak acid with F$^-$ as the conjugate base, or a weak base with Sn^{2+} as the conjugate acid. Those mixtures that result in roughly equal amounts of F$^-$ and HF will act as buffers. In all cases, the initial amount of F$^-$ is:

$$0.250 \, L \times \frac{0.150 \, mol. \, SnF_2}{1 \, L} \times \frac{2 \, mol. \, F^-}{1 \, mol \, SnF_2} = 0.0750 \, mol. \, F^-$$

a. The reaction to consider is: F$^-$(aq) + HCl(aq) → HF(aq) + Cl$^-$(aq)

$$0.100 \, mol. \, HCl \times \frac{1 \, mol. \, HCl}{1 \, mol \, F^-} = 0.100 \, mol. \, F^-$$

Since there is an excess of HCl, all the SnF_2 will be converted to to HF. Since no salt will remain, this mixture will not be a buffer.

b. $$0.060 \, mol. \, HCl \times \frac{1 \, mol. \, HCl}{1 \, mol \, F^-} = 0.060 \, mol. \, F^-$$

In this case, 0.060 mol. F$^-$ will react, giving 0.060 mol. HF, leaving 0.015 mol. F$^-$. Thus, both the acid (HF) and the conjugate base (F$^-$) will be present in solution, and this mixture will act as a buffer.

c. $$0.040 \, mol. \, HCl \times \frac{1 \, mol. \, HCl}{1 \, mol \, F^-} = 0.040 \, mol. \, F^-$$

In this case, 0.040 mol. F$^-$ will react, giving 0.040 mol. HF, leaving 0.035 mol. F$^-$. Thus, both the acid (HF) and the conjugate base (F$^-$) will be present in solution, and this mixture will act as a buffer.

270

d.   The reaction to consider is: $SnF_2(aq) + 2OH^-(aq) \rightarrow Sn(OH)_2(s) + 2F^-(aq)$

Since the $Sn(OH)_2$ is insoluble and precipitates from solution, this mixture cannot act as a buffer.

e.   There is no reaction to consider here. This simply produces a mixture with 0.040 mol HF (the weak acid) and 0.0750 mol. F^- (the conjugate base). Thus this mixture is a viable buffer.

Thus b, c and e will act as buffers.

---

34. *Refer to Section 14.1 and Table 13.2.*

Consider the products of the reaction of Na_3PO_4 and HCl. Determine the amounts of each, then calculate the pH.

$Na_3PO_4 + HCl \rightarrow NaCl + Na_2HPO_4$

$$5.60\,g\,Na_3PO_4 \times \frac{1\,mol.\,Na_3PO_4}{163.941\,g\,Na_3PO_4} = 0.0342\,mol.\,Na_3PO_4$$

$$0.0750\,L\,HCl \times \frac{0.225\,mol.\,HCl}{1\,L} = 0.0169\,mol.\,HCl$$

Since Na_3PO_4 and HCl react in a 1:1 ratio, HCl is the limiting reactant and is thus completely consumed. The amount of Na_2HPO_4 after the reaction is 0.0169 mol., and the amount of Na_3PO_4 is:

0.0342 mol. - 0.0169 mol. = 0.0173 mol. Na_3PO_4

$$[H^+] = K_a \times \frac{n_{HPO_4^{2-}}}{n_{PO_4^{3-}}} = 4.5 \times 10^{-13} \times \frac{0.0169}{0.0173} = 4.4 \times 10^{-13}\,M$$

pH = -log(4.4 x 10^{-13}) = 12.36

---

36. *Refer to Section 14.1.*

Consider the products of the reaction of $HC_4H_7O_2$ and NaOH. Determine the amounts of each, then calculate the pH.

$HC_4H_7O_2 + OH^- \rightarrow C_4H_7O_2^- + H_2O$

$$2.00\,g\,HC_4H_7O_2 \times \frac{1\,mol.\,HC_4H_7O_2}{88.106\,g\,HC_4H_7O_2} = 0.0227\,mol.\,HC_4H_7O_2$$

$$0.50\,\text{g NaOH} \times \frac{1\,\text{mol. NaOH}}{40.00\,\text{g NaOH}} = 0.0125\,\text{mol NaOH}$$

Since $HC_4H_7O_2$ and NaOH react in a 1:1 ratio, NaOH is the limiting reactant and is thus completely consumed. The amount of $C_4H_7O_2^-$ after the reaction is 0.0125 mol., and the amount of $HC_4H_7O_2$ is:

0.0227 mol. - 0.0125 mol. = 0.0102 mol. $HC_4H_7O_2$

$$[H^+] = K_a \times \frac{n_{HC_4H_7O_2}}{n_{C_4H_7O_2^-}} = 1.5 \times 10^{-5} \times \frac{0.0102}{0.0125} = 1.22 \times 10^{-5}\,M$$

pH = -log(1.22 x 10⁻⁵) = 4.91

---

38. Refer to Section 14.1.

a. $[H^+] = 10^{-pH} = 10^{-7.40} = 4.0 \times 10^{-8}\,M$

$$[H^+] = K_a \times \frac{[H_2PO_4^-]}{[HPO_4^{2-}]} \Rightarrow 4.0 \times 10^{-8} = 6.2 \times 10^{-8} \times \frac{[H_2PO_4^-]}{[HPO_4^{2-}]}$$

$$\frac{[H_2PO_4^-]}{[HPO_4^{2-}]} = 0.65$$

b. $[H^+] = 10^{-pH} = 10^{-6.80} = 1.6 \times 10^{-7}\,M$

$$[H^+] = K_a \times \frac{n_{H_2PO_4^-}}{n_{HPO_4^{2-}}} \Rightarrow 1.6 \times 10^{-7} = 6.2 \times 10^{-8} \times \frac{n_{H_2PO_4^-}}{n_{HPO_4^{2-}}}$$

$$\frac{n_{H_2PO_4^-}}{n_{HPO_4^{2-}}} = 2.6$$

To calculate the amount of HPO_4^{2-} ions that were converted, set $n_{HPO_4^{2-}} = 1$ in part (a) above. Then $n_{H_2PO_4^-} = 0.65$. Then set x = moles HPO_4^{2-} converted. Substituting gives:

$$2.6 = \frac{0.65 + x}{1.0 - x}$$

$x = 0.54$, thus 54% is converted.

c. $[H^+] = 10^{-pH} = 10^{-7.80} = 1.6 \times 10^{-8}\,M$

$$[H^+] = K_a \times \frac{n_{H_2PO_4^-}}{n_{HPO_4^{2-}}} \Rightarrow 1.6 \times 10^{-8} = 6.2 \times 10^{-8} \times \frac{n_{H_2PO_4^-}}{n_{HPO_4^{2-}}}$$

$$\frac{n_{H_2PO_4^-}}{n_{HPO_4^{2-}}} = 0.26$$

To calculate the amount of $H_2PO_4^{2-}$ ions that were converted, set $n_{H_2PO_4^-} = 1$ in part (a) above, then $n_{HPO_4^{2-}} = 1.5$. Then set x = moles HPO_4^{2-} converted. Substituting gives:

$$2.6 = \frac{1-x}{1.5-x}$$

$x = 0.48$, thus 48% is converted.

---

40. Refer to Sections 14.2 and 14.3, Example 14.7 and Figures 14.5 to 14.7.

For an indicator to be suitable for a given titration, the color change (end point) must occur at a pH corresponding to the equivalence point of the titration. At that point, small additions of the titrant result in dramatic changes in pH, and thus a sudden (definitive) color change. This occurs along the steep portion of the titration curves shown in Figures 14.5 -14.7.

a.   This is the titration of a weak acid with a strong base (see figure 14.6). An indicator that changes color between pH 8 to pH 10 would be ideal.
Phenolphthalein changes color in this range.

b.   This is the titration of a weak acid with a strong base (see figure 14.6). An indicator that changes color between pH 8 to pH 10 would be ideal.
Phenolphthalein changes color in this range.

c.   This is the titration of a strong acid with a strong base (see figure 14.5). An indicator that changes color between pH 4 to pH 10 would be ideal.
Any of the three indicators change color in this range.

d.   This is the titration of a weak base with a strong acid (see figure 14.7). An indicator that changes color between pH 2 to pH 6 would be ideal.
Methyl orange changes color in this range.

---

42. Refer to Section 14.2.

A color change occurs when pH $\approx$ pK_a

pK_a = -log K_a = -log(3.56 x 10^{-4}) = 3.45

The range will be from $\dfrac{[HIn]}{[In^-]} \geq 10$ to $\dfrac{[HIn]}{[In^-]} \leq 0.1$

$$[H^+] = K_a \times \frac{[HIn]}{[In^-]} \quad \Rightarrow \quad -\log[H^+] = -\log K_a - \log\frac{[HIn]}{[In^-]}$$

$$pH = pK_a - \log\frac{[HIn]}{[In^-]} = 3.45 - \log 10 = 2.45$$

$$pH = pK_a - \log\frac{[HIn]}{[In^-]} = 3.45 - \log 0.1 = 4.45$$

Thus the range is from pH 2.45 to 4.45.

At the pK_a, pH = 3.45. One would be half way between red and yellow, thus the color would be **orange**.

---

44. Refer to Section 14.3 and Example 14.5.

a. $2HNO_3(aq) + Ba(OH)_2(aq) \rightleftharpoons Ba(NO_3)_2(aq) + 2H_2O(l)$
   $2H^+(aq) + 2NO_3^-(aq) + Ba^{2+}(aq) + 2OH^-(aq) \rightleftharpoons Ba^{2+}(aq) + 2NO_3^-(aq) + 2H_2O(l)$
   $2H^+(aq) + 2OH^-(aq) \rightleftharpoons 2H_2O(l)$
   $H^+(aq) + OH^-(aq) \rightleftharpoons H_2O(l)$

b. At the equivalence point, all the base has reacted with acid. Thus, the species present are Ba^{2+}, NO_3^- and H_2O.

c. $0.05000\,L \times \dfrac{0.237\,mol.\,Ba(OH)_2}{1\,L} \times \dfrac{2\,mol.\,OH^-}{1\,mol.\,Ba(OH)_2} = 0.0237\,mol.\,OH^-$

   $0.0237\,mol.\,OH^- \times \dfrac{1\,mol.\,HNO_3}{1\,mol.\,OH^-} \times \dfrac{1\,L}{0.4000\,mol.\,HNO_3} = 0.0593\,L\,HNO_3$

d. Before HNO_3 is added, pH can be calculated from the pOH.

   $$\frac{0.237\,mol\,Ba(OH)_2}{1\,L} \times \frac{2\,mol\,OH^-}{1\,mol\,Ba(OH)_2} = 0.474\,M\,OH^-$$

   pOH = -log [OH⁻] = -log 0.474 = 0.324

   pH = 14.00 - pOH = 14.00 - 0.324 = 13.68

e. At half-way to the equivalence point, half of the base will remain, 0.0119 mol. of the original 0.0237 mol. Since half of the 0.0593 L of acid (calculated in part c) will have been added, the total volume will be 0.05000 L + ½(0.0593 L) = 0.0797 L.

   [OH⁻] = 0.0119 mol. / 0.0797 L = 0.149 M
   pOH = -log [OH⁻] = -log 0.149 = 0.83
   pH = 14.00 - pOH = 14.00 - 0.83 = 13.17

f. At the equivalence point, the only species present are the spectator ions and water, thus the pH will be 7.

---

46. Refer to Section 14.3 and Table 13.2.

---

a. $NaLac(aq) + HCl(aq) \rightleftharpoons NaCl(aq) + HLac(aq)$

$Na^+(aq) + Lac^-(aq) + H^+(aq) + Cl^-(aq) \rightleftharpoons Na^+(aq) + Cl^-(aq) + HLac(aq)$

$H^+(aq) + Lac^-(aq) \rightleftharpoons HLac(aq)$

b. At equivalence point, the lactate is converted to lactic acid. As a weak acid, however, a small amount will dissociate. Thus all the species in the ionic equation (H^+, Na^+, Cl^-, Lac^- and $HLac(aq)$) are present.

c. $0.05000\,L \times \dfrac{0.224\,mol.\ NaLac}{1\,L} \times \dfrac{1\,mol.\ Lac^-}{1\,mol.\ NaLac} = 0.0112\,mol.\ Lac^-$

$0.0112\,mol.\ Lac^- \times \dfrac{1\,mol.\ HCl}{1\,mol.\ Lac^-} \times \dfrac{1\,L}{0.1035\,mol.\ HCl} = 0.108\,L\ HCl$

d. At the beginning of the titration:

	Lac$^-$(aq)	+ H$_2$O(l)	$\rightleftharpoons$	HLac(aq)	+	OH$^-$(aq)
[ ]$_o$	0.224	---		0		0
Δ[ ]	-x	---		+x		+x
[ ]$_{eq}$	0.224 - x	---		x		x

$K_b = 7.1 \times 10^{-11} = \dfrac{[HLac][OH^-]}{[Lac^-]} = \dfrac{x^2}{0.224 - x}$

$K_b/b = 7.1 \times 10^{-11} / 0.224 = 3.2 \times 10^{-10}$

$K_b/b = = 3.2 \times 10^{-10} < 2.5 \times 10^{-3}$ (The 5% rule applies. See the explanation accompanying Chapter 13, Problem 38.)

$K_b = 7.1 \times 10^{-11} = \dfrac{x^2}{0.224}$

$x^2 = 1.6 \times 10^{-11}$

$x = [OH^-] = 4.0 \times 10^{-6}\,M$

$pOH = -\log\,(4.0 \times 10^{-6}) = 5.40$

$pH = 14.00 - pOH = 14.00 - 5.40 = 8.60$

e. At half-way to the equivalence point, half of the base will remain, 0.0056 mol. of the original 0.0112 mol. The other half will have been converted to HLac. Consequently, the system is a buffer.

$$[H^+] = K_a \times \frac{n_{HLac}}{n_{Lac^-}} = 1.4 \times 10^{-4} \times \frac{0.0056}{0.0056} = 1.4 \times 10^{-4}\ M$$

pH = -log(1.4×10^{-4}) = 3.85

f. At the equivalence point, all the Lac$^-$ has been converted to HLac, giving 0.0112 mol.

$$[HLac] = \frac{0.0112\ \text{mol. HLac}}{0.108\ L + 0.05000\ L} = 0.0709\ M$$

	HLac(aq)	$\rightleftharpoons$	H$^+$(aq)	+	Lac$^-$(aq)
[ ]$_o$	0.0709		0		0
Δ[ ]	-x		+x		+x
[ ]$_{eq}$	0.0709 - x		x		x

$$K_a = \frac{[H^+][Lac^-]}{[HLac]} \Rightarrow 1.4 \times 10^{-4} = \frac{x^2}{0.0709 - x}$$

K_a/a = 1.4 x 10^{-4} / 0.0709 = 2.0 x 10^{-3}
K_a/a = 2.0 x 10^{-3} < 2.5 x 10^{-3} (The 5% rule applies. See the explanation accompanying Chapter 13, Problem 38.)

$$K_a = \frac{[H^+][Lac^-]}{[HLac]} \Rightarrow 1.4 \times 10^{-4} = \frac{x^2}{0.0709}$$

x^2 = 9.9 x 10^{-6}

x = [H$^+$] = 3.2 x 10^{-3} M

pH = -log (3.2 x 10^{-3}) = 2.50

---

48. Refer to Section 14.3.

---

a. HClO is a weak acid. Calculate the molarity of the acid, and then, using K_a, [H$^+$].

$$30.0\ \text{mL} \times \frac{1.00\ g}{1.00\ \text{mL}} \times \frac{10.0\ g\ HClO}{100\ g\ \text{solution}} \times \frac{1\ \text{mol. HClO}}{52.46\ g} = 5.72 \times 10^{-2}\ \text{mol. HClO}$$

$$[HClO] = \frac{5.72 \times 10^{-2}\ \text{mol. HClO}}{0.0300\ L} = 1.91 M$$

	HClO(aq)	$\rightleftharpoons$	H$^+$(aq)	+	ClO$^-$(aq)
[ ]$_o$	1.91		0		0
Δ[ ]	-x		+x		+x
[ ]$_{eq}$	1.91 - x		x		x

$$K_a = \frac{[H^+][ClO^-]}{[HClO]} \Rightarrow 2.8 \times 10^{-8} = \frac{x^2}{1.91 - x}$$

$K_a/a = 2.8 \times 10^{-8} / 1.91 = 1.4 \times 10^{-8}$

$K_a/a = 1.4 \times 10^{-8} < 2.5 \times 10^{-3}$

(The 5% rule applies. See the explanation accompanying Chapter 13, Problem 38.)

$$K_a = \frac{[H^+][ClO^-]}{[HClO]} \Rightarrow 2.8 \times 10^{-8} = \frac{x^2}{1.91}$$

$x^2 = 5.3 \times 10^{-8}$

$x = [H^+] = 2.3 \times 10^{-4} M$

pH = -log (2.3 $\times 10^{-4}$) = 3.64

b. At half-way to the equivalence point, half of the HClO has been converted to ClO$^-$, giving 2.86 $\times 10^{-2}$ moles. Half remains as HClO. Consequently, the system is a buffer.

$$[H^+] = K_a \times \frac{n_{HClO}}{n_{ClO^-}} = 2.8 \times 10^{-8} \times \frac{2.86 \times 10^{-2}}{2.86 \times 10^{-2}} = 2.8 \times 10^{-8} M$$

pH = -log(2.8 $\times 10^{-8}$) = 7.55

c. At the equivalence point, all the HClO has been converted to ClO$^-$, giving 5.72 $\times 10^{-2}$ mol. The volume of the solution has increased, thus calculate the volume of base added and recalculate [ClO$^-$]. Then calculate [OH$^-$] using K_b, and from that, pOH and pH.

$$5.72 \times 10^{-2} \text{ mol. HClO} \times \frac{1 \text{ mol. KOH}}{1 \text{ mol. HClO}} \times \frac{1 \text{ L solution}}{0.419 \text{ mol. KOH}} = 0.137 \text{ L}$$

$$[ClO^-] = \frac{5.72 \times 10^{-2} \text{ mol. ClO}^-}{0.0300 \text{ L} + 0.137 \text{ L}} = 0.344 M$$

	ClO$^-$(aq)	H$_2$O(l)	$\rightleftharpoons$	HClO(aq)	+	OH$^-$(aq)
[ ]$_o$	0.344			0		0
Δ[ ]	-x			+x		+x
[ ]$_{eq}$	0.344 - x			x		x

$$K_b = \frac{[HClO][OH^-]}{[ClO^-]} \Rightarrow 3.6 \times 10^{-7} = \frac{x^2}{0.344 - x}$$

$K_b/b = 3.6 \times 10^{-7} / 0.344 = 1.0 \times 10^{-6}$

$K_b/b = 1.0 \times 10^{-6} < 2.5 \times 10^{-3}$ (The 5% rule applies. See the explanation accompanying Chapter 13, Problem 38.)

$$K_b = \frac{[HClO][OH^-]}{[ClO^-]} \Rightarrow 3.6 \times 10^{-7} = \frac{x^2}{0.344}$$

$x^2 = 1.24 \times 10^{-7}$

$x = [OH^-] = 3.52 \times 10^{-4} M$

$pOH = -\log(3.52 \times 10^{-4}) = 3.45$

$pH = 14.00 - pOH = 14.00 - 3.45 = 10.55$

---

50. Refer to Section 14.3.

a. Since the acid is monoprotic, 1 mol. HX reacts with 1 mol. NaOH.

$$0.0342 \text{ L} \times \frac{0.652 \text{ mol. NaOH}}{1 \text{ L}} \times \frac{1 \text{ mol. HX}}{1 \text{ mol. NaOH}} = = 0.0223 \text{ mol. HX}$$

$$\frac{2.500 \text{ g HX}}{0.0223 \text{ mol. HX}} = = 112 \text{ g/mol. HX}$$

b. As seen in problems (46e) and (48b) above, at halfway to the equivalence point, $pH = pK_a$. Thus: at 17.1 mL NaOH, $pH = pK_a = 3.17$.

$K_a = 10^{-pH} = 10^{-3.17} = 6.8 \times 10^{-4}$

c. $K_b = \dfrac{K_w}{K_a} = \dfrac{1.0 \times 10^{-14}}{6.8 \times 10^{-4}} = 1.5 \times 10^{-11}$

---

52. Refer to Section 10.3 and Chapter 13.

Use the osmotic pressure to calculate [HB]. Use pH to calculate $[H^+]$, which is also $[B^-]$. Then calculate K_a and finally K_b.

$\pi = MRT$

$0.878 \text{ atm} = (M)(0.0821 \text{ L·atm/mol.·K})(298 \text{ K})$

$M = 0.0359$ (this is [HB])

$$[H^+] = 10^{-pH} = 10^{-6.76} = 1.7 \times 10^{-7} \, M$$

$$K_a = \frac{[H^+][B^-]}{[HB]} = \frac{(1.7 \times 10^{-7})(1.7 \times 10^{-7})}{0.0359} = 8.1 \times 10^{-13}$$

$$K_b = \frac{K_w}{K_a} = \frac{1.0 \times 10^{-14}}{8.1 \times 10^{-13}} = 1.2 \times 10^{-2}$$

54. Refer to Section 14.3.

Each OH⁻ will react with one HX to give one X⁻ and a water molecule (not shown).

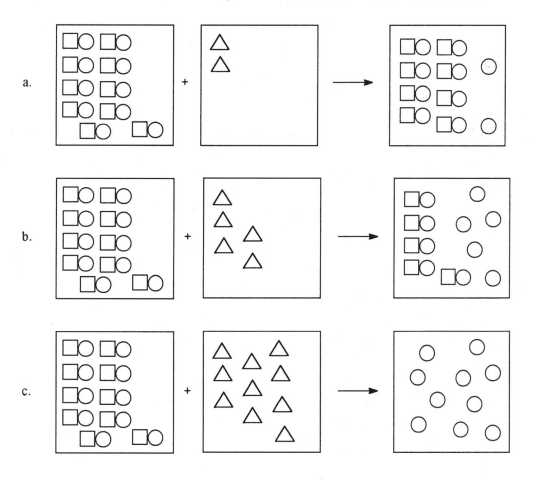

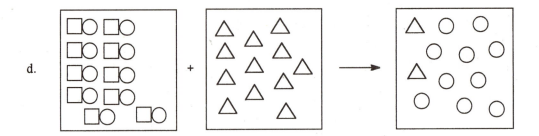

d.

b. above represents neutralization half-way to the equivalence point. Half the HX molecules have been converted to X⁻, and half remain as HX.

---

56. Refer to Section 12.5.

Acid in the blood reacts with the bicarbonate buffer, establishing the equilibrium:
$$H^+(aq) + HCO_3^-(aq) \rightleftharpoons H_2CO_3(aq) \qquad \text{eq. 1}$$

Carbonic acid also establishes an equilibrium, decomposing to carbon dioxide and water.
$$H_2CO_3(aq) \rightleftharpoons H_2O(l) + CO_2(g) \qquad \text{eq. 2}$$

According to Le Chatelier's principle, if the concentration of H^+ (a reactant) increases then the equilibrium will shift to produce more products, thus partially offsetting that increase in reactant concentration. Such an increase would result in an increase in the concentration of carbonic acid. This increase in $[H_2CO_3]$ would result in the equilibrium of eq. 2 also shifting to more products, thus producing more CO_2.

---

58. Refer to Section 14.1 and Chapter 13.

a. **False**. The formate ion concentration in $0.10\,M$ $HCHO_2$ is much less than the formate ion concentration in $0.10\,M$ $NaCHO_2$.
$HCHO_2$ is a weak acid, dissociating very little, while $NaCHO_2$ is a salt and dissociates completely.

b. **True**. Excessive amounts of acid will convert all the B⁻ to HB, changing the buffer solution to an acid solution.

c. **False**. A buffer is a mixture of roughly equal amounts of a weak acid and the conjugate base of that acid.

d. **False**. Since K_a for HCO_3^- is 4.7×10^{-11}, K_b for CO_3^{2-} is 2.1×10^{-4}.
$K_b = K_w / K_a$ for the conjugate base (B⁻) of the weak acid HB.

Write the equation for the reaction. Calculate [H$^+$] (from pH) and moles of acetate (from the reaction stoichiometry) and from that, moles of HB still in solution. Then calculate the moles of acetic acid added and finally, the volume of acetic acid.

$$HC_2H_3O_2(aq) + OH^-(aq) \rightleftharpoons C_2H_3O_2^-(aq) + H_2O(l)$$

$$[H^+] = 10^{-pH} = 10^{-4.20} = 6.31 \times 10^{-5}\,M$$

$$0.1000\,L \times \frac{1.25\,mol.\,NaOH}{1\,L} \times \frac{1\,mol.\,C_2H_3O_2^-}{1\,mol.\,NaOH} = 0.125\,mol.\,C_2H_3O_2^-$$

$$[H^+] = K_a \times \frac{n_{HC_2H_3O_2}}{n_{C_2H_3O_2^-}} \quad \Rightarrow \quad 6.31 \times 10^{-5} = 1.8 \times 10^{-5} \times \frac{n_{HC_2H_3O_2}}{0.125}$$

$$n_{HC_2H_3O_2} = 0.44\,mol.$$

This is the moles of acid present *after* reaction has occurred. The total moles of acid that must have been present *before* reaction is:

$$n_{total} = n_{HC_2H_3O_2} + n_{C_2H_3O_2^-} = 0.44\,mol. + 0.125\,mol. = 0.56\,mol.$$

Note that this solution contains roughly equal amounts of HB and B$^-$, and is therefore a buffer.

$$0.56\,mol. \times \frac{60.05\,g\,HC_2H_3O_2}{1\,mol.} \times \frac{100\,g\,solution}{98\,g\,HC_2H_3O_2} \times \frac{1\,mL}{1.0542\,g} = 33\,mL\,solution$$

Calculate the mole ratio of NH$_4^+$ to NH$_3$ needed to make a solution with the requisite pH. Consider the mole ratio and the definition of a buffer.

$$[H^+] = 10^{-pH} = 10^{-6.50} = 3.16 \times 10^{-7}\,M$$

$$[H^+] = K_a \times \frac{n_{NH_4^+}}{n_{NH_3}} \quad \Rightarrow \quad 3.16 \times 10^{-7} = 5.6 \times 10^{-10} \times \frac{n_{NH_4^+}}{n_{NH_3}}$$

$$\frac{n_{NH_4^+}}{n_{NH_3}} = 560$$

A buffer requires roughly equal amounts of a weak acid and its conjugate base. A 600:1 ratio does **not** meet this requirement.

62. *Refer to Sections 14.1 and 14.3.*

After solution A is titrated to its equivalence point, all the acid will have been converted to the conjugate base. Also, when solutions A and B are combined, [HB] = [B⁻], thus [HB] / [B⁻] = 1. Calculate the [H⁺] of this solution, and the K_a for the acid. From K_a and pH, of the original solution, one can calculate [HB] and thus the molar mass.

from the final solution,

$$[H^+] = 10^{-pH} = 10^{-4.26} = 5.50 \times 10^{-5}\,M$$

$$[H^+] = K_a \times \frac{[HB]}{[B^-]} \Rightarrow 5.50 \times 10^{-5} = K_a \times 1$$

$$K_a = 5.50 \times 10^{-5}$$

From the initial solution,

$$[H^+] = 10^{-pH} = 10^{-2.56} = 2.75 \times 10^{-3}\,M$$

$$K_a = \frac{[B^-][H^+]}{[HB]} \Rightarrow 5.50 \times 10^{-5} = \frac{(2.75 \times 10^{-3})^2}{[HB]}$$

$$[HB] = 0.138\,M$$

$$0.2500\,\text{L} \times \frac{0.138\,\text{mol. HB}}{1\,\text{L}} = 0.0345\,\text{mol. HB}$$

$$\frac{4.00\,\text{g}}{0.0345\,\text{mol. HB}} = 116\,\text{g/mol. HB}$$

63. *Refer to Sections 14.1 and 14.3 and Example 14.6.*

a. $$K_a = \frac{[NO_2^-][H^+]}{[HNO_2]} \Rightarrow 6.0 \times 10^{-4} = \frac{x^2}{1.000\,M}$$

$x^2 = 6.0 \times 10^{-4}$
$x = [H^+] = 0.024\,M$

pH = -log (0.0024) = 1.62

b. $$0.05000\,\text{L} \times \frac{1.000\,\text{mol.}}{1\,\text{L}} = 0.05000\,\text{mol. } HNO_2$$

At half neutralization, half of the HNO_2 will be converted to NO_2^-, thus:
$[NO_2^-] = [HNO_2] = 0.02500\,M$.

$$[H^+] = K_a \times \frac{n_{HNO_2}}{n_{NO_2^-}} = 6.0 \times 10^{-4} \times \frac{0.02500}{0.02500} = 6.0 \times 10^{-4} \, M$$

pH = -log (6.0×10^{-4}) = 3.22

c. At equivalence point, all the nitrous acid has been converted to nitrite ion and the solution is weakly basic.

$$NO_2^-(aq) + H_2O(l) \rightleftharpoons HNO_2(aq) + OH^-(aq)$$

Calculate the volume of base added and the total volume of the final solution. Then calculate [NO$_2^-$], [OH$^-$], pOH and finally pH.

$$0.0500 \, \text{mol. HNO}_2 \times \frac{1 \, \text{mol. NaOH}}{1 \, \text{mol. HNO}_2} \times \frac{1 \, L}{0.850 \, \text{mol. NaOH}} = 0.0588 \, L$$

V_{total} = 0.05000 L (HNO$_2$) + 0.0588 L (NaOH) = 0.1088 L

$$[NO_2^-] = \frac{0.05000 \, \text{mol.}}{0.1088 \, L} = 0.4596 \, M$$

$$K_b = \frac{[HNO_2][OH^-]}{[NO_2^-]} \Rightarrow 1.7 \times 10^{-11} = \frac{x^2}{0.4596 \, M}$$

$x^2 = 7.8 \times 10^{-12}$
$x = [OH^-] = 2.8 \times 10^{-6} \, M$

pOH = -log (2.8×10^{-6}) = 5.55
pH = 14.0 - pOH = 14.00 - 5.55 = 8.45

d. As calculated in part (c) above, 0.0588 L (58.8 mL) of NaOH solution were needed to reach equivalence. 0.10 mL less that volume is 0.0587 L (58.7 mL).

$$0.0587 \, L \times \frac{0.850 \, \text{mol. NaOH}}{1 \, L} \times \frac{1 \, \text{mol. HNO}_2}{1 \, \text{mol. NaOH}} = 0.0499 \, \text{mol. HNO}_2$$

0.05000 mol. HNO$_2$ - 0.0499 mol. HNO$_2$ = 0.0001 mol. HNO$_2$

$$[H^+] = K_a \times \frac{n_{HNO_2}}{n_{NO_2^-}} = 6.0 \times 10^{-4} \times \frac{0.0001}{0.0499} = 1.2 \times 10^{-6} \, M$$

pH = -log (1.2×10^{-6}) = 5.92

e. At this point, any NaOH added in addition to the 58.8 mL (from part (c) above) is excess. Calculate the moles of excess NaOH and [OH$^-$] in the resulting solution.

$$n_{NaOH} = 0.10 \, \text{mL} \times \frac{1 \, L}{1000 \, \text{mL}} \times \frac{0.850 \, \text{mol.}}{1 \, L} = 8.50 \times 10^{-5} \, \text{mol.} \qquad \text{(in excess)}$$

$$[OH^-] = \frac{8.50 \times 10^{-5} \text{ mol.}}{0.1089 \text{ L}} = 7.81 \times 10^{-4} \, M$$

Note that we can neglect the OH⁻ formed by the nitrate (see part (c) above) since its contribution is so small, considering that K_b has only two significant figures.

pOH = -log (7.81 x 10⁻⁴) = 3.11
pH = 14.00 - pOH = 14.00 - 3.11 = 10.89

f. Consider one more data point, with 10.00 mL NaOH in excess, giving a total volume of 68.8 mL.

$$n_{\text{NaOH}} = 10.00 \text{ mL} \times \frac{1 \text{ L}}{1000 \text{ mL}} \times \frac{0.850 \text{ mol.}}{1 \text{ L}} = 8.50 \times 10^{-3} \text{ mol.} \qquad \text{(in excess)}$$

$$[OH^-] = \frac{8.50 \times 10^{-3} \text{ mol.}}{0.0688 \text{ L}} = 0.124 \, M$$

pOH = -log (0.124) = 0.907
pH = 14.00 - pOH = 14.00 - 0.838 = 13.09

pH	NaOH (mL)
1.62	0.00
3.22	29.4
5.92	58.7
8.45	58.8
10.89	58.9
13.09	68.8

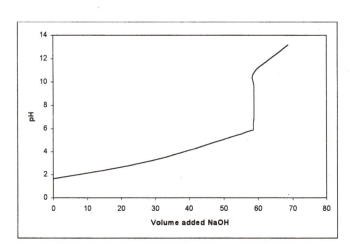

---

64. *Refer to Section 14.2.*

The indicator will be in the midst of changing color when [HIn] = [In⁻]. Calculate the [H⁺] at this point. Determine the amount of $C_2H_3O_2^-$ that is present, and thus the amount of NaOH used to produce this amount of $C_2H_3O_2^-$.

$$[H^+] = K_a \times \frac{[HIn]}{[In^-]} \quad \Rightarrow \quad [H^+] = 1.0 \times 10^{-5} \times 1 = 1.0 \times 10^{-5} \, M$$

284

$$[H^+] = K_a \times \frac{n_{HC_2H_3O_2}}{n_{C_2H_3O_2^-}} \quad \Rightarrow \quad 1.0 \times 10^{-5} = 1.8 \times 10^{-5} \times \frac{n_{HC_2H_3O_2}}{n_{C_2H_3O_2^-}}$$

$$\frac{n_{HC_2H_3O_2}}{n_{C_2H_3O_2^-}} = 0.56 \quad \Rightarrow \quad n_{HC_2H_3O_2} = 0.56 n_{C_2H_3O_2^-} \qquad \text{eq. 1}$$

Before titration:

$$0.05000 \text{ L} \times \frac{1.00 \text{ mol.}}{1 \text{ L}} = 0.0500 \text{ mol. } HC_2H_3O_2$$

After titration:

$$n_{HC_2H_3O_2} = x, \text{ and } n_{C_2H_3O_2^-} = 0.0500 - x \qquad \text{Substituting into eq. 1,}$$

$$x = 0.56(0.0500 - x)$$
$$x = [HC_2H_3O_2] = 0.018 \, M$$

$$n_{C_2H_3O_2^-} = 0.0500 - x = 0.032 \text{ mol.}$$

$$0.032 \text{ mol. } C_2H_3O_2^- \times \frac{1 \text{ mol. NaOH}}{1 \text{ mol. } C_2H_3O_2^-} \times \frac{1 \text{ L}}{1.00 \text{ mol. NaOH}} = 0.032 \text{ L NaOH}$$

A better indicator would be phenolphthalein (see problem 40a above).

---

65. Refer to Section 14.1 and Appendix 3.

$$[H^+] = K_a \times \frac{[HB]}{[B^-]}$$

Take the -log of both sides.

$$-\log[H^+] = -\log\left(K_a \times \frac{[HB]}{[B^-]}\right) = -\log K_a - \log\left(\frac{[HB]}{[B^-]}\right)$$

$$pH = pK_a - \log\left(\frac{[HB]}{[B^-]}\right) \quad \Rightarrow \quad pH = pK_a + \log\left(\frac{[B^-]}{[HB]}\right)$$

---

66. Refer to Section 13.4 and Example 13.9.

---

a. $H_2SO_4(aq) \rightleftharpoons H^+(aq) + HSO_4^-(aq)$
    This is a strong acid and ionizes completely.

$[H^+] = 0.1500 \, M$

$pH = -\log(0.1500) = 0.8239$

b. For the second ionization: $HSO_4^-(aq) \rightleftharpoons H^+(aq) + SO_4^{2-}(aq)$

$$[SO_4^{2-}] = x$$
$$[HSO_4^-] = 0.1500 - x$$
$$[H^+] = 0.1500 + x$$

$$K_a = \frac{[SO_4^{2-}][H^+]}{[HSO_4^-]} \Rightarrow 1.1 \times 10^{-2} = \frac{(0.1500 + x)(x)}{0.1500 - x}$$

$K_a/a = 1.1 \times 10^{-2} / 0.1500 = 7.3 \times 10^{-2}$
$K_a/a = 7.3 \times 10^{-2} > 2.5 \times 10^{-3}$ (The 5% rule does **not** apply. See the explanation accompanying Chapter 13, Problem 38.)

$$K_a = \frac{[SO_4^{2-}][H^+]}{[HSO_4^-]} \Rightarrow 1.1 \times 10^{-2} = \frac{(0.1500 + x)(x)}{0.1500 - x} = \frac{0.1500x + x^2}{0.1500 - x}$$

$$0.00165 - 0.011x = 0.1500x + x^2$$
$$x^2 + 0.161x - 0.00165 = 0$$

$$x = \frac{-b \pm \sqrt{b^2 - 4ac}}{2a} = \frac{-0.161 \pm \sqrt{(0.161)^2 - 4(1)(-0.00165)}}{2(1)} = 0.0097$$

$$[H^+] = 0.1500 + 0.0097 = 0.1597\, M$$

$$pH = -\log(0.1597) = 0.7968$$

---

67. Refer to Section 14.3.

Since 21.0 mL is needed for neutralization, the titration with 7.00 mL represents ⅓ (7/21) neutralization. Thus ⅓ of the base (B) has been converted to conjugate acid (HB⁺), with ⅔ remaining as B.

Consequently, $\dfrac{n_{HB^+}}{n_B} = \dfrac{1/3}{2/3} = 0.5$

$$[H^+] = 10^{-pH} = 10^{-8.95} = 1.1 \times 10^{-9}\, M$$

$$[H^+] = K_a \times \frac{[HB^+]}{[B]} \Rightarrow 1.1 \times 10^{-9} = K_a \times 0.5$$

$$K_a = 2.2 \times 10^{-9}$$

$$K_b = \frac{K_w}{K_a} = \frac{1.0 \times 10^{-14}}{2.2 \times 10^{-9}} = 4.5 \times 10^{-6}$$

Chapter 15: Complex Ions

---

2. *Refer to Section 15.1 and Example 15.1.*

    a. $[Ni(H_2O)_2Cl_2(OH)_2]^{2-}$ has three different ligands:
        H_2O: Water is a molecule and has no charge.
        Cl^-:   Chloride is an anion with a -1 charge.
        OH^-:   Hydroxide is an anion with a -1 charge.

    b. metal oxidation number = charge of the complex - the sum of the charges of the ligands.
        $M_{o.n.} = (-2) - (2(0) + 2(-1) + 2(-1)) = +2$.

    c. Sodium is a cation with a +1 charge, thus 2 Na^+ are needed to balance the charge on the Ni
        complex. $Na_2[Ni(H_2O)_2Cl_2(OH)_2]$

---

4. *Refer to Section 15.1.*

Put the metal and the ligands together and add up the charges to get the overall charge.

    a. $Pt^{2+} + 2NH_3 + C_2O_4^{2-} \rightarrow [Pt(NH_3)_2(C_2O_4)]$

    b. $Pt^{2+} + 2NH_3 + SCN^- + Br^- \rightarrow [Pt(NH_3)_2(SCN)Br]$

    c. $Pt^{2+} + en + 2NO_2^- \rightarrow [Pt(en)(NO_2)_2]$
          * *en* = ethylene diamine

---

6. *Refer to Section 15.1.*

The coordination number is the total number of atoms bonded to the metal.

    a. $[Co(en)_2(SCN)Cl]^+$ (recall that (*en*) is a chelating ligand, taking up 2 coordination
                                sites on the Co)
    $2(2en) + 1SCN^- + 1Cl^- = 6$, thus the coordination number is **6**.

    b. $[Zn(en)(C_2O_4)]$ (recall that (*en*) and ($C_2O_4^{2-}$) are both chelating ligands)
    $2(1en) + 2(1C_2O_4^{2-}) = 4$, thus the coordination number is **4**.

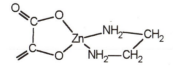

---

c. $Ag(NH_3)Cl$
    $1NH_3 + 1Cl^- = 2$, thus the coordination number is **2**.
d. $[Cu(H_2O)_4]^{2+}$
    $4H_2O$, thus the coordination number is **4**.

---

8. *Refer to Section 15.1.*

Use the coordination number of the metal ion to determine the number of ligands, and add the charges of the metal ion and the ligands to get the charge of the complex.

a. Ag^+ has a coordination number of 2, and H_2O has no charge, thus
    $[Ag(H_2O)_2]^+$

b. Pt^{2+} has a coordination number of 4, and H_2O has no charge, thus
    $[Pt(H_2O)_4]^{2+}$

c. Pd^{2+} has a coordination number of 4, and Br^- has a -1 charge, thus
    $[PdBr_4]^{2-}$

d. Fe^{3+} has a coordination number of 6, and $C_2O_4^{2-}$ has a -2 charge and is chelating (each oxalate occupying 2 coordination sites), thus
    $[Fe(C_2O_4)_3]^{3-}$

---

10. *Refer to Section 15.1 and Chapter 3.*

Since chloride has a -1 charge, 2 moles of the Cl^- associate with one mole of complex cation: $[Cr(H_2O)_5(OH)]Cl_2$

$$M = 51.996 + 5(18.02) + 1(17.01)] + 2(35.45) = 230.0 \text{ g/mol.}$$

$$\text{mass \%} = \frac{\text{mass of Cr}}{\text{mass of salt}} \times 100\% = \frac{51.996 \text{ g}}{230.0 \text{ g}} \times 100\% = 22.61\%$$

---

12. *Refer to Section 15.1, Problem 3.85 (p. 82) and Problem 10 (above).*

$$\text{mass \%} = \frac{\text{mass of Fe}}{\text{mass of hemoglobin}} \times 100\%$$

$$0.35\% = \frac{4(55.85 \text{ g/mol.})}{M_{\text{hemoglobin}}} \times 100\%$$

$$M_{\text{hemoglobin}} = 6.4 \times 10^4 \text{ g/mol.}$$

Note that the atom actually bonded to the metal is that ligand atom which is capable of acting as a Lewis base. Also note that $OH_2 = H_2O$; it is written in this fashion to emphasize that it is the oxygen that is bonded to the metal.

a.

b.

c.

en =

d.

e.

16. Refer to Sections 15.1 and 15.2.

The dech ligand will chelate to the Pd in much the same manner as en.

a. As in Figure 15.5, there are 4 of one ligand (SCN^-), and 2 of another (NH_3). There are two possible geometric isomers (NH_3 *cis*, and NH_3 *trans*).

cis   trans

b. As in Figure 15.6, there are 3 of one ligand (NO_2^-), and 3 of another (NH_3). There are two possible geometric isomers (NH_3 in a *facial* , and NH_3 in an *meridial* arrangement).

facial   *meridial*

c. This complex complex has 3 NH_3, 2 H_2O and 1 OH^-. The NH_3 can be distributed in either *facial* or *meridial* fashion. For the *meridial* isomer, the H_2O can be either *cis* or *trans*.

facial   *meridial, trans* H_2O *meridial, cis* H_2O

This complex complex has 3 NH_3, 2 Cl^- and 1 Br^-. The NH_3 can be distributed in either *facial* or *meridial* fashion. For the *meridial* isomer, the Cl^- can be either *cis* or *trans*.

facial   *meridial, trans* Cl *meridial, cis* Cl

22. Refer to Section 15.3 and Chapter 6.

When forming a cation, the electrons are removed first from the **s** orbital then the **d** orbital.

a. Ti: $[Ar]4s^2 3d^2$
    Ti^{3+}: $[Ar]3d^1$

b. Cr: $[Ar]4s^2 3d^4$
    Cr^{2+}: $[Ar]3d^4$

c. Ru: $[Kr]5s^2 4d^6$
    Ru^{4+}: $[Kr]4d^4$

d. Pd: $[Kr]5s^2 4d^8$
    Pd^{2+}: $[Kr]4d^8$

e. Mo: $[Kr]5s^2 4d^4$
    Mo^{3+}: $[Kr]4d^3$

24. Refer to Section 15.3 and Chapter 6.

a. $[Ar]3d^1$:   (↑)( )( )( )( )
    1 unpaired electrons

b. $[Ar]3d^4$:   (↑)(↑)(↑)(↑)( )
    4 unpaired electrons

c. $[Kr]4d^4$:   (↑)(↑)(↑)(↑)( )
    4 unpaired electrons

d. $[Kr]4d^8$:   (↑↓)(↑↓)(↑↓)(↑)(↑)
    2 unpaired electrons

e. $[Kr]4d^3$:   (↑)(↑)(↑)( )( )
    3 unpaired electrons

a. Fe: $[Ar]4s^23d^6$
    Fe^{2+}: $[Ar]3d^6$

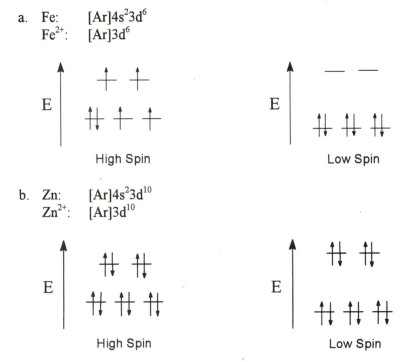

b. Zn: $[Ar]4s^23d^{10}$
    Zn^{2+}: $[Ar]3d^{10}$

Note that the number of unpaired electrons does not change.

V: $[Ar]4s^23d^3$
V^{3+}: $[Ar]3d^2$

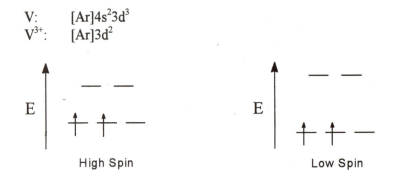

The high and low spin systems have the same electron distribution.

Both complexes contain Co^{3+} ($[Ar]3d^6$). To be diamagnetic, $[Co(NH_3)_6]^{3+}$ must have all electrons paired, which would make this a low spin complex. To be paramagnetic, $[CoF_6]^{3-}$ must have unpaired electrons, which would make this a high spin complex. This is consistent with the fact that NH_3 is a stronger field ligand than F^-.

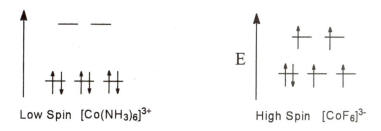

Low Spin $[Co(NH_3)_6]^{3+}$ High Spin $[CoF_6]^{3-}$

For strong field ligands, Δ_0 is large, resulting in low spin complexes.

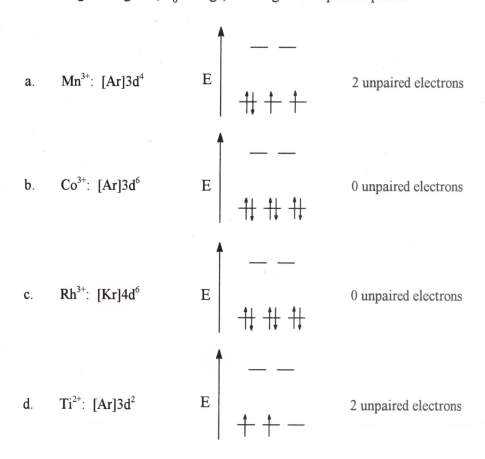

a. Mn^{3+}: $[Ar]3d^4$ E   2 unpaired electrons

b. Co^{3+}: $[Ar]3d^6$ E   0 unpaired electrons

c. Rh^{3+}: $[Kr]4d^6$ E   0 unpaired electrons

d. Ti^{2+}: $[Ar]3d^2$ E   2 unpaired electrons

e. Mo^{2+}: [Kr]4d^4

2 unpaired electrons

---

34. Refer to Section 15.3 (Color) and Chapter 6.

$$E = \frac{hc}{\lambda} = \frac{(6.626 \times 10^{-34} \text{ J} \cdot \text{s})(2.998 \times 10^8 \text{ m/s})}{399 \times 10^{-9} \text{ m}} = 4.98 \times 10^{-19} \text{ J}$$

$$E = 4.98 \times 10^{-19} \text{ J} \times \frac{1 \text{ kJ}}{1000 \text{ J}} \times \frac{6.022 \times 10^{23}}{1 \text{ mol.}} = 3.00 \times 10^2 \text{ kJ/mol.}$$

---

36. Refer to Section 15.3, Figure 15.12 and Table 15.3.

The color observed is the complimentary color of that absorbed by the complex. Since red is observed, the complimentary color (from the color wheel) is green-blue. Thus the wavelength of maximum absorption is approximately **500 nm**.

---

38. Refer to Section 15.4 and Example 15.6.

Calculate pOH and [OH⁻]. Use [OH⁻] and K_f to calculate $[Al^{3+}]/[Al(OH)_4^-]$

At pH 2.0:   pOH = 14.00 - 2.00 = 12.00
               [OH⁻] = 10^{-pOH} = 10$^{-12.00}$ = 1.0 × 10^{-12} M

$$Al^{3+} + 4OH^- \rightleftharpoons [Al(OH)_4^-]$$

$$K_f = 1 \times 10^{33} = \frac{[Al(OH)_4^-]}{[Al^{3+}][OH^-]^4} = \frac{[Al(OH)_4^-]}{[Al^{3+}](1 \times 10^{-12})^4}$$

$$\frac{[Al(OH)_4^-]}{[Al^{3+}]} = 1 \times 10^{-15} \quad \Rightarrow \quad \frac{[Al^{3+}]}{[Al(OH)_4^-]} = 1 \times 10^{15}$$

At pH 7.0:   pOH = 14.00 - 7.00 = 7.00
               [OH⁻] = 10^{-pOH} = 10$^{-7.00}$ = 1.0 × 10^{-7} M

$$K_f = 1 \times 10^{33} = \frac{[Al(OH)_4^-]}{[Al^{3+}](1 \times 10^{-7})^4}$$

$$\frac{[Al(OH)_4^-]}{[Al^{3+}]}=1\times10^5 \;\Rightarrow\; \frac{[Al^{3+}]}{[Al(OH)_4^-]}=\mathbf{1\times10^{-5}}$$

At pH 11.0: pOH = 14.00 - 11.00 = 3.00

$$[OH^-] = 10^{-pOH} = 10^{-3.00} = 1.0\times10^{-3}\,M$$

$$K_f = 1\times10^{33} = \frac{[Al(OH)_4^-]}{[Al^{3+}](1\times10^{-3})^4}$$

$$\frac{[Al(OH)_4^-]}{[Al^{3+}]}=1\times10^{21} \;\Rightarrow\; \frac{[Al^{3+}]}{[Al(OH)_4^-]}=\mathbf{1\times10^{-21}}$$

---

40. Refer to Section 15.4, Example 15.6 and Table 15.4.

a. $K_f = \dfrac{[Cd(NH_3)_4^{2+}]}{[Cd^{2+}][NH_3]^4} = 2.8\times10^7$

When $[Cd(NH_3)_4^{2+}] = [Cd^{2+}]$, then $K_f = \dfrac{1}{[NH_3]^4} = 2.8\times10^7$

$[NH_3]^4 = 3.6\times10^{-8}$

$[NH_3] = 1.4\times10^{-2}\,M$

b. $K_f = \dfrac{[Co(NH_3)_6^{2+}]}{[Co^{2+}][NH_3]^6} = 1\times10^5$

When $[Co(NH_3)_6^{2+}] = [Co^{2+}]$, then $K_f = \dfrac{1}{[NH_3]^6} = 1\times10^5$

$[NH_3]^6 = 1\times10^{-5}$

$[NH_3] = 1\times10^{-1}\,M = 0.1\,M$

---

42. Refer to Section 15.1 and Chapter 3.

a. Assume 100 g of sample.

$$22.0\,g\,Co \times \frac{1\,mol.}{58.93\,g} = 0.373\,mol.\,Co$$

$$31.4\,g\,N \times \frac{1\,mol.}{14.01\,g} = 2.24\,mol.\,N$$

$$6.78 \text{ g H} \times \frac{1 \text{ mol.}}{1.008 \text{ g}} = 6.73 \text{ mol. H}$$

$$39.8 \text{ g Cl} \times \frac{1 \text{ mol.}}{35.45 \text{ g}} = 1.12 \text{ mol. Cl}$$

0.373 / 0.373 = 1 mol. Co
2.24 / 0.373 = 6 mol. N
6.73 / 0.373 = 18 mol. H
1.12 / 0.373 = 3 mol. Cl

Thus the simplest formula is: $CoN_6H_{18}Cl_3$.

b. The salt most likely contains $6NH_3$ molecules and $3Cl^-$ ions.

$$[Co(NH_3)_6]Cl_3(s) \rightleftharpoons [Co(NH_3)_6]^{2+}(aq) + 3Cl^-(aq)$$

---

44. Refer to "Chemistry Beyond the Classroom: Chelates."

In venous blood, the iron is not coordinated to oxygen (hemoglobin, blue), while in arterial blood, the iron is coordinated to oxygen (oxyhemoglobin, red). When there is a lot of oxygen around (as in the lungs), oxyhemoglobin is formed. When there is a deficiency of oxygen, the oxygen of the oxyhemoglobin dissociates from the iron, forming hemoglobin.

$$\text{hemoglobin} + O_2 \rightleftharpoons \text{oxyhemoglobin} + H_2O$$

It is because of this equilibrium that hemoglobin is able to transport oxygen from the lungs to the cells.

---

46. Refer to Chapters 7 and 15.

a. Rust contains iron in the form of Fe^{3+}, which will form a complex with the oxalate ion, $[Fe(C_2O_4)_3]^{3-}$. This complex is water soluble, while rust is not.

b. There are no geometric isomers of tetrahedral complexes because any two ligands are 109.5° apart. Thus all spatial orientations are identical. Contrast this with square planar or octahedral, in which two ligands could be either 90° or 180° apart.

c. Co^{2+} has empty orbitals (such as the 4s), which can accept a pair of electrons from a Lewis base.

d.

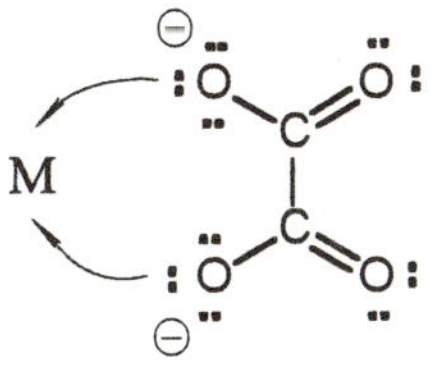

It should be clear from the Lewis structure that $C_2O_4{}^{2-}$ has two atoms with unshared electron pairs, which can act as Lewis bases. Also, those atoms are held in the correct position for both of them to bond to the metal simultaneously.

e.

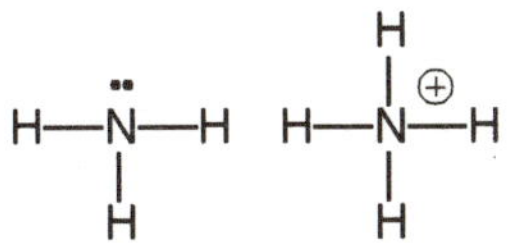

Ammonia has an unshared pair of electrons, which allows it to bond to metals. By contrast, the ammonium ion does not.

---

48. *Refer to Section 15.4 and Chapter 12.*

---

a. **False**. As learned in Chapter 12, greater K_f corresponds to more product at equilibrium. Thus larger K_f's, also correspond to more stable products.

b. **True**. $[Ag(NH_3)_2]^+$ undergoes the following reaction:
$$[Ag(NH_3)_2]^+ \rightleftharpoons Ag^+ + 2NH_3 \qquad \text{eq. 1}$$
The NH_3 can then react with HNO_3 to form $NH_4{}^+$.
$$NH_3 + HNO_3 \rightleftharpoons NH_4{}^+ + NO_3{}^- \qquad \text{eq. 2}$$
At this point, the NH_3 is effectively removed from solution. With no NH_3, equation 1 cannot reverse and reform $[Ag(NH_3)_2]^+$.

c. **False**. $[AgBr_2]^-$ undergoes the following reaction:
$$[AgBr_2]^- \rightleftharpoons Ag^+ + 2Br^-$$
Br^- is a spectator (neutral) anion and does not react with HNO_3. The nitric acid, therefore, does not cause $[AgBr_2]^-$ to dissociate.

d. **True**. AgI undergoes the following reactions in the presence of CN^-:
$$AgI \rightleftharpoons Ag^+ + I^-$$
$$Ag^+ + 2CN^- \rightleftharpoons [Ag(CN)_2]^-$$
AgI is insoluble, thus the equilibrium lies far to the left (reactant). The reaction with cyanide forms the soluble $[Ag(CN)_2]^-$ complex. NaCN is a salt, which dissociates completely to Na^+ and CN^-. HCN is a weak acid, thus only a small amount dissociates to form CN^-. Consequently, less NaCN would be needed.

e. **True**. The reactions of interest here are:
$$Ag^+ + 2Br^- \rightleftharpoons [AgBr_2]^-$$
$$Ag^+ + 2CN^- \rightleftharpoons [Ag(CN)_2]^-$$
The ratio of products to reactants is given by the formation constant:
$$K_f = \frac{[AgL_2]^-}{[Ag^+][L^-]^2}, \text{ where } L^- = Br^- \text{ or } CN^-$$
Since $[Br^-] = [CN^-]$, the species with the greater K_f will have the greater ratio of $[AgL_2]^-/Ag^+$.

49. Refer to Section 15.1, Figure 15B and "Chemistry Beyond the Classroom: Chelates."

Calculate the amount of Pb that was consumed, and then the amount of EDTA needed to bind the lead. Recall that EDTA forms a 1:1 complex with Pb.

$$10.0\,\text{g paint} \times \frac{5.0\,\text{g Pb}}{100\,\text{g paint}} \times \frac{1\,\text{mol. Pb}}{207.2\,\text{g}} = 2.41 \times 10^{-3}\,\text{mol. Pb}$$

$$Na_4(EDTA) = Na_4(C_{10}H_{12}N_2O_8)$$

$$M_{EDTA} = 380.2\,\text{g/mol.}$$

$$2.41 \times 10^{-3}\,\text{mol. Pb} \times \frac{1\,\text{mol. EDTA}}{1\,\text{mol. Pb}} \times \frac{1\,\text{mol. Na}_4\text{EDTA}}{1\,\text{mol. EDTA}} \times \frac{380.2\,\text{g}}{1\,\text{mol. Na}_4\text{EDTA}} = 0.92\,\text{g EDTA}$$

50. Refer to Section 15.1 and Chapter 3.

Calculate the molar mass of the simplest formula, then determine the molecular formula. From that, determine the ligand(s) present and cation and anion.

$$M\,(PtN_2H_6Cl_2) = 300.2\,\text{g/mol.}$$

600 is 2 x 300, thus the molecular formula is $(Pt_{1x2}N_{2x2}H_{6x2}Cl_{2x2}) = Pt_2N_4H_{12}Cl_4$

Since we have a complex cation and a complex anion, it is likely one Pt is with the cation, and the other is with the anion. Also N_4H_{12} is likely to be $4NH_3$. According to Table 15.2, Pt has a +2 charge and a coordination number of 4. Thus a likely candidate is:

$$[Pt(NH_3)_4]^{2+}\ [PtCl_4]^{2-} \Rightarrow [Pt(NH_3)_4][PtCl_4]$$

A second candidate is:

$$[Pt(NH_3)_3Cl]^+\ [Pt(NH_3)Cl_3]^- \Rightarrow [Pt(NH_3)_3Cl][Pt(NH_3)Cl_3]$$

51. Refer to Section 15.1, Problem 46 above and Chapter 3.

Assume a 100 g sample.

$$20.25\,\text{g Cu} \times \frac{1\,\text{mol.}}{63.55\,\text{g}} = 0.3186\,\text{mol. Cu}$$

$$15.29\,\text{g C} \times \frac{1\,\text{mol.}}{12.01\,\text{g}} = 1.273\,\text{mol. C}$$

$$7.07\,\text{g H} \times \frac{1\,\text{mol.}}{1.008\,\text{g}} = 7.01\,\text{mol. H}$$

$$26.86\,\text{g N} \times \frac{1\,\text{mol.}}{14.01\,\text{g}} = 1.917\,\text{mol. N}$$

$$20.39\,\text{g O} \times \frac{1\,\text{mol.}}{16.00\,\text{g}} = 1.274\,\text{mol. O}$$

$$10.23\,\text{g S} \times \frac{1\,\text{mol.}}{32.07\,\text{g}} = 0.3190\,\text{mol. S}$$

0.3186 / 0.3186 = 1 mol. Cu
1.273 / 0.3186 = 4 mol. C
7.07 / 0.3186 = 22 mol. H
1.917 / 0.3186 = 6 mol. N
1.274 / 0.3186 = 4 mol. O
0.3190 / 0.3186 = 1 mol. S

Thus the simplest formula is: $CuC_4H_{22}N_6O_4S$.

b.   To deduce the structural formula, one must determine the possible ligands that are present. The O_4S could be the sulfate anion. The -2 charge of the sulfate anion would balance the +2 charge that accompanies most copper cations, which would indicate that the remaining ligands are neutral. A ligand containing C, H and N mentioned in this chapter is ethylenediamine (*en*). If we subtract two en units from the simplest formula, we are left with H_6N_2, which is likely $2NH_3$. This gives: $[Cu(C_2H_8N_2)_2(NH_3)_2]^{2+}\ SO_4^{2-}$

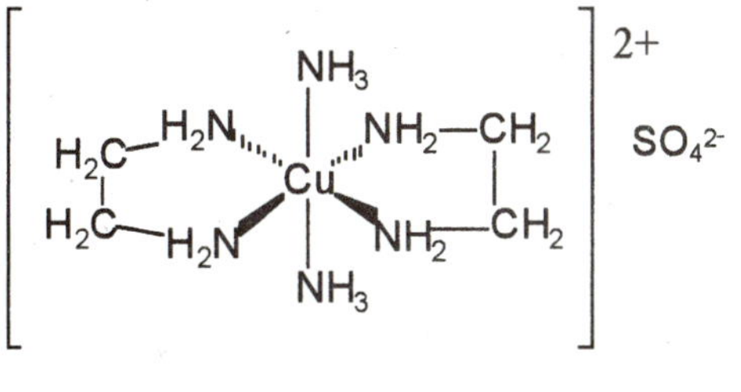

NH_3 *trans*

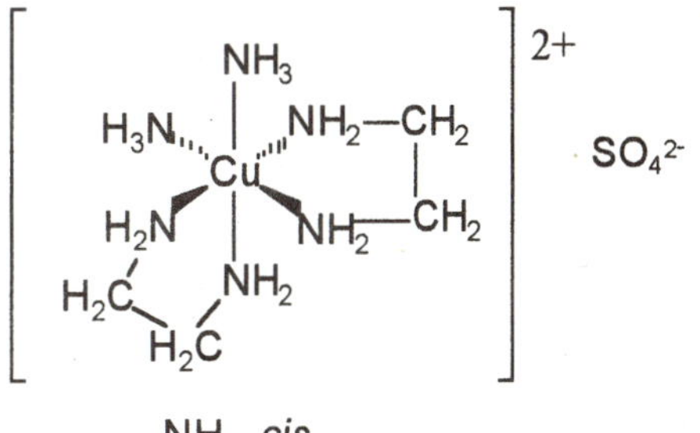

NH_3 *cis*

Calculate the energy per atom (or photon) in Joules and then calculate the wavelength of light with that energy. The observed color will be the complement of the absorbed color.

$\Delta_0 = E = 55$ kcal/mol

$$\frac{55\,\text{kcal}}{1\,\text{mol.}} \times \frac{4.184\,\text{kJ}}{1\,\text{kcal}} \times \frac{1000\,\text{J}}{1\,\text{kJ}} \times \frac{1\,\text{mol.}}{6.02 \times 10^{23}\,\text{photons}} = 3.8 \times 10^{-19} \text{ J/photon}$$

$$E = \frac{hc}{\lambda} \implies 3.8 \times 10^{-19}\,\text{J} = \frac{(6.63 \times 10^{-34}\,\text{J} \cdot \text{s})(3.00 \times 10^{8}\,\text{m/s})}{\lambda}$$

$\lambda = 5.2 \times 10^{-7}$ m = 520 nm

This corresponds to green. If the absorbed color is green, the observed color is **red-violet**.

Chapter 16: Precipitation Equilibria

2. Refer to Section 16.1 and Example 16.1.

a. $Co_2S_3(s) \rightleftharpoons 2Co^{3+}(aq) + 3S^{2-}(aq)$ $\qquad K_{sp} = [Co^{3+}]^2[S^{2-}]^3$

b. $PbCl_2(s) \rightleftharpoons Pb^{2+}(aq) + 2Cl^-(aq)$ $\qquad K_{sp} = [Pb^{2+}][Cl^-]^2$

c. $Zn_2P_2O_7(s) \rightleftharpoons 2Zn^{2+}(aq) + P_2O_7^{4-}(aq)$ $\qquad K_{sp} = [Zn^{2+}]^2[P_2O_7^{4-}]$

d. $Sc(OH)_3(s) \rightleftharpoons Sc^{3+}(aq) + 3OH^-(aq)$ $\qquad K_{sp} = [Sc^{3+}][OH^-]^3$

4. Refer to Section 16.1 and Example 16.1.

The ions are those of the product side of the reaction, and the exponents in the K_{sp} expression are the coefficients in the equilibrium expression.

a. $CaCO_3(s) \rightleftharpoons Ca^{2+}(aq) + CO_3^{2-}(aq)$

b. $Co(OH)_3(s) \rightleftharpoons Co^{3+}(aq) + 3OH^-(aq)$

c. $Ag_2S(s) \rightleftharpoons 2Ag^+(aq) + S^{2-}(aq)$

d. $PbCl_2(s) \rightleftharpoons Pb^{2+}(aq) + 2Cl^-(aq)$

6. Refer to Section 16.1 and Example 16.2.

Write the balanced net ionic equation and the expression for K_{sp}. Solve for the missing variable.

	$[Ag^+]$	$[CO_3^{2-}]$
a.	$1 \times 10^{-4}\ M$	$7 \times 10^{-4}\ M$
b.	$0.003\ M$	$9 \times 10^{-7}\ M$
c.	$3 \times 10^{-4}\ M$	$\frac{1}{2}[Ag^+]$
d.	$12[CO_3^{2-}]$	$4 \times 10^{-5}\ M$

$Ag_2CO_3(s) \rightleftharpoons 2Ag^+(aq) + CO_3^{2-}(aq)$ $\qquad K_{sp} = [Ag^+]^2[CO_3^{2-}]$

a. $8 \times 10^{-12} = [Ag^+]^2(7 \times 10^{-4})$
   $1 \times 10^{-8} = [Ag^+]^2$
   $[Ag^+] = 1 \times 10^{-4}\ M$

b. $8 \times 10^{-12} = (0.003)^2[CO_3^{2-}]$
   $[CO_3^{2-}] = 9 \times 10^{-7} M$

c. $8 \times 10^{-12} = [Ag^+]^2(\frac{1}{2}[Ag^+])$
   $8 \times 10^{-12} = \frac{1}{2}[Ag^+]^3$
   $1.6 \times 10^{-11} = [Ag^+]^3$
   $[Ag^+] = 3 \times 10^{-4} M$

d. $8 \times 10^{-12} = (12[CO_3^{2-}])^2[CO_3^{2-}]$
   $8 \times 10^{-12} = 144[CO_3^{2-}]^3$
   $6 \times 10^{-14} = [CO_3^{2-}]^3$
   $[CO_3^{2-}] = 4 \times 10^{-5} M$

---

8. Refer to Section 16.1 and Example 16.2.

Write the balanced net ionic equation and the K_{sp} expression. Then solve for the ion concentration.

a. $PbBr_2(s) \rightleftharpoons Pb^{2+}(aq) + 2Br^-(aq)$ $\qquad\qquad K_{sp} = [Pb^{2+}][Br^-]^2$
   $K_{sp} = 6.6 \times 10^{-6} = [Pb^{2+}](0.019)^2$
   $[Pb^{2+}] = 1.8 \times 10^{-2} M$

b. $Hg_2Br_2(s) \rightleftharpoons Hg_2^{2+}(aq) + 2Br^-(aq)$ $\qquad\qquad K_{sp} = [Hg_2^{2+}][Br^-]^2$
   $K_{sp} = 6 \times 10^{-23} = [Hg_2^{2+}](0.019)^2$
   $[Hg_2^{2+}] = 2 \times 10^{-19} M$

c. $AgBr(s) \rightleftharpoons Ag^+(aq) + Br^-(aq)$ $\qquad\qquad K_{sp} = [Ag^+][Br^-]$
   $K_{sp} = 5 \times 10^{-13} = [Ag^+](0.019)$
   $[Ag^+] = 3 \times 10^{-11} M$

---

10. Refer to Section 16.1 and Example 16.3.

---

a. A precipitate will begin to form when $P = K_{sp}$.
   $Ca_3(PO_4)_2(s) \rightleftharpoons 3Ca^{2+}(aq) + 2PO_4^{3-}(aq)$ $\qquad\qquad K_{sp} = [Ca^{2+}]^3[PO_4^{3-}]^2$
   $K_{sp} = 1 \times 10^{-33} = [Ca^{2+}]^3(0.15)^2$
   $[Ca^{2+}] = 4 \times 10^{-11} M$

b. 15% of the original $[PO_4^{3-}]$ is: $0.15(0.15 M) = 0.023 M$
   $K_{sp} = 1 \times 10^{-33} = [Ca^{2+}]^3(0.023)^2$
   $[Ca^{2+}] = 1 \times 10^{-10} M$

12. *Refer to Section 16.1 and Examples 16.3 and 16.4.*

Calculate the molar concentrations of Pb^{2+} and CrO_4^{2-} in solution, then calculate P.

$$\frac{0.50\,\text{mg Pb(NO}_3)_2}{1\,\text{L}} \times \frac{1\,\text{g}}{1000\,\text{mg}} \times \frac{1\,\text{mol. Pb(NO}_3)_2}{331.22\,\text{g}} \times \frac{1\,\text{mol. Pb}^{2+}}{1\,\text{mol. Pb(NO}_3)_2} = 1.5 \times 10^{-6}\,M\;\text{Pb}^{2+}$$

$$\frac{0.020\,\text{mg K}_2\text{CrO}_4}{1\,\text{L}} \times \frac{1\,\text{g}}{1000\,\text{mg}} \times \frac{1\,\text{mol. K}_2\text{CrO}_4}{194.2\,\text{g K}_2\text{CrO}_4} \times \frac{1\,\text{mol. CrO}_4^{2-}}{1\,\text{mol. K}_2\text{CrO}_4} = 1.0 \times 10^{-7}\,M\;\text{CrO}_4^{2-}$$

$P = [Pb^{2+}][CrO_4^{2-}] = (1.5 \times 10^{-6})(1.0 \times 10^{-7}) = 1.5 \times 10^{-13}$

$P > K_{sp}$ ($= 2 \times 10^{-14}$), therefore **a precipitate will form**.

To just start a precipitate:
$K_{sp} = [Pb^{2+}][CrO_4^{2-}] \;\Rightarrow\; 2 \times 10^{-14} = [Pb^{2+}](1.0 \times 10^{-7})$
$[Pb^{2+}] = 2.0 \times 10^{-7}\,M$

---

14. *Refer to Section 16.1 and Examples 16.3 and 16.4.*

a. Calculate P for Hg_2Cl_2 (HNO_3 will dissociate completely and can be disregarded).

$Hg_2(NO_3)_2(aq) + 2HCl(aq) \rightarrow Hg_2Cl_2(s) + 2H^+(aq) + 2NO_3^-(aq)$
$Hg_2Cl_2(s) \rightleftharpoons Hg_2^{2+}(aq) + 2Cl^-(aq)$

$$0.01300\,\text{L} \times \frac{0.0021\,\text{mol. Hg}_2^{2+}}{1\,\text{L}} = 2.7 \times 10^{-5}\,\text{mol. Hg}_2^{2+}$$

$$[\text{Hg}_2^{2+}] = \frac{2.7 \times 10^{-5}\,\text{mol. Hg}_2^{2+}}{(0.01300\,\text{L} + 0.0250\,\text{L})} = 7.2 \times 10^{-4}\,M\;\text{Hg}_2^{2+}$$

$$0.0250\,\text{L} \times \frac{0.015\,\text{mol. Cl}^-}{1\,\text{L}} = 3.8 \times 10^{-4}\,\text{mol. Cl}^-$$

$$[\text{Cl}^-] = \frac{3.8 \times 10^{-4}\,\text{mol. Cl}^-}{(0.01300\,\text{L} + 0.0250\,\text{L})} = 1.0 \times 10^{-2}\,M\;\text{Cl}^-$$

$P = [Hg_2^+][Cl^-]^2 = (7.2 \times 10^{-4})(1.0 \times 10^{-2})^2 = 7.0 \times 10^{-8}$

$P > K_{sp}$ ($= 1 \times 10^{-18}$), therefore **a precipitate will form**.

b. NO_3^- is a spectator ion, so all the NO_3^- introduced initially will remain.

$$[\text{NO}_3^-] = \frac{7.2 \times 10^{-4}\,\text{mol. Hg}_2(\text{NO}_3)_2}{1\,\text{L}} \times \frac{2\,\text{mol. NO}_3^-}{1\,\text{mol. Hg}_2(\text{NO}_3)_2} = 1.4 \times 10^{-3}\,M\;\text{NO}_3^-$$

Since [Cl$^-$] is more than twice that of [Hg$_2^{2+}$], it is safe to assume (given the very small K_{sp}) that all the Hg$_2^{2+}$ precipitates and that the only Cl$^-$ left in solution is the excess Cl$^-$.

$$\frac{7.2 \times 10^{-4} \text{ mol. Hg}_2^{2+}}{1 \text{ L}} \times \frac{2 \text{ mol. Cl}^-}{1 \text{ mol. Hg}_2^{2+}} = 1.4 \times 10^{-3} \ M \text{ Cl}^- \quad \text{(precipitated as Hg}_2\text{Cl}_2\text{)}$$

[Cl$^-$] = 1.0 x 10^{-2} M (initial) - 1.4 x 10^{-3} M (precipitated)
[Cl$^-$] = 8.6 x 10^{-3} M (remaining in solution)

The [Hg$_2^+$] is calculated from the K_{sp} and [Cl$^-$].

K_{sp} = [Hg$_2^+$][Cl$^-$]^2 $\Rightarrow$ 1 x 10^{-18} = [Hg$_2^+$](8.6 x 10^{-3})^2

[Hg$_2^+$] = 1 x 10^{-14} M

---

16. Refer to Section 16.1.

Write the balanced net ionic equation, calculate the concentration of the ions and then determine K_{sp}.

$$\text{Zn(CN)}_2(s) \rightleftharpoons \text{Zn}^{2+}(aq) + 2\text{CN}^-(aq)$$

$$\frac{1.5 \text{ mg Zn(CN)}_2}{0.1000 \text{ L}} \times \frac{1 \text{ g}}{1000 \text{ mg}} \times \frac{1 \text{ mol. Zn(CN)}_2}{117.43 \text{ g Zn(CN)}_2} \times \frac{1 \text{ mol. Zn}^{2+}}{1 \text{ mol. Zn(CN)}_2} = 1.3 \times 10^{-4} \ M \text{ Zn}^{2+}$$

$$\frac{1.5 \text{ mg Zn(CN)}_2}{0.1000 \text{ L}} \times \frac{1 \text{ g}}{1000 \text{ mg}} \times \frac{1 \text{ mol. Zn(CN)}_2}{117.43 \text{ g Zn(CN)}_2} \times \frac{2 \text{ mol. CN}^-}{1 \text{ mol. Zn(CN)}_2} = 2.6 \times 10^{-4} \ M \text{ CN}^-$$

K_{sp} = [Zn^{2+}][CN$^-$]^2 = (1.3 x 10^{-4})(2.6 x 10^{-4})^2 = 8.8 x 10^{-12}

---

18. Refer to Section 16.1, Examples 16.5 and 16.6 and Table 16.1.

Write the balanced net ionic equation, set up the expression for K_{sp}. Calculate the concentration of Mg^{2+}, and from that the mass of Mg(OH)$_2$.

$$\text{Mg(OH)}_2(s) \rightleftharpoons \text{Mg}^{2+}(aq) + 2\text{OH}^-(aq) \qquad\qquad K_{sp} = [\text{Mg}^{2+}][\text{OH}^-]^2$$

a. K_{sp} = 6 x 10^{-12} = (x)(2x)^2 = 4x^3
   x = 1 x 10^{-4} M Mg^{2+}

$$\frac{1 \times 10^{-4} \text{ mol. Mg}^{2+}}{1 \text{ L}} \times \frac{1 \text{ mol. Mg(OH)}_2}{1 \text{ mol. Mg}^{2+}} \times \frac{58.3 \text{ g Mg(OH)}_2}{1 \text{ mol. Mg(OH)}_2} = 6 \times 10^{-3} \text{ g/L Mg(OH)}_2$$

b. [OH$^-$] = 2(0.041) M = 0.082 M (We can ignore the small contribution from Mg(OH)$_2$)

   K_{sp} = 6 x 10^{-12} = (x)(0.082)^2
   x = 9 x 10^{-10} M Mg^{2+}

$$\frac{9 \times 10^{-10} \text{ mol. Mg}^{2+}}{1\,\text{L}} \times \frac{1\,\text{mol. Mg(OH)}_2}{1\,\text{mol. Mg}^{2+}} \times \frac{58.3\,\text{g Mg(OH)}_2}{1\,\text{mol. Mg(OH)}_2} = 5 \times 10^{-8} \text{ g/L Mg(OH)}_2$$

c. $[Mg^{2+}] = 0.0050\,M$ $\qquad$ (We can ignore the small contribution from $Mg(OH)_2$)

$K_{sp} = 6 \times 10^{-12} = (0.0050)(2x)^2$
$x^2 = 3 \times 10^{-10}\,M\,OH^-$
$x = 2 \times 10^{-5}\,M\,OH^-$

$$\frac{2 \times 10^{-5} \text{ mol. OH}^-}{1\,\text{L}} \times \frac{1\,\text{mol. Mg(OH)}_2}{2\,\text{mol. (OH)}^-} \times \frac{58.3\,\text{g Mg(OH)}_2}{1\,\text{mol. Mg(OH)}_2} = 5 \times 10^{-4} \text{ g/L Mg(OH)}_2$$

---

20. Refer to Section 16.1, Example 16.5 and Table 16.1.

$PbCl_2(s) \rightleftharpoons Pb^{2+}(aq) + 2Cl^-(aq)$ $\qquad\qquad$ $K_{sp} = 1.7 \times 10^{-5} = [Pb^{2+}][Cl^-]^2$

$$\frac{1.0\,\text{g PbCl}_2}{1\,\text{L}} \times \frac{1\,\text{mol. PbCl}_2}{278.1\,\text{g PbCl}_2} = 0.0036\,M\,PbCl_2$$

$P = [Pb^{2+}][Cl^-]^2 = (x)(2x)^2 = 4x^3 = 4(0.0036)^3 = 1.9 \times 10^{-7}$

$P < K_{sp}$, therefore all the $PbCl_2$ will remain in solution (**none will precipitate**).

---

22. Refer to Section 16.2 and Example 16.8.

a. $Cu_2S(s) + 2H^+(aq) \rightarrow 2Cu^+(aq) + H_2S(aq)$

b. $Hg_2Cl_2(s) + H^+(aq) \rightarrow Hg_2Cl_2(s) + H^+(aq)$
$\qquad$ Thus no reaction occurs

c. $SrCO_3(s) + 2H^+(aq) \rightleftharpoons Sr^{2+}(aq) + H_2CO_3(aq)$
$\quad\underline{H_2CO_3(aq) \rightleftharpoons H_2O(l) + CO_2(g)}$
$\quad SrCO_3(s) + 2H^+(aq) \rightarrow Sr^{2+}(aq) + H_2O(l) + CO_2(g)$

d. $Cu(NH_3)_4^{2+}(aq) + 4H^+(aq) \rightarrow Cu^{2+}(aq) + 4NH_4^+(aq)$

e. $Ca(OH)_2(s) + 2H^+(aq) \rightarrow Ca^{2+}(aq) + 2H_2O(l)$

---

24. Refer to Section 16.2 and Table 16.2.

a. $Cu(OH)_2(s) + 4NH_3(aq) \rightleftharpoons Cu(NH_3)_4^{2+}(aq) + 2OH^-(aq)$

b. $Cd^{2+}(aq) + 4NH_3(aq) \rightleftharpoons Cd(NH_3)_4^{2+}(aq)$

c. The lead must be combining with and anion to give an electrically neutral precipitate. A likely candidate is OH^- formed by the reaction of ammonia and water.

$$Pb^{2+}(aq) + 2NH_3(aq) + 2H_2O(l) \rightleftharpoons Pb(OH)_2(s) + 2NH_4^+(aq)$$

---

26. Refer to Section 16.2, Example 16.11 and Table 16.2.

a. $Sb^{3+}(aq) + 3OH^-(aq) \rightleftharpoons Sb(OH)_3(s)$

b. $Sb(OH)_3(s) + OH^-(aq) \rightleftharpoons Sb(OH)_4^-(aq)$

c. $Sb^{3+}(aq) + 4OH^-(aq) \rightleftharpoons Sb(OH)_4^-(aq)$

---

28. Refer to Section 16.2 and Example 16.9.

$Al(OH)_3(s) \rightleftharpoons Al^{3+}(aq) + 3OH^-(aq)$ $\qquad$ $K_1 = K_{sp} = 2 \times 10^{-31}$
$3[OH^-(aq) + H^+(aq) \rightleftharpoons H_2O(l)]$ $\qquad$ $K_2 = (1/K_w)^3 = (1 \times 10^{14})^3$
$\overline{Al(OH)_3(s) + 3H^+(aq) \rightleftharpoons Al^{3+}(aq) + 3H_2O(l)}$ $\quad$ $K = (K_{sp})(1/K_w)^3 = 2 \times 10^{11}$

---

30. Refer to Section 16.2 and Example 16.9.

a. $Zn(OH)_2(s) \rightleftharpoons Zn^{2+}(aq) + 2OH^-(aq)$ $\qquad$ $K_{sp} = 4 \times 10^{-17}$
$Zn(CN)_2(s) \rightleftharpoons Zn^{2+}(aq) + 2CN^-(aq)$ $\qquad$ $K_{sp} = 8.0 \times 10^{-12}$

$Zn(OH)_2(s) \rightleftharpoons Zn^{2+}(aq) + 2OH^-(aq)$ $\qquad$ $K_1 = K_{sp} = 4 \times 10^{-17}$
$Zn^{2+}(aq) + 2CN^-(aq) \rightleftharpoons Zn(CN)_2(s)$ $\qquad$ $K_2 = 1/K_{sp} = 1.3 \times 10^{11}$
$\overline{Zn(OH)_2(s) + 2CN^-(aq) \rightleftharpoons Zn(CN)_2(s) + 2OH^-(aq)}$ $\quad$ $K = (K_1)(K_2) = 5 \times 10^{-6}$

b. The K_{sp} of $Zn(CN)_2$ is greater than the K_{sp} of $Zn(OH)_2$, indicating that $Zn(CN)_2$ is more soluble. Therefore, if NaCN is added to a saturated solution of $Zn(OH)_2$, $Zn(CN)_2$ will form, but will **not precipitate**.

---

32. Refer to Section 16.2 and Example 16.9.

a. $Cu(OH)_2(s) \rightleftharpoons Cu^{2+}(aq) + 2OH^-(aq)$ $\qquad$ $K_{sp} = 2 \times 10^{-19}$
$Cu^{2+}(aq) + 4NH_3(aq) \rightleftharpoons Cu(NH_3)_4^{2+}(aq)$ $\qquad$ $K_f = 2 \times 10^{12}$
$\overline{Cu(OH)_2(s) + 4NH_3(aq) \rightleftharpoons Cu(NH_3)_4^{2+}(aq) + 2OH^-(aq)}$ $\quad$ $K = K_{sp} \times K_f = 4 \times 10^{-7}$

b. $K = \dfrac{[Cu(NH_3)_4{}^{2+}][OH^-]^2}{[NH_3]^4}$ and $2[Cu(NH_3)_4{}^{2+}] = [OH^-]$

$$K = 4 \times 10^{-7} = \dfrac{(x)(2x)^2}{(4.5\,M)^4} = \dfrac{4x^3}{410}$$

$x^3 = 4 \times 10^{-5}$

$x = 3 \times 10^{-2}\,M$

---

34. Refer to Sections 16.1 and 16.2 and Example 16.9.

Calculate K for the reaction. Then calculate $[Pb(OH)_3]^-$ from the equilibrium expression. Bear in mind that solids do not factor into the equation.

$PbCl_2(s) \rightleftharpoons Pb^{2+}(aq) + 2Cl^-(aq)$	$K_{sp} = 1.7 \times 10^{-5}$
$Pb^{2+}(aq) + 3OH^-(aq) \rightleftharpoons [Pb(OH)_3]^-(aq)$	$K_f = 3.8 \times 10^{14}$
$PbCl_2(s) + 3OH^-(aq) \rightleftharpoons 2Cl^-(aq) + [Pb(OH)_3]^-(aq)$	$K = K_{sp} \times K_f = 6.5 \times 10^{9}$

	$PbCl_2(s)$	$+$	$3OH^-(aq)$	$\rightleftharpoons$	$2Cl^-(aq)$	$+$	$[Pb(OH)_3]^-(aq)$
$[\ ]_0$	--		2.0		0		0
$\Delta[\ ]$	--		$-3x$		$+2x$		$+x$
$[\ ]_{eq}$	--		$2.0 - 3x$		$2x$		x

$$K = \dfrac{[Cl^-]^2[Pb(OH)_3{}^-]}{[OH^-]^3} \Rightarrow 6.5 \times 10^9 = \dfrac{(2x)^2(x)}{(2.0 - 3x)^3} = \dfrac{4x^3}{(2.0 - 3x)^3}$$

$$\sqrt[3]{6.5 \times 10^9} = \sqrt[3]{\dfrac{4x^3}{(2.0 - 3x)^3}} \Rightarrow 1.9 \times 10^3 = \dfrac{1.6x}{2.0 - 3x}$$

$3.8 \times 10^3 - 5.7 \times 10^3 x = 1.6x$

$5.7 \times 10^3 x = 3.8 \times 10^3$

$x = 0.67\,M$

---

36. Refer to Section 16.2 and Example 16.9.

a. Calculate K for the reaction, recalling that when you add 2 equilibrium equations, you must multiply the equilibrium constants.

$CdC_2O_4(s) \rightleftharpoons Cd^{2+}(aq) + C_2O_4{}^{2-}(aq)$	$K_{sp} = 1.5 \times 10^{-8}$
$Cd^{2+}(aq) + 4NH_3(aq) \rightleftharpoons Cd(NH_3)_4{}^{2+}(aq)$	$K_f = 2.8 \times 10^{7}$
$CdC_2O_4(s) + 4NH_3(aq) \rightleftharpoons Cd(NH_3)_4{}^{2+}(aq) + C_2O_4{}^{2-}(aq)$	$K = K_{sp} \times K_f = 4.2 \times 10^{-1}$

b. Calculate $[C_2O_4^{2-}]$ and $[Cd(NH_3)_4^{2+}]$, substitute the values in the equilibrium constant expression and solve for $[NH_3]$.

$$\frac{2.00\,g}{1\,L} \times \frac{1\,mol.\,CdC_2O_4}{200.42\,g} \times \frac{1\,mol.\,C_2O_4^{2-}}{1\,mol.\,CdC_2O_4} = 0.00998\,M\,C_2O_4^{2-}$$

$$K = \frac{[Cd(NH_3)_4^{2+}][C_2O_4^{2-}]}{[NH_3]^4} \quad and \quad [Cd(NH_3)_4^{2+}] = [C_2O_4^{2-}]$$

$$K = 0.42 = \frac{(0.00998)^2}{x^4}$$

$$x^4 = 2.4 \times 10^{-4}$$
$$x = 0.12\,M\,NH_3$$

---

38. Refer to Section 16.3 and Table 16.3.

	Cation	Analytical Group	Precipitating Agent	Precipitate Formed
a.	Ag^+	**Gp I**	$6\,M$ HCl	AgCl
b.	Bi^{3+}	**Gp II**	$0.1\,M\,H_2S$ at pH = 0.5	Bi_2S_3
c.	Co^{2+}	**Gp III**	$0.1\,M\,H_2S$ at pH = 9	CoS
d.	Mg^{2+}	**Gp IV**	$0.2\,M\,(NH_4)_2CO_3$ at pH = 9.5	$MgCO_3$

a. $Ag^+(aq) + Cl^-(aq) \rightleftharpoons AgCl(s)$

b. $2Bi^{3+}(aq) + 3S^{2-}(aq) \rightleftharpoons Bi_2S_3(s)$

c. $Co^{2+}(aq) + S^{2-}(aq) \rightleftharpoons CoS(s)$

d. $Mg^{2+}(aq) + CO_3^{2-}(aq) \rightleftharpoons MgCO_3(s)$

---

40. Refer to Section 16.3.

That a precipitate forms with HCl indicates that one or more of the group I metals (Ag^+, Pb^{2+} and/or Hg_2^{2+}) is present. Precipitate formation with $(NH_4)_2CO_3$ indicates that Mg^{2+}, Ca^{2+} and/or Ba^{2+} is present.

42. Refer to Section 16.3.

$$Cd^{2+}(aq) + H_2S(aq) \rightleftharpoons CdS(s) + 2H^+(aq) \qquad K = 1/K = 1 \times 10^9$$
$$Mn^{2+}(aq) + H_2S(aq) \rightleftharpoons MnS(s) + 2H^+(aq) \qquad K = 1/K = 2 \times 10^{-7}$$

The sulfide with the largest K (**CdS**) will precipitate first.

44. Refer to Section 16.1.

Recall that 2 ppm indicate that there are 2 grams F^- per 1 million total grams. Determine $[F^-]$ and P. Compare P to K_{sp} to determine if a precipitate forms.

$$Ca^{2+}(aq) + 2F^-(aq) \rightleftharpoons CaF_2(s) \qquad\qquad K_{sp} = 1.5 \times 10^{-10}$$

$$[F^-] = \frac{2.0\,g\,F^-}{1 \times 10^6\,g\,solution} \times \frac{1\,mol.\,F^-}{19\,g\,F^-} \times \frac{1\,g\,solution}{1\,mL\,solution} \times \frac{1000\,mL}{1\,L} = 1.1 \times 10^{-4}\,mol./L$$

$$P = [Ca^{2+}][F^-]^2 = (3.5 \times 10^{-4})(1.1 \times 10^{-4})^2 = 4.2 \times 10^{-12}$$

$K_{sp} > P$, therefore a **precipitate will not form**.

46. Refer to Sections 16.1 and 13.5.

a. $\dfrac{17.5\,g\,NH_4Cl}{1.50\,L} \times \dfrac{1\,mol.\,NH_4Cl}{53.49\,g\,NH_4Cl} = 0.218\,M\,NH_4Cl$

	$NH_3(aq)$	+	$H_2O(l)$	$\rightleftharpoons$	$NH_4^+(aq)$	+	$OH^-(aq)$
$[\ ]_0$	3.75		---		0.218		0
$\Delta[\ ]$	$-x$		---		$+x$		$+x$
$[\ ]_{eq}$	$3.75 - x$		---		$0.218 + x$		x

$$K_b = 1.8 \times 10^{-5} = \frac{[NH_4^+][OH^-]}{[NH_3]} = \frac{(0.218 + x)(x)}{(3.75 - x)}$$

Assuming the 5% rule is valid for both $[NH_4^+]$ and $[OH^-]$, then

$$K_b = 1.8 \times 10^{-5} = \frac{(0.218)(x)}{3.75}$$

$$x = [OH^-] = 3.1 \times 10^{-4}\,M$$

b. $Mg^{2+}(aq) + 2OH^-(aq) \rightleftharpoons Mg(OH)_2(s)$ $K_{sp} = 6 \times 10^{-12}$

$$[Mg^{2+}] = \frac{5.0 \text{ g MgCl}_2}{1.50 \text{ L}} \times \frac{1 \text{ mol. MgCl}_2}{95.2 \text{ g MgCl}_2} \times \frac{1 \text{ mol. Mg}^{2+}}{1 \text{ mol. MgCl}_2} = 0.0350 \, M \text{ Mg}^{2+}$$

$P = [Mg^{2+}][OH^-]^2 = (0.0350)(3.1 \times 10^{-4})^2 = 3.4 \times 10^{-9}$

$K_{sp} < P$, Therefore a **precipitate will form**.

c. $K_{sp} = [Mg^{2+}][OH^-]^2$

     $6 \times 10^{-12} = [Mg^{2+}](3.1 \times 10^{-4})^2$

     $[Mg^{2+}] = 6 \times 10^{-5} \, M$

---

48. Refer to Section 16.1.

Since the top box in the problem represents a saturated solution, we know that 4 moles of MX_2 will dissolve in one liter. For the subsequent boxes, dissolve up to 4 MX_2's, and leave any remainder undissolved.

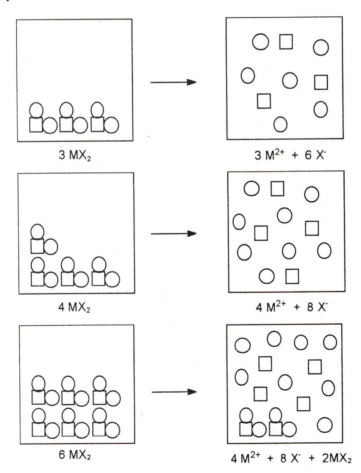

3 MX_2 3 M^{2+} + 6 X^-

4 MX_2 4 M^{2+} + 8 X^-

6 MX_2 4 M^{2+} + 8 X^- + 2MX_2

50. *Refer to Section 16.2.*

The lower the pH, the higher [H^+]. Acid rain has a lower pH and therefore a higher [H^+] than normal rain. By LeChatelier's principle, increasing [H^+], one of the reactants, will cause the equilibrium to shift to more products, thus dissolving more $CaCO_3$.

52. *Refer to Sections 16.2 and 12.5.*

Consider the equilibrium equations with "heat" as a product (exothermic) or reactant (endothermic) and apply LeChatelier's principle.

exothermic: $M(s) \rightleftharpoons M(aq) + \text{heat}$
endothermic: $M(s) + \text{heat} \rightleftharpoons M(aq)$

In the former case, adding heat to the system would shift the reaction to the left (more reactants), which would produce more undissolved material. This corresponds to a decrease in solubility with increasing temperature.

In the latter case, adding heat to the system would shift the reaction to the fight (more products), which would produce more dissolved material. This corresponds to an increase in solubility with increasing temperature.

Dissolving this material is an endothermic process.

53. *Refer to Sections 16.1 and 16.2 and Examples 16.5 and 16.9.*

Calculate the solubility of $Mg(OH)_2$ in water using K_{sp}. Then calculate K for the reaction given. Using K, calculate the solubility of $Mg(OH)_2$ in 0.2 M NH_4Cl and compare the two solubilities.

Solubility in water:
$$K_{sp} = 6 \times 10^{-12} = [Mg^{2+}][OH^-]^2 = (x)(2x)^2 = 4x^3$$
$$x^3 = 1.5 \times 10^{-12}$$
$$x = 1 \times 10^{-4} M$$

Solubility in 0.2 M NH_4Cl:

$Mg(OH)_2(s) \rightleftharpoons Mg^{2+}(aq) + 2OH^-(aq)$	$K_{sp} = 6 \times 10^{-12}$
$2[\ NH_4^+(aq) \rightleftharpoons H^+(aq) + NH_3(aq)\ ]$	$K_1 = (K_a)^2 = 3.1 \times 10^{-19}$
$\underline{2[\ H^+(aq) + OH^-(aq) \rightleftharpoons H_2O(l)\ ]}$	$K_2 = (1/K_w)^2 = 1 \times 10^{28}$
$Mg(OH)_2(s) + NH_4^+(aq) \rightleftharpoons Mg^{2+}(aq) + 2NH_3(aq) + 2H_2O(l)$	$K = K_{sp} \times K_1 \times K_2 = 0.02$

$$K = \frac{[Mg^{2+}][NH_3]^2}{[NH_4^+]^2} \qquad \text{set } [Mg^{2+}] = x, \text{ then } [NH_3] = 2x \text{ and } [NH_4^+] = 0.2 - x$$

$$K = 0.02 = \frac{(x)(2x)^2}{(0.2-x)^2} \quad \Rightarrow \quad 0.0008 - 0.008x + x^2 = 4x^3$$

Using successive approximations: $\left(x \approx \sqrt[3]{0.0008/4}\right)$

$x = 0.06\ M$

$Mg(OH)_2$ is approximately 600 times more soluble in $0.2\ M\ NH_4Cl$ than in "pure" water.

---

54. Refer to Section 16.2 and Chapter 14.

Calculate K for the given reaction. Determine $[H^+]$ using the equation for buffers, and with that, calculate $[Ca^{2+}]$.

$CaF_2(s) \rightleftharpoons Ca^{2+}(aq) + 2F^-(aq)$	$K_{sp} = 1.5 \times 10^{-10}$
$2[\ H^+(aq) + F^-(aq) \rightleftharpoons HF(aq)\ ]$	$K_1 = (1/K_a)^2 = 2.1 \times 10^6$
$CaF_2(s) + 2H^+(aq) \rightleftharpoons Ca^{2+}(aq) + 2HF(aq)$	$K = K_{sp} \times K_1 = 3.2 \times 10^{-4}$

$$[H^+] = K_a \times \frac{[HCHO_2]}{[CHO_2^-]} = 1.9 \times 10^{-4} \times \frac{0.30}{0.20} = 2.9 \times 10^{-4}\ M$$

$$K = \frac{[Ca^{2+}][HF]^2}{[H^+]^2} \quad \Rightarrow \quad 3.2 \times 10^{-4} = \frac{(x)(2x)^2}{(2.9 \times 10^{-4})^2} = \frac{4x^3}{8.4 \times 10^{-8}}$$

$x^3 = 6.8 \times 10^{-12}$

$x = 1.9 \times 10^{-4}\ M$

---

55. Refer to Section 16.1.

Since AgI has the smaller K_{sp}, it will precipitate first. Thereafter, this is a common ion problem. Calculate the $[Ag^+]$ in solution at the point at which AgCl will just begin to precipitate. Using that concentration and K_{sp} (AgI), calculate $[I^-]$.

K_{sp} (AgCl) $= 1.8 \times 10^{-10} = [Ag^+][Cl^-] = (x)(0.020\ M)$
    $x = 9.0 \times 10^{-9}\ M\ Ag^+$

K_{sp} (AgI) $= 1 \times 10^{-16} = [Ag^+][I^-] = (9.0 \times 10^{-9}\ M)(x)$
    $x = 1 \times 10^{-8}\ M\ I^-$

---

56. Refer to Section 16.1.

a. $K_{sp} = 6 \times 10^{-12} = [Mg^{2+}][OH^-]^2 = (0.056)(x)^2$
    $x^2 = 1 \times 10^{-10}$
    $x = 1 \times 10^{-5}\ M$

b. Na^+:   No. NaOH is extremely soluble.

   Ca^{2+}: $P = [Ca^{2+}][OH^-]^2 = (0.01)(1 \times 10^{-5})^2 = 1 \times 10^{-12}$
            $P < K_{sp}$, therefore **Ca^{2+} does not precipitate**.

   Al^{3+}: $P = [Al^{3+}][OH^-]^3 = (4 \times 10^{-7})(1 \times 10^{-5})^3 = 4 \times 10^{-22}$
            $P > K_{sp}$, therefore **Al^{3+} does precipitate**.

   Fe^{3+}: $P = [Fe^{3+}][OH^-]^3 = (2 \times 10^{-7})(1 \times 10^{-5})^3 = 2 \times 10^{-22}$
            $P > K_{sp}$, therefore **Fe^{3+} does precipitate**.

c. Calculate $[OH^-]$ when half the Mg^{2+} is precipitated, then calculate $[Al^{3+}]$ and $[Fe^{3+}]$ in solution and the percent precipitated.

$K_{sp} = 6 \times 10^{-12} = [Mg^{2+}][OH^-]^2 = (0.028)(x)^2$
$x = 1.5 \times 10^{-5} M$

$K_{sp}\{Al(OH)_3\} = [Al^{3+}][OH^-]^3 \implies 2 \times 10^{-31} = (x)(1.5 \times 10^{-5})^3$
$x = 6 \times 10^{-17} M$

$$\text{percent in solution} = \frac{6 \times 10^{-17}}{4 \times 10^{-7}} \times 100\% = 1 \times 10^{-8}\%$$

Essentially all the Al^{3+} will have precipitated.

$K_{sp}\{Fe(OH)_3\} = [Fe^{3+}][OH^-]^3 \implies 3 \times 10^{-39} = (x)(1.5 \times 10^{-5})^3$
$x = 9 \times 10^{-25} M$

$$\text{percent in solution} = \frac{9 \times 10^{-25}}{2 \times 10^{-7}} \times 100\% = 4 \times 10^{-16}\%$$

Essentially all the Fe^{3+} will have precipitated.

d. Calculate and add up the masses of the species that precipitate.

$$0.028 \text{ mol. } Mg(OH)_2 \times \frac{58.32 \text{ g}}{1 \text{ mol.}} = 1.6 \text{ g}$$

$$4 \times 10^{-7} \text{ mol. } Al(OH)_3 \times \frac{70.00 \text{ g}}{1 \text{ mol.}} = 3 \times 10^{-5} \text{ g}$$

$$2 \times 10^{-7} \text{ mol. } Fe(OH)_3 \times \frac{106.9 \text{ g}}{1 \text{ mol.}} = 2 \times 10^{-5} \text{ g}$$

Total mass = 1.6 g.

a. $Zn(NH_3)_4^{2+}(aq) \rightleftharpoons Zn^{2+}(aq) + 4NH_3(aq)$ $K_1 = 1/K_f = 2.8 \times 10^{-9}$

    $Zn^{2+}(aq) + 4OH^-(aq) \rightleftharpoons Zn(OH)_4^{2-}(aq)$ $K_f = 3 \times 10^{14}$

$Zn(NH_3)_4^{2+}(aq) + 4OH^-(aq) \rightleftharpoons Zn(OH)_4^{2-}(aq) + 4NH_3(aq)$ $K = K_1 \times K_f = 8 \times 10^5$

b. Calculate the $[OH^-]$ in a $1.0\,M$ NH_3 solution, then calculate the ratio of the two complexes using the equilibrium concentration expression.

$$NH_3(aq) + H_2O \rightleftharpoons NH_4^+(aq) + OH^-(aq)$$

$$K_b = \frac{[NH_4^+][OH^-]}{[NH_3]} \quad \Rightarrow \quad 1.8 \times 10^{-5} = \frac{x^2}{1.0}$$

$x^2 = 1.8 \times 10^{-5}$

$x = 0.0042\,M\ OH^-$

$$K = \frac{[Zn(OH)_4^{2-}][NH_3]^4}{[Zn(NH_3)_4^{2+}][OH^-]^4} \quad \Rightarrow \quad 8 \times 10^5 = \frac{[Zn(OH)_4^{2-}]}{[Zn(NH_3)_4^{2+}]} \times \frac{(1.0)^4}{(0.0042)^4}$$

$$\frac{[Zn(OH)_4^{2-}]}{[Zn(NH_3)_4^{2+}]} = 2 \times 10^{-4}$$

$$\frac{[Zn(NH_3)_4^{2+}]}{[Zn(OH)_4^{2-}]} = 4 \times 10^3$$

$Ag(NH_3)_2^+(aq) \rightleftharpoons Ag^+(aq) + 2NH_3(aq)$ $K_1 = 1/K_f = 5.9 \times 10^{-8}$

$2[\,H^+(aq) + NH_3(aq) \rightleftharpoons NH_4^+(aq)\,]$ $K_2 = (1/K_a)^2 = 3.2 \times 10^{18}$

$Ag^+(aq) + Cl^-(aq) \rightleftharpoons AgCl(s)$ $K_3 = 1/K_{sp} = 5.6 \times 10^9$

$Ag(NH_3)_2^+(aq) + 2H^+(aq) + Cl^-(aq) \rightleftharpoons AgCl(s) + NH_4^+(aq)$ $K = K_1 \times K_2 \times K_3$

                             $K = 1.0 \times 10^{21}$

Chapter 17: Spontaneity of Reaction

---

2. Refer to Section 17.1.

    a. Non-spontaneous. Cards do not assemble themselves into houses-of-cards (although they do collapse spontaneously). Human intervention is required to build such a structure.

    b. Non-spontaneous. Cleaning up a desk reduces disorder, thus ΔS is negative and the process is not spontaneous (unfortunately).

    c. Non-spontaneous. Clothes do not fold themselves, but require human intervention (which can involve an enormous energy of activation).

    d. Spontaneous. Wind scattering leaves increases disorder, thus ΔS is positive and the process is spontaneous.

---

4. Refer to Section 17.1.

    a. Spontaneous. This particular example is not as familiar as the others, but zinc metal, when placed in acid, does react to give hydrogen gas. Gases are more random than liquids or solids.

    b. Non-spontaneous. This example becomes obvious when one recognizes that $CaCO_3$ is limestone. That limestone has survived the rains and has not been converted to $Ca(OH)_2$ and carbonic acid indicates that this is not a spontaneous process.

    c. Spontaneous. Methane (CH_4) is the prime constituent in natural gas and definitely burns in air, hence its use in cooking and home heating.

    d. Spontaneous. We know from Chapter 16 that AgCl is an insoluble salt that forms when Ag^+ and Cl^- are mixed. If you have done qualititative analysis, you have seen this first hand.

---

6. Refer to Section 17.2 and Example 17.2.

    a. ΔS (-) Entropy generally decreases changing from a liquid to a solid (freezing).

    b. ΔS (-) Precipitating $PbCl_2$ from Pb^{2+} and $2Cl^-$ involves converting 3 moles of aqueous ions to 1 mole of solid, thus entropy decreases.

c. ΔS (+) During the process of combustion, a liquid (candle wax) is converted to gases (water and CO_2). Entropy increases as liquids are converted to gases.

d. ΔS (-) Weeding a garden results in a neater, more ordered garden.

---

8. *Refer to Section 17.2 and Example 17.2.*

a. ΔS (+) In this reaction, 1 mole of solid is converted to 1 mole of another solid (no entropy change) and 2 moles of liquid are converted to one mole of gas. The number of moles decreased (smaller entropy change) and a liquid was converted to a gas (larger entropy change), so overall, entropy increases.

b. ΔS (-) In this reaction, 2 moles of solid and 1 mole of gas are being converted to 2 moles of solid. The total number of moles is decreased (entropy decreases) and a gas is being converted to a solid (entropy decreases).

c. ΔS (+) In this reaction, 2 moles of gas are converted to 4 moles of gas. The total moles of gas increases, thus the entropy increases.

d. ΔS (+) In this reaction, 2 moles of liquid and 3 moles of gas are being converted to 6 moles of gas. The total number of moles increased (entropy increases) and liquid is converted to gas (entropy increases).

---

10. *Refer to Section 17.2 and Example 17.2.*

a. ΔS (+) In this reaction, 1 mole of gas is converted to 2 moles of gas. The total moles of gas increases, thus the entropy increases.

b. ΔS (-) In this reaction, 2 moles of gas are converted to 1 mole of gas. The total moles of gas decreases, thus the entropy decreases.

c. ΔS (-) One mole of solid and 5 moles of liquid are converted to one mole of solid. Since the total number of moles decreases, entropy decreases.

---

12. *Refer to Section 17.2 and Example 17.3.*

$\Delta S^{\circ} = \Sigma\ S^{\circ}_{(products)} - \Sigma\ S^{\circ}_{(reactants)}$

a. $\Delta S^{\circ} = [(4\ mol.)(240.0\ J/mol.\cdot K) + (6\ mol.)(188.7\ J/mol.\cdot K)]$
  $- [(4\ mol.)(192.3\ J/mol.\cdot K) + (7\ mol.)(205.0\ J/mol.\cdot K)]$
$\Delta S^{\circ} = -112.0\ J/K$

b. $\Delta S° = [(1 \text{ mol.})(191.5 \text{ J/mol.·K}) + (4 \text{ mol.})(188.7 \text{ J/mol.·K})]$
     $- [(2 \text{ mol.})(109.6 \text{ J/mol.·K}) + (1 \text{ mol.})(121.2 \text{ J/mol.·K})]$
   $\Delta S° = 605.9 \text{ J/K}$

c. $\Delta S° = [(1 \text{ mol.})(213.6 \text{ J/mol.·K})] - [(1 \text{ mol.})(5.7 \text{ J/mol.·K}) + (1 \text{ mol.})(205.0 \text{ J/mol.·K})]$
   $\Delta S° = 2.9 \text{ J/K}$

d. $\Delta S° = [(1 \text{ mol.})(201.7 \text{ J/mol.·K}) + (3 \text{ mol.})(186.8 \text{ J/mol.·K})]$
     $- [(1 \text{ mol.})(186.2 \text{ J/mol.·K}) + (3 \text{ mol.})(223.0 \text{ J/mol.·K})]$
   $\Delta S° = -93.1 \text{ J/K}$

---

14. *Refer to Section 17.2 and Example 17.3.*

$\Delta S° = \Sigma\, S°_{\text{(products)}} - \Sigma\, S°_{\text{(reactants)}}$

a. $\Delta S° = [(1 \text{ mol.})(-112.1 \text{ J/mol.·K}) + (1 \text{ mol.})(130.6 \text{ J/mol.·K})]$
     $- [(1 \text{ mol.})(41.6 \text{ J/mol.·K}) + (2 \text{ mol.})(0.0 \text{ J/mol.·K})]$
   $\Delta S° = -23.1 \text{ J/K}$

b. $\Delta S° = [(1 \text{ mol.})(69.9 \text{ J/mol.·K})] - [(1 \text{ mol.})(0.0 \text{ J/mol.·K}) + (1 \text{ mol.})(-10.8 \text{ J/mol.·K})]$
   $\Delta S° = 80.7 \text{ J/K}$

c. $\Delta S° = [(1 \text{ mol.})(113.4 \text{ J/mol.·K}) + (1 \text{ mol.})(-10.8 \text{ J/mol.·K})]$
     $- [(1 \text{ mol.})(192.3 \text{ J/mol.·K}) + (1 \text{ mol.})(69.9 \text{ J/mol.·K})]$
   $\Delta S° = -159.6 \text{ J/K}$

---

16. *Refer to Section 17.2 and Example 17.3.*

$\Delta S° = \Sigma\, S°_{\text{(products)}} - \Sigma\, S°_{\text{(reactants)}}$

a. $\Delta S° = [(2 \text{ mol.})(188.7 \text{ J/mol.·K}) + (2 \text{ mol.})(248.1 \text{ J/mol.·K})]$
     $- [(2 \text{ mol.})(205.7 \text{ J/mol.·K}) + (3 \text{ mol.})(205.0 \text{ J/mol.·K})]$
   $\Delta S° = -152.8 \text{ J/K}$

b. $\Delta S° = [(1 \text{ mol.})(72.7 \text{ J/mol.·K}) + (1 \text{ mol.})(69.9 \text{ J/mol.·K}) + (1 \text{ mol.})(240.0 \text{ J/mol.·K})]$
     $- [(1 \text{ mol.})(42.6 \text{ J/mol.·K}) + (2 \text{ mol.})(0.0 \text{ J/mol.·K}) + (1 \text{ mol.})(146.4 \text{ J/mol.·K})]$
   $\Delta S° = 193.6 \text{ J/K}$

c. $\Delta S° = [(1 \text{ mol.})(-128.9 \text{ J/mol.·K}) + (1 \text{ mol.})(248.1 \text{ J/mol.·K}) + (2 \text{ mol.})(69.9 \text{ J/mol.·K})]$
     $- [(1 \text{ mol.})(20.1 \text{ J/mol.·K}) + (4 \text{ mol.})(0.0 \text{ J/mol.·K}) + (1 \text{ mol.})(29.9 \text{ J/mol.·K})]$
   $\Delta S° = 209.0 \text{ J/K}$

$\Delta G^\circ = \Delta H^\circ - T\Delta S^\circ$ (Convert T to Kelvin and ΔS° to kJ/K)

a. $\Delta G^\circ = -79.6$ kJ $- (318$ K$)(0.4331$ kJ/K$)$
   $\Delta G^\circ = -217$ kJ

b. $\Delta G^\circ = 837.4$ kJ $- (318$ K$)(0.1738$ kJ/K$)$
   $\Delta G^\circ = 782.1$ kJ

c. $\Delta G^\circ = -34$ kJ $- (318$ K$)(-0.039$ kJ/K$)$
   $\Delta G^\circ = -22$ kJ

---

20. *Refer to Section 17.4, Example 17.4 and Problem 12 above.*

Calculate ΔH° and ΔS° (see Problem 12 above) for the reactions. Then calculate ΔG°. If ΔG° is negative, the reaction is spontaneous.

a. $\Delta H^\circ = \Sigma \Delta H_f^\circ{}_{(products)} - \Sigma \Delta H_f^\circ{}_{(reactants)}$
   $\Delta H^\circ = [(4$ mol.$)(33.2$ kJ/mol.$) + (6$ mol.$)(-241.8$ kJ/mol.$)]$
        $- [(4$ mol.$)(-46.1$ kJ/mol.$) + (7$ mol.$)(0.0$ kJ/mol.$)]$
   $\Delta H^\circ = -1133.6$ kJ

   $\Delta G^\circ = \Delta H^\circ - T\Delta S^\circ$
   $\Delta G^\circ = -1133.6$ kJ $- (415$ K$)(-0.1120$ kJ/K$)$
   $\Delta G^\circ = -1087.1$ kJ;   **Spontaneous.**

b. $\Delta H^\circ = \Sigma \Delta H_f^\circ{}_{(products)} - \Sigma \Delta H_f^\circ{}_{(reactants)}$
   $\Delta H^\circ = [(1$ mol.$)(0.0$ kJ/mol.$) + (4$ mol.$)(-241.8$ kJ/mol.$)]$
        $- [(2$ mol.$)(-187.8$ kJ/mol.$) + (1$ mol.$)(50.6$ kJ/mol.$)]$
   $\Delta H^\circ = -642.2$ kJ

   $\Delta G^\circ = \Delta H^\circ - T\Delta S^\circ$
   $\Delta G^\circ = -642.2$ kJ $- (415$ K$)(0.6059$ kJ/K$)$
   $\Delta G^\circ = -894$ kJ;   **Spontaneous.**

c. $\Delta H^\circ = \Sigma \Delta H_f^\circ{}_{(products)} - \Sigma \Delta H_f^\circ{}_{(reactants)}$
   $\Delta H^\circ = [(1$ mol.$)(-393.5$ kJ/mol.$)] - [(1$ mol.$)(0.0$ kJ/mol.$) + (1$ mol.$)(0.0$ kJ/mol.$)]$
   $\Delta H^\circ = -393.5$ kJ

   $\Delta G^\circ = \Delta H^\circ - T\Delta S^\circ$
   $\Delta G^\circ = -393.5$ kJ $- (415$ K$)(0.0029$ kJ/K$)$
   $\Delta G^\circ = -394.7$ kJ;   **Spontaneous.**

d. $\Delta H° = \Sigma\ \Delta H_f°_{(products)} - \Sigma\ \Delta H_f°_{(reactants)}$

$\Delta H° = [(1\ mol.)(-134.5\ kJ/mol.) + (3\ mol.)(-92.3\ kJ/mol.)]$
$\qquad - [(1\ mol.)(-74.8\ kJ/mol.) + (3\ mol.)(0.0\ kJ/mol.)]$

$\Delta H° = -336.6\ kJ$

$\Delta G° = \Delta H° - T\Delta S°$

$\Delta G° = -336.6\ kJ - (415\ K)(-0.0931\ kJ/K)$

$\Delta G° = -298.0\ kJ;$ **Spontaneous.**

---

22. Refer to Section 17.4

$\Delta G° = \Sigma\ \Delta G_f°_{(products)} - \Sigma\ \Delta G_f°_{(reactants)}$

a. $\Delta G° = [(1\ mol.)(-147.1\ kJ/mol.) + (1\ mol.)(0.0\ kJ/mol.)]$
$\qquad - [(1\ mol.)(0.0\ kJ/mol.) + (2\ mol.)(0.0\ kJ/mol.)]$

$\Delta G° = -147.1\ kJ$

b. $\Delta G° = [(1\ mol.)(-237.2\ kJ/mol.)]$
$\qquad - [(1\ mol.)(0.0\ kJ/mol.) + (1\ mol.)(-157.2\ kJ/mol.)]$

$\Delta G° = -80.0\ kJ$

c. $\Delta G° = [(1\ mol.)(-79.3\ kJ/mol.) + (1\ mol.)(-157.2\ kJ/mol.)]$
$\qquad - [(1\ mol.)(-16.5\ kJ/mol.) + (1\ mol.)(-237.2\ kJ/mol.)]$

$\Delta G° = 17.2\ kJ$

---

24. Refer to Section 17.4 and Example 17.5.

Determine $\Delta H_f°$ from the table in Appendix 1. Write the equation for the formation of the compounds from their elements. Calculate $\Delta S°$ and then $\Delta G_f°$.

a. $Ca(s) + C(s) + \tfrac{3}{2}O_2(g)\ \rightarrow\ CaCO_3(s)$

$\Delta H_{reaction} = \Delta H_f° = -1206.9\ kJ$

$\Delta S° = \Sigma\ S°_{products} - \Sigma\ S°_{reactants}$

$\Delta S° = [(1\ mol.)(0.0929\ kJ/mol.\cdot K)] - [(1\ mol.)(0.0414\ kJ/mol.\cdot K)$
$\qquad + (1\ mol.)(0.0057\ kJ/mol.\cdot K) + (\tfrac{3}{2}\ mol.)(0.2050 kJ/mol.\cdot K)]$

$\Delta S° = -0.2617\ kJ/K$

$\Delta G_{reaction} = \Delta G_f° = \Delta H° - T\Delta S°$

$\Delta G_f° = -1206.9\ kJ - (298\ K)(-0.2617\ kJ/K) = -1128.9\ kJ$

Note that $\Delta G_f°$ is defined for 1 mole, thus $\Delta G_f° = -1128.9\ kJ/mol.$

319

b. $Mg(s) + \frac{1}{2}O_2(g) \rightarrow MgO(s)$

$\Delta H_{reaction} = \Delta H_f^\circ = -601.7$ kJ

$\Delta S^\circ = \Sigma \; S^\circ_{products} - \Sigma \; S^\circ_{reactants}$
$\Delta S^\circ = [(1 \; mol.)(0.0269 \; kJ/mol.\cdot K)]$
$\qquad\qquad - [(1 \; mol.)(0.0327 \; kJ/mol.\cdot K) + (\frac{1}{2} \; mol.)(0.2050 \; kJ/mol.\cdot K)]$
$\Delta S^\circ = -0.1083$ kJ/K

$\Delta G_{reaction} = \Delta G_f^\circ = \Delta H^\circ - T\Delta S^\circ$

$\Delta G_f^\circ = -601.7$ kJ $- (298 \; K)(-0.1083 \; kJ/K)$

$\Delta G_f^\circ = -569.4$ kJ

Note that ΔG_f° is defined for 1 mole, thus $\Delta G_f^\circ = -569.4$ kJ/mol.

c. $\frac{1}{4}P_4(s) + \frac{5}{2}Cl_2(g) \rightarrow PCl_5(g)$

$\Delta H_{reaction} = \Delta H_f^\circ = -374.9$ kJ

$\Delta S^\circ = \Sigma \; S^\circ_{(prod)} - \Sigma \; S^\circ_{(react.)}$
$\Delta S^\circ = [(1 \; mol.)(0.3645 \; kJ/mol.\cdot K)]$
$\qquad\qquad - [(\frac{1}{4} \; mol.)(0.1644 \; kJ/mol.\cdot K) + (\frac{5}{2} \; mol.)(0.2230 \; kJ/mol.\cdot K)]$
$\Delta S^\circ = -0.2341$ kJ/K

$\Delta G_{reaction} = \Delta G_f^\circ = \Delta H^\circ - T\Delta S^\circ$

$\Delta G_f^\circ = -374.9$ kJ $- (298 \; K)(-0.2341 \; kJ/K)$

$\Delta G_f^\circ = -305.1$ kJ

Note that ΔG_f° is defined for 1 mole, thus $\Delta G_f^\circ = -305.1$ kJ/mol.

---

26. *Refer to Section 17.4.*

$CH_3OH(l) \rightarrow CH_4(g) + \frac{1}{2}O_2(g)$

$\Delta G^\circ = \Sigma \; \Delta G_f^\circ{}_{(products)} - \Sigma \; \Delta G_f^\circ{}_{(reactants)}$

$\Delta G^\circ = [(1 \; mol.)(-50.7 \; kJ/mol.) + (\frac{1}{2} \; mol.)(0.0 \; kJ/mol.)] - [(1 \; mol.)(-166.3 \; kJ/mol.)]$
$\Delta G^\circ = 115.6$ kJ

ΔG° is positive, so the reaction is not spontaneous and the reaction is **not feasible**.

$Mg(s) + 2H_2O(l) \rightarrow Mg(OH)_2(s) + H_2(g)$

$\Delta G° = \Sigma \, \Delta G_f°_{(products)} - \Sigma \, \Delta G_f°_{(reactants)}$

$\Delta G° = [(1 \text{ mol.})(-833.6 \text{ kJ/mol.}) + (1 \text{ mol.})(0.0 \text{ kJ/mol.})]$
$\qquad\qquad - [(1 \text{ mol.})(0.0 \text{ kJ/mol.}) + (2 \text{ mol.})(-237.2 \text{ kJ/mol.})]$

$\Delta G° = -359.2 \text{ kJ}$ $\qquad\qquad$ (at 25°C)

To calculate $\Delta G°$ at 15°C, calculate $\Delta H°$ and $\Delta S°$, then use $\Delta G° = \Delta H° - T\Delta S°$

$\Delta H° = \Sigma \, \Delta H_f°_{(products)} - \Sigma \, \Delta H_f°_{(reactants)}$
$\Delta H° = [(1 \text{ mol.})(-924.5 \text{ kJ/mol.}) + (1 \text{ mol.})(0.0 \text{ kJ/mol.})]$
$\qquad\quad - [(1 \text{ mol.})(0.0 \text{ kJ/mol.}) + (2 \text{ mol.})(-285.5 \text{ kJ/mol.})]$
$\Delta H° = -352.9 \text{ kJ}$

$\Delta S° = \Sigma \, S°_{(products)} - \Sigma \, S°_{(reactants)}$
$\Delta S° = [(1 \text{ mol.})(0.0632 \text{ kJ/mol.·K}) + (1 \text{ mol.})(0.1306 \text{ kJ/mol.·K})]$
$\qquad\quad - [(1 \text{ mol.})(0.0327 \text{ kJ/mol.·K}) + (2 \text{ mol.})(0.0699 \text{ kJ/mol.·K})]$
$\Delta S° = 0.0213 \text{ kJ/K}$

$\Delta G° = \Delta H° - T\Delta S°$
$\Delta G° = -352.9 \text{ kJ} - (288 \text{ K})(0.0213 \text{ kJ/K})$
$\Delta G° = -359.0 \text{ kJ}$ $\qquad\qquad$ (at 15°C)

a. $\Delta G° = \Delta H° - T\Delta S°$
   $-148.4 \text{ kJ} = -109.0 \text{ kJ} - (298 \text{ K})(\Delta S°)$
   $\Delta S° = 0.132 \text{ kJ/K}$
   The sign of $\Delta S°$ is reasonable since Δn_{gas} is positive.

b. $\Delta S° = \Sigma \, S°_{(products)} - \Sigma \, S°_{(reactants)}$
   $0.132 \text{ kJ/K} = [(4 \text{ mol.})(0.0645 \text{ kJ/mol.·K}) + (1 \text{ mol.})(0.2050 \text{ kJ/mol.·K})]$
   $\qquad\qquad - [(2 \text{ mol.})S°(N_2O_2(s)) + (2 \text{ mol.})(0.0699 \text{ kJ/mol.·K})]$
   $(2 \text{ mol.})S°(N_2O_2(s)) = 0.191 \text{ kJ/K}$
   $S°(N_2O_2(s)) = 0.0955 \text{ kJ/mol.·K}$

c. $\Delta H°_{(reaction)} = \Sigma \, \Delta H_f°_{(products)} - \Sigma \, \Delta H_f°_{(reactants)}$
   $-109.0 \text{ kJ} = [(4 \text{ mol.})(-425.6 \text{ kJ/mol.}) + (1 \text{ mol.})(0.0 \text{ kJ/mol.})]$
   $\qquad\qquad - [(2 \text{ mol.})\Delta H_f°(N_2O_2(s)) + (2 \text{ mol.})(-285.8 \text{ kJ/mol.})]$
   $(2 \text{ mol.})\Delta H_f°(N_2O_2(s)) = -1021.8 \text{ kJ}$
   $\Delta H_f°(N_2O_2(s)) = -510.9 \text{ kJ/mol.}$

Calculate $\Delta S°_{(reaction)}$ using the Gibbs-Helmholtz equation. Then use $\Delta S°_{(reaction)}$ to solve for $S°(Co^{2+})$

a. $\Delta G° = \Delta H° - T\Delta S°$
    -452.4 kJ $= -353.2$ kJ $- (298$ K$)(\Delta S°)$
    $\Delta S° = 0.333$ kJ/K $= 333$ J/K

b. $\Delta S° = \Sigma\, S°_{(products)} - \Sigma\, S°_{(reactants)}$
    0.333 kJ/K $= [(2$ mol.$)(0.1868$ kJ/mol.$\cdot$K$) + (2$ mol.$)S°(COCl_2(s))]$
    $\qquad\qquad - [(2$ mol.$)(0.2017$ kJ/mol.$\cdot$K$) + (1$ mol.$)(0.2050$ kJ/mol.$\cdot$K$)]$
    $(2$ mol.$)S°(COCl_2(s)) = 0.568$ kJ/K
    $S°(COCl_2(s)) = 0.284$ kJ/mol.$\cdot$K $= 284$ J/mol.$\cdot$K

c. $\Delta H°_{(reaction)} = \Sigma\, \Delta H_f°_{(products)} - \Sigma\, \Delta H_f°_{(reactants)}$
    -353.2 kJ $= [(2$ mol.$)(-92.3$ kJ/mol.$) + (2$ mol.$)(\Delta H_f°(COCl_2(s)))]$
    $\qquad\qquad - [(2$ mol.$)(-134.5$ kJ/mol.$) + (1$ mol.$)(0$ kJ/mol.$)]$
    $(2$ mol.$)\Delta H_f°(COCl_2(s)) = -437.6$ kJ
    $\Delta H_f°(COCl_2(s)) = -218.8$ kJ/mol.

Set up the equation for calculating $\Delta G°$. Then examine the impact changing T would have on the sign of $\Delta G°$. Specifically, would raising or lowering T change the sign of $\Delta G°$?

a. $\Delta G° = \Delta H° - T\Delta S°$
    $\Delta G° = 851.5$ kJ $- (T)(0.0385$ kJ/K$)$
    $\quad \Delta G° = 0$ when $T = 2.21 \times 10^4$ K,
    $\quad \Delta G°$ is (+) when $T < 2.21 \times 10^4$ K and
    $\quad \Delta G°$ is (-) when $T > 2.21 \times 10^4$ K.
    Therefore, varying T can alter the spontaneity of the reaction.

b. $\Delta G° = \Delta H° - T\Delta S°$
    $\Delta G° = -50.6$ kJ $- (T)(0.3315$ kJ/K$)$
    $\quad \Delta G° = 0$ when $T = -153$ K,
    Therefore, $\Delta G°$ is always negative and changing T will not alter the spontaneity of the reaction.

c. $\Delta G° = \Delta H° - T\Delta S°$

$\Delta G° = 98.9$ kJ - $(T)(0.0939$ kJ/K$)$

$\Delta G° = 0$ when $T = 1050$ K,

$\Delta G°$ is (+) when $T < 1050$ K and

$\Delta G°$ is (-) when $T > 1050$ K.

Therefore, varying T can alter the spontaneity of the reaction.

---

36. Refer to Section 17.5, Example 17.7 and Problem 34 above.

Write the Gibbs-Helmholtz equation for the reaction, set $\Delta G° = 0$ and solve for T. Those temperatures at which $\Delta G° < 0$, represent the temperatures at which the reaction is spontaneous. See problem 34 above.

a. Spontaneous for $T > 2.21 \times 10^4$ K.

b. Spontaneous at all temperatures.

c. Spontaneous for $T > 1050$ K.

---

38. Refer to Sections 17.4 and 17.5 and Example 17.7.

Calculate $\Delta H°_{reaction}$ and $\Delta S°_{reaction}$. Using the Gibbs-Helmholtz equation, set $\Delta G° = 0$, and solve for T.

$\Delta H° = \Sigma \; \Delta H_f°_{products} - \Sigma \; \Delta H_f°_{reactants}$

$\Delta H° = [(1$ mol.$)(-46.1$ kJ/mol.$) + (1$ mol.$)(-92.3$ kJ/mol.$)] - [(1$ mol.$)(-314.4$ kJ/mol.$)]$

$\Delta H° = +176.0$ kJ

$\Delta S° = \Sigma \; S°_{products} - \Sigma \; S°_{reactants}$

$\Delta S° = [(1$ mol.$)(0.1923$ kJ/mol.$\cdot$K$) + (1$ mol.$)(0.1868$ kJ/mol.$\cdot$K$)]$

$\qquad - [(1$ mol.$)(0.0946$ kJ/mol.$\cdot$K$)]$

$\Delta S° = 0.2845$ kJ/K

$\Delta G° = \Delta H° - T\Delta S°$

$0 = 176.0$ kJ - $(T)(0.2845$ kJ/K$)$

$T = 618.6$ K. Since $\Delta H°$ and $\Delta S°$ are positive, $\Delta G°$ is negative (and the reaction is spontaneous) at temperatures above 618.6 K (345.5°C).

a. Calculate $\Delta H^\circ_{reaction}$ and $\Delta S^\circ_{reaction}$. Use these values and $T = 100$ K to 500 K to calculate ΔG°.

$\Delta H^\circ = \Sigma \Delta H_f^\circ{}_{products} - \Sigma \Delta H_f^\circ{}_{reactants}$

$\Delta H^\circ = [(4 \text{ mol.})(0.0 \text{ kJ/mol.}) + (3 \text{ mol.})(-393.5 \text{ kJ/mol.})]$
$\qquad - [(2 \text{ mol.})(-824.2 \text{ kJ/mol.}) + (3 \text{ mol.})(0.0 \text{ kJ/mol.})]$

$\Delta H^\circ = 467.9 \text{ kJ}$

$\Delta S^\circ = \Sigma S^\circ{}_{products} - \Sigma S^\circ{}_{reactants}$

$\Delta S^\circ = [(4 \text{ mol.})(0.0273 \text{ kJ/mol.}\cdot\text{K}) + (3 \text{ mol.})(0.2136 \text{ kJ/mol.}\cdot\text{K})]$
$\qquad - [(2 \text{ mol.})(0.0874 \text{ kJ/mol.}\cdot\text{K}) + (3 \text{ mol.})(0.0057 \text{ kJ/mol.}\cdot\text{K})]$

$\Delta S^\circ = 0.5581 \text{ kJ/K}$

$\Delta G^\circ = \Delta H^\circ - T\Delta S^\circ$

$\Delta G^\circ = 467.9 \text{ kJ} - (T)(0.5581 \text{ kJ/K})$

T (K)	100	200	300	400	500
ΔG° (kJ)	412.1	356.3	300.5	244.7	188.9

b. $\Delta G^\circ = 0 = 467.9 \text{ kJ} - (T)(0.5581 \text{ kJ/K})$

$T = 838.4 \text{ K}$

ΔG° becomes negative (and the reaction becomes spontaneous) at temperatures above 838.4 K, therefore this is the lowest temperature at which smelting is possible.

Calculate $\Delta H^\circ_{reaction}$ and $\Delta S^\circ_{reaction}$. Then calculate T for $\Delta G^\circ = 0$ to determine the lowest temperature at which the reaction is spontaneous.

a. $SnO_2(s) \rightarrow Sn(s) + O_2(g)$

$\Delta H^\circ = \Sigma \Delta H_f^\circ{}_{products} - \Sigma \Delta H_f^\circ{}_{reactants}$

$\Delta H^\circ = [(1 \text{ mol.})(0.0 \text{ kJ/mol.}) + (1 \text{ mol.})(0.0 \text{ kJ/mol.})] - [(1 \text{ mol.})(-580.7 \text{ kJ/mol.})]$

$\Delta H^\circ = 580.7 \text{ kJ}$

$\Delta S^\circ = \Sigma S^\circ{}_{products} - \Sigma S^\circ{}_{reactants}$

$\Delta S^\circ = [(1 \text{ mol.})(0.0516 \text{ kJ/mol.}\cdot\text{K}) + (1 \text{ mol.})(0.2050 \text{ kJ/mol.}\cdot\text{K})]$
$\qquad - [(1 \text{ mol.})(0.0523 \text{ kJ/mol.}\cdot\text{K})]$

$\Delta S^\circ = 0.2043 \text{ kJ/K}$

$\Delta G° = \Delta H° - T\Delta S°$

$\Delta G° = 0 = 580.7 \text{ kJ} - (T)(0.2043 \text{ kJ/K})$

$\Delta G° = 0$ when $T = 2842$ K

b. $SnO_2(s) + 2H_2(g) \rightarrow Sn(s) + 2H_2O(g)$

$\Delta H° = \Sigma \Delta H_f°_{products} - \Sigma \Delta H_f°_{reactants}$
$\Delta H° = [(1 \text{ mol.})(0.0 \text{ kJ/mol.}) + (2 \text{ mol.})(-241.8 \text{ kJ/mol.})]$
$\qquad - [(1 \text{ mol.})(-580.7 \text{ kJ/mol.}) + (2 \text{ mol.})(0.0 \text{ kJ/mol.})]$
$\Delta H° = 97.1$ kJ

$\Delta S° = \Sigma S°_{products} - \Sigma S°_{reactants}$
$\Delta S° = [(1 \text{ mol.})(0.0516 \text{ kJ/mol.·K}) + (2 \text{ mol.})(0.1887 \text{ kJ/mol.·K})]$
$\qquad - [(1 \text{ mol.})(0.0523 \text{ kJ/mol.·K}) + (2 \text{ mol.})(0.1306 \text{ kJ/mol.·K})]$
$\Delta S° = 0.1155$ kJ/K

$\Delta G° = \Delta H° - T\Delta S°$
$\Delta G° = 0 = 97.1 \text{ kJ} - (T)(0.1155 \text{ kJ/K})$

$\Delta G° = 0$ when $T = 841$ K

c. $SnO_2(s) + C(s) \rightarrow Sn(s) + CO_2(g)$

$\Delta H° = \Sigma \Delta H_f°_{products} - \Sigma \Delta H_f°_{reactants}$
$\Delta H° = [(1 \text{ mol.})(0.0 \text{ kJ/mol.}) + (1 \text{ mol.})(-393.5 \text{ kJ/mol.})]$
$\qquad - [(1 \text{ mol.})(-580.7 \text{ kJ/mol.}) + (1 \text{ mol.})(0.0 \text{ kJ/mol.})]$
$\Delta H° = 187.2$ kJ

$\Delta S° = \Sigma S°_{products} - \Sigma S°_{reactants}$
$\Delta S° = [(1 \text{ mol.})(0.0516 \text{ kJ/mol.·K}) + (1 \text{ mol.})(0.2136 \text{ kJ/mol.·K})]$
$\qquad - [(1 \text{ mol.})(0.0523 \text{ kJ/mol.·K}) + (1 \text{ mol.})(0.0057 \text{ kJ/mol.·K})]$
$\Delta S° = 0.2072$ kJ/K

$\Delta G° = \Delta H° - T\Delta S°$
$\Delta G° = 0 = 187.2 \text{ kJ} - (T)(0.2072 \text{ kJ/K})$

$\Delta G° = 0$ when $T = 903.5$ K

Reaction (b) is spontaneous at the lowest temperature.

---

44. *Refer to Section 17.5.*

Calculate $\Delta H°_{reaction}$ and $\Delta S°_{reaction}$. $\Delta G° = 0$ at equilibrium, so calculate T for $\Delta G° = 0$ to determine the temperature at which the reaction is at equilibrium.

$P_{red}(s) \rightleftharpoons P_{white}(s)$

$\Delta H^{\circ} = \Sigma \Delta H_f^{\circ}{}_{products} - \Sigma \Delta H_f^{\circ}{}_{reactants}$
$\Delta H^{\circ} = [(1 \text{ mol.})(0.0 \text{ kJ/mol.})] - [(1 \text{ mol.})(-17.6 \text{ kJ/mol.})]$
$\Delta H^{\circ} = 17.6 \text{ kJ}$

$\Delta S^{\circ} = \Sigma S^{\circ}{}_{products} - \Sigma S^{\circ}{}_{reactants}$
$\Delta S^{\circ} = [(1 \text{ mol.})(0.04109 \text{ kJ/mol.·K})] - [(1 \text{ mol.})(0.02280 \text{ kJ/mol.·K})]$
$\Delta S^{\circ} = 0.01829 \text{ kJ/K}$

$\Delta G^{\circ} = \Delta H^{\circ} - T\Delta S^{\circ}$
$\Delta G^{\circ} = 0 = 17.6 \text{ kJ} - (T)(0.01829 \text{ kJ/K})$
$T = 962 \text{ K}$

The system is at equilibrium when $T = 962$ K

---

46. *Refer to Section 17.5.*

Calculate $\Delta H^{\circ}{}_{reaction}$ and $\Delta S^{\circ}{}_{reaction}$. $\Delta G^{\circ} = 0$ at equilibrium, so calculate T for $\Delta G^{\circ} = 0$ to determine the temperature at which the reaction is at equilibrium.

$C_{graphite}(s) \rightleftharpoons C_{diamond}(s)$

$\Delta H^{\circ} = \Sigma \Delta H_f^{\circ}{}_{products} - \Sigma \Delta H_f^{\circ}{}_{reactants}$
$\Delta H^{\circ} = [(1 \text{ mol.})(1.9 \text{ kJ/mol.})] - [(1 \text{ mol.})(0 \text{ kJ/mol.})]$
$\Delta H^{\circ} = 1.9 \text{ kJ}$

$\Delta S^{\circ} = \Sigma S^{\circ}{}_{products} - \Sigma S^{\circ}{}_{reactants}$
$\Delta S^{\circ} = [(1 \text{ mol.})(0.0024 \text{ kJ/mol.·K})] - [(1 \text{ mol.})(0.0057 \text{ kJ/mol.·K})]$
$\Delta S^{\circ} = -0.0033 \text{ kJ/K}$

$\Delta G^{\circ} = \Delta H^{\circ} - T\Delta S^{\circ}$
$\Delta G^{\circ} = 0 = 1.9 \text{ kJ} - (T)(-0.0033 \text{ kJ/K})$
$T = -576 \text{ K}$

The system is at equilibrium when $T = -576$ K. Since the temperature can never be negative, there is no temperature at which graphite and diamond are in equilibrium (for $P = 1$ atm).

---

48. *Refer to Section 17.5.*

Calculate $\Delta H^{\circ}{}_{reaction}$ and $\Delta S^{\circ}{}_{reaction}$. Calculate ΔG° for the given pressure, then calculate T for that value of ΔG°.

$I_2(s) \rightleftharpoons I_2(g)$

$\Delta G^{\circ} = -RT \ln(P_{(I_2)}) = -RT \ln(1 \text{ atm}) = 0.$

$\Delta H^{\circ} = \Sigma \Delta H_f^{\circ}{}_{products} - \Sigma \Delta H_f^{\circ}{}_{reactants}$

$\Delta H° = [(1 \text{ mol.})(62.4 \text{ kJ/mol.})] - [(1 \text{ mol.})(0 \text{ kJ/mol.})] = 62.4 \text{ kJ}$

$\Delta S° = \Sigma\ S°_{products} - \Sigma\ S°_{reactants}$
$\Delta S° = [(1 \text{ mol.})(0.2606 \text{ kJ/mol.·K})] - [(1 \text{ mol.})(0.1161 \text{ kJ/mol.·K})] = 0.1445 \text{ kJ/K}$

$\Delta G° = \Delta H° - T\Delta S°$
$\Delta G° = 0 = 62.4 \text{ kJ} - (T)(0.1445 \text{ kJ/K})$
$T = 432 \text{ K}$

At 1 atm, I_2 sublimes when $T = 432 \text{ K}$.

---

50. Refer to Section 17.5.

Calculate Q for the concentrations listed, calculate $\Delta G°$ using the values in Appendix 1, then calculate $\Delta G_{reaction}$.

a. $Q = \dfrac{[H^+][F^-]}{[HF]} = \dfrac{(1)(1)}{(1)} = 1$

$\Delta G = \Delta G° + RT \ln Q$
$\Delta G = 18.0 \text{ kJ} + (8.31 \text{ J/mol.·K})(298 \text{ K})(0.001 \text{ kJ/J})(\ln 1.0)$
$\Delta G = 18.0 \text{ kJ} + 0 \text{ kJ} = 18.0 \text{ kJ}$
ΔG is positive, so the reaction is **non-spontaneous**.

b. $Q = \dfrac{[H^+][F^-]}{[HF]} = \dfrac{(1.0 \times 10^{-3})(1.0 \times 10^{-3})}{(1.0)} = 1.0 \times 10^{-6}$

$\Delta G = \Delta G° + RT \ln Q$
$\Delta G = 18.0 \text{ kJ/mol.} + (8.31 \text{ J/mol.·K})(298 \text{ K})(0.001 \text{ kJ/J})(\ln 1.0 \times 10^{-6})$
$\Delta G = 18.0 \text{ kJ} - 34.2 \text{ kJ} = -16.2 \text{ kJ}$
ΔG is negative, so the reaction is **spontaneous**.

---

52. Refer to Sections 17.4 and 17.5 and Example 17.8.

Calculate $\Delta H°_{reaction}$ and $\Delta S°_{reaction}$. Calculate $\Delta G°$ for the given pressure, then calculate T for that value of $\Delta G°$.

a. $\Delta G° = \Sigma\ \Delta G_f°{}_{products} - \Sigma\ \Delta G_f°{}_{reactants}$
$\Delta G° = [(1 \text{ mol.})(0.0 \text{ kJ/mol.}) + (1 \text{ mol.})(0.0 \text{ kJ/mol.}) + (2 \text{ mol.})(-157.2 \text{ kJ/mol.})]$
$\qquad - [(2 \text{ mol.})(-237.2 \text{ kJ/mol.}) + (2 \text{ mol.})(-131.2 \text{ kJ/mol.})]$
$\Delta G° = 422.4 \text{ kJ}$

b. $[H^+] = 10^{-pH} = 10^{-11.98} = 1.05 \times 10^{-12}\,M$

$$[OH^-] = \frac{K_w}{[H^+]} = \frac{1.00 \times 10^{-14}}{1.05 \times 10^{-12}} = 9.6 \times 10^{-3}\,M$$

$$Q = \frac{(P_{H_2})(P_{Cl_2})[OH^-]^2}{[Cl^-]^2} = \frac{(0.250)(0.250)(9.6 \times 10^{-3})^2}{(0.335)^2} = 5.1 \times 10^{-5}$$

$\Delta G = \Delta G° + RT \ln Q$
$\Delta G = 422.4\text{ kJ} + (8.31\text{ J/mol.·K})(0.001\text{ kJ/J})(298\text{ K})(\ln 5.1 \times 10^{-5})$
$\Delta G = 397.9\text{ kJ}$

---

54. Refer to Sections 17.4 and 17.5 and Example 17.8.

a. $\Delta G° = \Sigma\,\Delta G_f°_{\text{products}} - \Sigma\,\Delta G_f°_{\text{reactants}}$
$\Delta G° = [(1\text{ mol.})(77.1\text{ kJ/mol.}) + (1\text{ mol.})(-131.2\text{ kJ/mol.})] - [(1\text{ mol.})(-109.8\text{ kJ/mol.})]$
$\Delta G° = 55.7\text{ kJ}$

b. Assume 25°C as in part (a) above. Set $\Delta G = 0$, solve the equation for $[Ag^+]$ and $[Cl^-]$. Note that $[Ag^+] = [Cl^-] = x$.

$\Delta G = \Delta G° + RT \ln Q$
$Q = [Ag^+][Cl^-]$
$-1.0\text{ kJ} = 55.7\text{ kJ} + (8.31\text{ J/mol.·K})(0.001\text{ kJ/J})(298\text{ K})(\ln [Ag^+][Cl^-])$
$\ln [Ag^+][Cl^-] = -22.9$
$[Ag^+][Cl^-] = x^2 = 1.15 \times 10^{-10}$
$x = [Ag^+] = [Cl^-] = 1.07 \times 10^{-5}\,M$

c. $K_{sp} = [Ag^+][Cl^-] = 1.15 \times 10^{-10}$
Actual $K_{sp} = 1.8 \times 10^{-10}$

Since the calculated value and K_{sp} are approximately equal, the answer in (b) is reasonable.

---

56. Refer to Section 17.7.

Manipulate the equations so that their sum is the desired equation. Remember that reversing the equation changes the sign of $\Delta G°$, and any factor applied to the equation must also be applied to $\Delta G°$.

$2[FeCl_2(s) \rightarrow Fe(s) + Cl_2(g)]$	$\Delta G° = -(2)(-302.3\text{ kJ})$
$2[Fe(s) + {}^3/_2Cl_2(g) \rightarrow FeCl_3(s)]$	$\Delta G° = (2)(-334.0\text{ kJ})$

$2FeCl_2(s) + 2Fe(s) + 3Cl_2(g) \rightarrow 2FeCl_3(s) + 2Fe(s) + 2Cl_2(g)$
$2FeCl_2(s) + Cl_2(g) \rightarrow 2FeCl_3(s)$ $\Delta G° = -63.4\text{ kJ}$

58. Refer to Sections 17.4 and 17.7 and Example 17.10.

a. $\Delta G° = \Sigma \, \Delta G_f°_{products} - \Sigma \, \Delta G_f°_{reactants}$

$\Delta G° = [(1 \text{ mol.})(-32.9 \text{ kJ/mol.}) + (1 \text{ mol.})(0.0 \text{ kJ/mol.})] - [(2 \text{ mol.})(-50.7 \text{ kJ/mol.})]$

$\Delta G° = 68.5 \text{ kJ}$

$\Delta G°$ is positive, this reaction is not spontaneous and therefore **not feasible** (at 25°C).

b.
$$2CH_4(g) \rightarrow C_2H_6(g) + H_2(g) \qquad \Delta G° = 68.5 \text{ kJ}$$
$$\underline{H_2(g) + \tfrac{1}{2}O_2(g) \rightarrow H_2O(g) \qquad \Delta G° = -228.6 \text{ kJ}}$$
$$2CH_4(g) + H_2(g) + \tfrac{1}{2}O_2(g) \rightarrow H_2O(g) + C_2H_6(g) + H_2(g)$$
$$2CH_4(g) + \tfrac{1}{2}O_2(g) \rightarrow H_2O(g) + C_2H_6(g) \qquad \Delta G° = -160.1 \text{ kJ}$$

This reaction is spontaneous, and therefore **feasible** at 25°C

60. Refer to Section 17.7 and Example 17.10.

Calculate the moles of ADP required to give $\Delta G°_{reaction} = -390 \text{ kJ}$.

$\Delta G°_{reaction} = (x \text{ mol.})(\Delta G°_{(ADP \, reaction)}) + \Delta G°_{(glucose \, reaction)}$

$-390 \text{ kJ} = (x)(31 \text{ kJ}) + (-2870 \text{ kJ})$

$x = 80 \text{ mol.}$ (Multiply the ADP reaction by 80 and add to the $C_6H_{12}O_6$ reaction.)

$$80[ADP(aq) + HPO_4^{2-}(aq) + 2H^+(aq) \rightarrow ATP(aq) + H_2O(l)] \qquad \Delta G° = 80(31 \text{ kJ})$$
$$\underline{C_6H_{12}O_6(aq) + 6O_2(g) \rightarrow 6CO_2(g) + 6H_2O(l) \qquad \Delta G° = -2870 \text{ kJ}}$$
$$80ADP(aq) + 80HPO_4^{2-}(aq) + 160H^+(aq) + C_6H_{12}O_6(aq) + 6O_2(g)$$
$$\rightarrow 6CO_2(g) + 6H_2O(l) + 80ATP(aq) + 80H_2O(l)$$

$$80ADP(aq) + 80HPO_4^{2-}(aq) + 160H^+(aq) + C_6H_{12}O_6(aq) + 6O_2(g)$$
$$\rightarrow 6CO_2(g) + 86H_2O(l) + 80ATP(aq) \qquad \Delta G° = -390 \text{ kJ}$$

62. Refer to Sections 17.4 and 17.6.

a. $\Delta G° = \Sigma \, \Delta G_f°_{products} - \Sigma \, \Delta G_f°_{reactants}$

$\Delta G° = [(1 \text{ mol.})(0.0 \text{ kJ/mol.}) + (1 \text{ mol.})(-157.2 \text{ kJ/mol.})] - [(1 \text{ mol.})(-237.2 \text{ kJ/mol.})]$

$\Delta G° = +80.0 \text{ kJ}$

b. $\Delta G° = -RT\ln K$

$80.0 \text{ kJ} = -(8.31 \text{ J/mol·K})(298 \text{ K})(0.001 \text{ kJ/J})(\ln K_w)$

$\ln K_w = -32.3$

$K_w = e^{-32.3} = 9.4 \times 10^{-15}$

64. *Refer to Section 17.6 and Example 17.9.*

a. $\Delta G° = -RT \ln K$
    $\Delta G° = -(8.31 \text{ J/mol·K})(0.001 \text{ kJ/J})(298 \text{ K}) \ln (1.0 \times 10^{-37})$
    $\Delta G° = 2.11 \times 10^2 \text{ kJ}$

b. $\Delta G° = \Sigma \Delta G_f°_{products} - \Sigma \Delta G_f°_{reactants}$
    $2.11 \times 10^2 \text{ kJ} = (2 \text{ mol.})(\Delta G°_f(Cl_2)) - [(1 \text{ mol.})(0.0 \text{ kJ/mol.})]$
    $\Delta G°_f(Cl_2) = 1.05 \times 10^2 \text{ kJ/mol.}$

66. *Refer to Section 17.5.*

$$K = \frac{(P_{O_2})(P_O)}{P_{O_3}}$$

$\Delta G° = -RT \ln K$
$-6.86 \times 10^4 \text{ J} = -(8.31 \text{ J/mol·K})(298 \text{ K}) \ln K$
$\ln K = 27.7$
$K = 1.1 \times 10^{12}$

$\Delta G° = -RT \ln K$
$-4.63 \times 10^4 \text{ J} = -(8.31 \text{ J/mol·K})(1200 \text{ K}) \ln K$
$\ln K = 4.64$
$K = 1.0 \times 10^2$

68. *Refer to Section 17.6.*

Calculate $\Delta H°$ and $\Delta S°$. Then calculate $\Delta G°$ and K_a at each of the temperatures.

$$HF(aq) \rightleftharpoons H^+(aq) + F^-(aq)$$

$\Delta H° = \Sigma \Delta H_f°_{products} - \Sigma \Delta H_f°_{reactants}$
$\Delta H° = [(1 \text{ mol.})(0.0 \text{ kJ/mol.}) + (1 \text{ mol.})(-332.6 \text{ kJ/mol.})] - [(1 \text{ mol.})(-320.1 \text{ kJ/mol.})]$
$\Delta H° = -12.5 \text{ kJ}$

$\Delta S° = \Sigma S°_{products} - \Sigma S°_{reactants}$
$\Delta S° = [(1 \text{ mol.})(0.0 \text{ kJ/mol·K}) + (1 \text{ mol.})(-0.0138 \text{ kJ/mol·K})] - [(1 \text{ mol.})(0.0887 \text{ kJ/mol·K.})]$
$\Delta S° = -0.1025 \text{ kJ/K}$

At 25°C:
   $\Delta G° = \Delta H° - T\Delta S°$
   $\Delta G° = -12.5 \text{ kJ} - (298 \text{ K})(-0.1025 \text{ kJ/K})$
   $\Delta G° = 18.0 \text{ kJ} = 1.80 \times 10^4 \text{ J}$

$\Delta G° = -RT \ln K$

$1.80 \times 10^4 \text{ J} = -(8.31 \text{ J/mol·K})(298 \text{ K}) \ln K$

$\ln K = -7.27$

$K = 7.0 \times 10^{-4}$

---

70. Refer to Section 17.6 and Chapter 13.

Calculate $[H^+]$ (and $[B^-]$) from the pH. Then calculate K_a, and from that, $\Delta G°$. Note that K_a is for the reverse of the desired reaction, therefore the sign of $\Delta G°$ must also be reversed.

$[B^-] = [H^+] = 10^{-pH} = 10^{-3.71} = 2.0 \times 10^{-4}$

	HB(aq)	$\rightleftharpoons$	H$^+$(aq)	+	B$^-$(aq)
$[\ ]_0$	0.13		0		0
$\Delta[\ ]$	-2.0×10^{-4}		$+2.0 \times 10^{-4}$		$+2.0 \times 10^{-4}$
$[\ ]_{eq}$	$0.31 - 2.0 \times 10^{-4}$		2.0×10^{-4}		2.0×10^{-4}

$$K_a = \frac{[H^+][B^-]}{[HB]} = \frac{(2.0 \times 10^{-4})(2.0 \times 10^{-4})}{0.13 - (2.0 \times 10^{-4})} = 3.1 \times 10^{-7}$$

$\Delta G° = -RT \ln K$

$\Delta G° = -(8.314 \text{ J/mol·K})(0.001 \text{ kJ/J})(298 \text{ K}) \ln(3.1 \times 10^{-7})$

$\Delta G° = +37.1 \text{ kJ}$ [for: HB(aq) $\rightleftharpoons$ H$^+$(aq) + B$^-$(aq)]

$\Delta G° = -37.1 \text{ kJ}$ [for: H$^+$(aq) + B$^-$(aq) $\rightleftharpoons$ HB(aq)]

---

72. Refer to Sections 17.4 and 17.6 and Chapter 11.

a. $\Delta G° = \Sigma \Delta G_f°{}_{products} - \Sigma \Delta G_f°{}_{reactants}$

    $\Delta G° = [(1 \text{ mol.})(-368.57 \text{ kJ/mol.}) + (3 \text{ mol.})(-237.2 \text{ kJ/mol.})]$

        $- [(2 \text{ mol.})(-50.7 \text{ kJ/mol.}) + (1 \text{ mol.})(-16.5 \text{ kJ/mol.}) + (^5/_2 \text{ mol.})(0.0 \text{ kJ/mol.})]$

    $\Delta G° = -962.3 \text{ kJ}$

    $\Delta G°$ is negative, this **reaction will proceed spontaneously** (at 25°C).

b. $\Delta G° = -RT \ln K$

    $-962.3 \text{ kJ} = -(8.314 \text{ J/mol·K})(0.001 \text{ kJ/J})(298 \text{ K}) \ln K$

    $\ln K = 388.4 \text{ kJ}$

    $K = 4.8 \times 10^{168}$

    The positive value for K indicates that at equilibrium, **products are favored over reactants**.

c.	The calculations in (a) and (b) above only tell us that the reaction is spontaneous and lies far to the right at equilibrium, but **tell us nothing about how rapidly the reaction proceeds**. Rates of reaction (kinetics) is covered in chapter 11.

74. *Refer to Sections 17.1 - 17.3.*

a.	**False**. An exothermic reaction is *usually* spontaneous, but not always. One must also consider entropy changes, which could render an exothermic reaction non-spontaneous.

b.	**False**. When $\Delta G°$ is positive, the reaction is not spontaneous under the standard conditions (25°C and 1 atm), but may be spontaneous under different conditions.

c.	**False**. $\Delta S°$ is positive for a reaction in which there is an increase in the moles of *gas*.

d.	**False**. If $\Delta H°$ and $\Delta S°$ are both negative, $\Delta G°$ will be negative at low temperatures and positive at high temperatures.

76. *Refer to Sections 17.2, 17.4, 17.5 and 17.6.*

a.	$\Delta H°$ and $\Delta G°$ become equal at **0 K**.
$\Delta G° = \Delta H° + T\Delta S°$, if $T = 0$ K, then
$\Delta G° = \Delta H°$

b.	$\Delta G°$ and ΔG are equal when $Q = 1$
$\Delta G = \Delta G° + RT \ln Q$, when $Q = 1$, then $\ln Q = 0$ and
$\Delta G = \Delta G°$

c.	$S°$ for steam is **greater** than $S°$ for water. Entropy is always higher for gases than for liquids.

78. *Refer to Section 17.2.*

a.	Gases have a higher entropy than solids and liquids. Therefore, if the moles of gas decrease in a reaction, the **entropy decreases** and $\Delta S°$ is negative.

b.	**Entropy changes due to temperature are small** and may be ignored. The exception is when the temperature change also involves a phase change.

c.	The entropy of solids is lower than the entropy of liquids because **solids have an ordered three dimensional structure**, which largely inhibit molecular motion, while liquids do not have an ordered three dimensional structure and have significant molecular motion.

80. Refer to Sections 17.1, 17.2 and 17.5.

The reaction is exothermic. This indicates that ΔH is negative.

$\Delta n_g < 0$. This indicates that ΔS is negative.

The system is at equilibrium, and $K = 1$ at 300°C. This indicates that $\Delta G° = 0$ at $T = 300$ K.

The three above points taken together indicate that $\Delta G°$ is negative below 300 K and positive above 300 K.

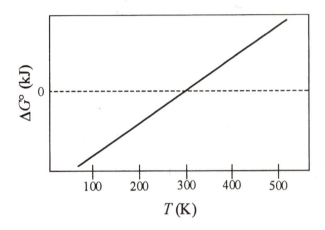

81. Refer to Section 17.7 and Chapter 9.

For the equation $C_2H_5OH(l) \rightleftharpoons C_2H_5OH(g)$, $\Delta G° = 0$ since this is a phase change (see problem 44 above).

$$\Delta G° = \Delta H° - T \Delta S° \quad \Rightarrow \quad \Delta H° = T \Delta S°$$

Calculate ΔS, ΔH_{vap} and the use the Claysius-Clapeyron equation in Chapter 9.

$\Delta S° = \Sigma \, S°_{products} - \Sigma \, S°_{reactants}$
$\Delta S° = [(1 \text{ mol.})(0.2827 \text{ kJ/mol.·K})] - [(1 \text{ mol.})(0.1607 \text{ kJ/mol.·K})]$
$\Delta S° = 0.1220 \text{ kJ/K}$

$\Delta H° = T \Delta S°$
$\Delta H° = (351.6 \text{ K})(0.1220 \text{ kJ/K})$
$\Delta H° = \Delta H_{vap} = 42.89 \text{ kJ}$

$$\ln\left(\frac{P_2}{P_1}\right) = \frac{+\Delta H_{vap}}{R}\left(\frac{1}{T_1} - \frac{1}{T_2}\right)$$

$$\ln\left(\frac{760 \text{ mm Hg}}{357 \text{ mm Hg}}\right) = \frac{42890 \text{ J/mol.}}{8.31 \text{ J/mol.} \cdot \text{K}}\left(\frac{1}{T_1} - \frac{1}{351.6 \text{ K}}\right)$$

$$0.756 = (5161 \text{ K})\left(\frac{1}{T_1} - 0.00285\right) \quad \Rightarrow \quad \frac{1}{T_1} = 0.00300 \text{ K}^{-1}$$

$T_1 = 333 \text{ K} = 60°\text{C}$

---

82. *Refer to Sections 17.4, 17.5 and 17.6.*

Calculate $\Delta S°$ and $\Delta H°$, then $\Delta G°$. Use $\Delta G°$ to calculate K. Finally, calculate the final pressures of the gases using the equilibrium expression.

$\Delta S° = \Sigma\, S°_{products} - \Sigma\, S°_{reactants}$
$\Delta S° = [(1 \text{ mol.})(0.1306 \text{ kJ/mol.} \cdot \text{K}) + (1 \text{ mol.})(0.2607 \text{ kJ/mol.} \cdot \text{K})]$
$\qquad - [(2 \text{ mol.})(0.2065 \text{ kJ/mol.} \cdot \text{K})]$
$\Delta S° = -0.0217 \text{ kJ/K}$

$\Delta H° = \Sigma\, \Delta H_f°_{products} - \Sigma\, \Delta H_f°_{reactants}$
$\Delta H° = [(1 \text{ mol.})(0.0 \text{ kJ/mol.}) + (1 \text{ mol.})(62.4 \text{ kJ/mol.})] - [(2 \text{ mol.})(26.5 \text{ kJ/mol.})]$
$\Delta H° = 9.4 \text{ kJ}$

$\Delta G° = \Delta H° - T \Delta S°$
$\Delta G° = 9.4 \text{ kJ} - (773 \text{ K})(-0.0217 \text{ kJ/K})$
$\Delta G° = 26.2 \text{ kJ}$

$\Delta G° = -RT \ln K$
$26.2 \text{ kJ} = -(8.31 \text{ J/mol.} \cdot \text{K})(0.001 \text{ kJ/J})(773 \text{ K}) \ln K$
$\ln K = -4.08$
$K = 0.0169$

Since $\Delta G°$ is positive, the reaction will proceed in the reverse direction, giving:
$\quad [H_2] = [I_2] = 0.200 - x \quad \text{and} \quad [HI] = 0.200 + 2x$

$$K = \frac{(P_{H_2})(P_{I_2})}{(P_{HI})^2} \quad \Rightarrow \quad 0.0169 = \frac{(0.200 - x)^2}{(0.200 + 2x)^2}$$

$$\sqrt{0.0169} = \frac{0.200 - x}{0.200 + 2x} \quad \Rightarrow \quad 0.0260 + 0.260x = 0.200 - x$$

$x = 0.138 \text{ atm}$

$[H_2] = [I_2] = 0.200 - x = 0.062 \text{ atm}$
$[HI] = 0.200 + 2x = 0.476 \text{ atm}$

a. Assume the phase change is at equilibrium.

$$\Delta H° = \Delta H_{fusion} = \frac{333\,J}{1\,g} \times \frac{18.02\,g}{1\,mol.} = 6.00 \times 10^3 \ J/mol. = 6.00\,kJ/mol.$$

b. $\Delta G° = -RT \ln K$. Since neither solids nor liquids appear in the expression for K (both are one), we get: $\Delta G° = -RT \ln 1 = 0$.

c. $\Delta G° = \Delta H° - T\,\Delta S°$
$0 = 6.00\,kJ - (273\,K)(\Delta S°)$
$\Delta S° = 0.0220\,kJ/K$

d. $\Delta G° = \Delta H° - T\,\Delta S°$
$\Delta G° = 6.00\,kJ - (253\,K)(0.0220\,kJ/K)$
$\Delta G° = 0.43\,kJ$ Thus, ice does not melt spontaneously at -20°C.

e. $\Delta G° = \Delta H° - T\,\Delta S°$
$\Delta G° = 6.00\,kJ - (293\,K)(0.0220\,kJ/K)$
$\Delta G° = -0.45\,kJ$ But ice does melt spontaneously at +20°C

a. Calculate $\Delta S°$ at 25°C using the Gibbs-Helmholtz equation, then calculate $\Delta G°$ at 37°C. Calculate the amount of energy from one gram sugar with 25% efficiency.

$\Delta G° = \Delta H° - T\,\Delta S°$
$-5790\,kJ = -5650\,kJ - (298\,K)(\Delta S°)$
$\Delta S° = 0.470\,kJ/K$

$\Delta G° = \Delta H° - T\,\Delta S°$
$\Delta G° = -5650\,kJ - (310\,K)(0.470\,kJ/K)$
$\Delta G° = -5796\,kJ$

$$\frac{-5796\,kJ}{mol.} \times \frac{1\,mol.}{342.0\,g} \times 1.00\,g \times 0.25 = -4.2\,kJ$$

Thus one gram of sugar provides 4.2 kJ of work.

b. $m = 120 \text{ lb} \times \dfrac{453.6 \text{ g}}{1 \text{ lb}} \times \dfrac{1 \text{ kg}}{1000 \text{ g}} = 54.4 \text{ kg}$

$w = (9.79 \times 10^{-3})(54.4 \text{ kg})(4158 \text{ m}) = 2210 \text{ kJ}$

$2210 \text{ kJ} \times \dfrac{1.00 \text{ g sugar}}{4.2 \text{ kJ}} = 530 \text{ g sugar}$

Thus, little more than one pound of sugar will provide enough energy for a 120 lb person to climb a 2.5 mile high mountain.

---

85. *Refer to Section 17.5.*

$CaH_2(s) \rightarrow Ca(s) + H_2(g)$

$\Delta H^\circ = -\Delta H_f^\circ = +186.2 \text{ kJ}$

$\Delta S^\circ = \Sigma\, S^\circ_{products} - \Sigma\, S^\circ_{reactants}$
$\Delta S^\circ = [(1 \text{ mol.})(0.0414 \text{ kJ/mol.·K}) + (1 \text{ mol.})(0.1306 \text{ kJ/mol.·K})]$
$\qquad - [(1 \text{ mol.})(0.0420 \text{ kJ/mol.·K})]$
$\Delta S^\circ = 0.130 \text{ kJ/K}$

$\Delta G^\circ = -RT \ln K = -RT \ln (1 \text{ atm } H_2) = 0 \text{ kJ}$
$\Delta G^\circ = \Delta H^\circ - T\,\Delta S^\circ$
$0 \text{ kJ} = 186.2 \text{ kJ} - (T)(0.130 \text{ kJ/K})$
$T = 1432 \text{ K} = 1159°C$

This is not a feasible source of H_2 fuel.

---

86. *Refer to Sections 17.4, 17.5 and 17.7.*

$Cu(s) + \frac{1}{2}O_2(g) \rightarrow CuO(s)$

$\Delta G^\circ = \Sigma\, \Delta G^\circ_{f\ products} - \Sigma\, \Delta G^\circ_{f\ reactants}$
$\Delta G^\circ = [(1 \text{ mol.})(-129.7 \text{ kJ/mol.})] - [(1 \text{ mol.})(0.0 \text{ kJ/mol.}) + (\frac{1}{2} \text{ mol.})(0.0 \text{ kJ/mol.})]$
$\Delta G^\circ = -129.7 \text{ kJ}$
ΔG° is negative, so the black CuO forms spontaneously at room temperature.

$2CuO(s) \rightarrow Cu_2O(s) + \frac{1}{2}O_2(g)$

$\Delta H^\circ = \Sigma\, \Delta H^\circ_{f\ products} - \Sigma\, \Delta H^\circ_{f\ reactants}$
$\Delta H^\circ = [(1 \text{ mol.})(-168.6 \text{ kJ/mol.}) + (\frac{1}{2} \text{ mol.})(0.0 \text{ kJ/mol.})] - [(2 \text{ mol.})(-157.3 \text{ kJ/mol.})]$
$\Delta H^\circ = 146.0 \text{ kJ}$

$\Delta S^\circ = \Sigma \ S^\circ_{products} - \Sigma \ S^\circ_{reactants}$

$\Delta S^\circ = [(1 \text{ mol.})(0.0931 \text{ kJ/mol.·K}) + (\frac{1}{2} \text{ mol.})(0.2050 \text{ kJ/mol.·K})]$
$\qquad - [(2 \text{ mol.})(0.0426 \text{ kJ/mol.·K})]$

$\Delta S^\circ = 0.1104 \text{ kJ/K}$

The reaction becomes spontaneous when $\Delta G^\circ < 0$, giving

$\Delta G^\circ = \Delta H^\circ - T\Delta S^\circ$

$0 = 146.0 \text{ kJ} - (T)(0.1104 \text{ kJ/K})$

$T = 1322 \text{ K} = 1049°C$ (Reaction becomes spontaneous above 1049°C)

Thus the black CuO converts to red Cu_2O when the temperature exceeds 1049°C.

$Cu_2O(s) \ \rightarrow \ 2Cu(s) \ + \frac{1}{2}O_2(g)$

$\Delta H^\circ = \Sigma \ \Delta H^\circ_{f \ products} - \Sigma \ \Delta H^\circ_{f \ reactants}$

$\Delta H^\circ = [(2 \text{ mol.})(0.0 \text{ kJ/mol.}) + (\frac{1}{2} \text{ mol.})(0.0 \text{ kJ/mol.})] - [(1 \text{ mol.})(-168.6 \text{ kJ/mol.})]$

$\Delta H^\circ = 168.6 \text{ kJ}$

$\Delta S^\circ = \Sigma \ S^\circ_{products} - \Sigma \ S^\circ_{reactants}$

$\Delta S^\circ = [(2 \text{ mol.})(0.0332 \text{ kJ/mol.}) + (\frac{1}{2} \text{ mol.})(0.2050 \text{ kJ/mol.})] - [(1 \text{ mol.})(0.0931 \text{ kJ/mol.})]$

$\Delta S^\circ = 0.0758 \text{ kJ}$

$\Delta G^\circ = \Delta H^\circ - T\Delta S^\circ$

The reaction becomes spontaneous when $\Delta G^\circ < 0$, giving

$0 = 168.6 \text{ kJ} - (T)(0.0758 \text{ kJ/K})$

$T = 2220 \text{ K} = 1950°C$ (Reaction becomes spontaneous above 1950°C)

Thus the red Cu_2O reverts to metallic Cu when the temperature exceeds 1950°C.

Chapter 18: Electrochemistry

---

2. Refer to Section 18.1 and Chapter 4.

Remember that the oxidation is always shown on the left side of the cell notation. For parts (a) and (c), note that the Pt serves as an inert electrode and is not part of the chemical equation.

a. $H_2(g) \rightarrow 2H^+(aq) + 2e^-$ oxidation half-reaction

    $Fe^{3+}(aq) + e^- \rightarrow Fe^{2+}(aq)$ reduction half-reaction

Multiply the reduction half-reaction by two, then add the two half-reactions together.

$2Fe^{3+}(aq) + 2e^- + H_2(g) \rightarrow 2Fe^{2+}(aq) + 2H^+(aq) + 2e^-$

$2Fe^{3+}(aq) + H_2(g) \rightarrow 2Fe^{2+}(aq) + 2H^+(aq)$

b. $Cd(s) \rightarrow Cd^{2+}(aq) + 2e^-$ oxidation half-reaction

    $Ni^{2+}(aq) + 2e^- \rightarrow Ni(s)$ reduction half-reaction

    $Cd(s) + Ni^{2+}(aq) + 2e^- \rightarrow Ni(s) + Cd^{2+}(aq) + 2e^-$

    $Cd(s) + Ni^{2+}(aq) \rightarrow Ni(s) + Cd^{2+}(aq)$

c. Note: The equation cannot be balanced as written. The cell notation should be:

        $Pt \mid Cl^- \mid Cl_2 \parallel MnO_4^-, H^+ \mid Mn^{2+} \mid Pt$

$2Cl^-(aq) \rightarrow Cl_2(g) + 2e^-$ oxidation half-reaction

$MnO_4^-(aq) + 5e^- \rightarrow Mn^{2+}(aq)$ reduction half-reaction

$MnO_4^-(aq) + 5e^- + 8H^+(aq) \rightarrow Mn^{2+}(aq)$

$MnO_4^-(aq) + 5e^- + 8H^+(aq) \rightarrow Mn^{2+}(aq) + 4H_2O(l)$

Balance electrons by multiplying the oxidation equation by 5, and the reduction equation by 2.

$5[2Cl^-(aq) \rightarrow Cl_2(g) + 2e^-]$

$2[MnO_4^-(aq) + 5e^- + 8H^+(aq) \rightarrow Mn^{2+}(aq) + 4H_2O(l)]$

$10Cl^-(aq) \rightarrow 5Cl_2(g) + 10e^-$

$2MnO_4^-(aq) + 10e^- + 16H^+(aq) \rightarrow 2Mn^{2+}(aq) + 8H_2O(l)$

$2MnO_4^-(aq) + 10e^- + 16H^+(aq) + 10Cl^-(aq) \rightarrow 5Cl_2(g) + 10e^- + 2Mn^{2+}(aq) + 8H_2O(l)$

$2MnO_4^-(aq) + 16H^+(aq) + 10Cl^-(aq) \rightarrow 5Cl_2(g) + 2Mn^{2+}(aq) + 8H_2O(l)$

a. Sn(*s*) is oxidized at the anode to $Sn^{2+}(aq)$,
   $Ag^+(aq)$ is reduced at the cathode to Ag(*s*).

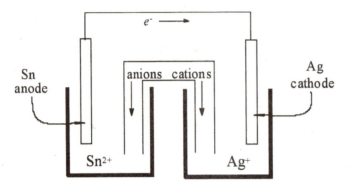

b. $H_2(g)$ is oxidized at the anode to $H^+(aq)$,
   $Hg_2^{2+}(Hg_2Cl_2)$ is reduced at the cathode to Hg(*l*).

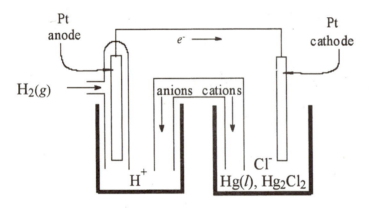

c. Pb(*s*) is oxidized at the anode to $Pb^{2+}(PbSO_4(s))$,
   $Pb^{4+}(PbO_2(s))$ is reduced at the cathode to $Pb^{2+}(PbSO_4(s))$.

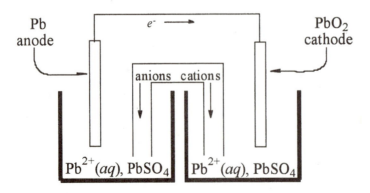

6. *Refer to Section 18.1 and Example 18.1.*

$$3[\ Mn^0 \rightarrow Mn^{2+} + 2e^-]$$
$$\underline{2[\ Cr^{3+} + 3e^- \rightarrow Cr^0\ ]}$$
$$2Cr^{3+} + 6e^- + 3Mn^0 \rightarrow 3Mn^{2+} + 6e^- + 2Cr^0$$
$$2Cr^{3+}(aq) + 3Mn(s) \rightarrow 3Mn^{2+}(aq) + 2Cr(s)$$

oxidation half-reaction: anode
reduction half-reaction: cathode

$$Mn \mid Mn^{2+} \parallel Cr^{3+} \mid Cr$$

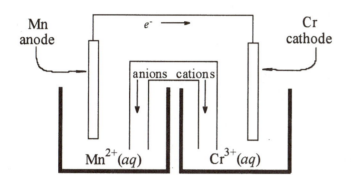

8. *Refer to Section 18.2 and Example 18.2.*

An oxidizing agent is reduced in the process of oxidizing another species. The more effective oxidizing agent is the one that is more easily reduced; thus the species with the higher reduction potential is the better oxidizing agent.

a. $Br_2(l) + 2e^- \rightarrow 2Br^-(aq)$ $\qquad E^\circ_{red} = 1.077\ V$

$H_2O_2(aq) + 2H^+(aq) + 2e^- \rightarrow H_2O(l)$ $\qquad E^\circ_{red} = 1.763\ V$

H_2O_2 is the better oxidizing agent.

b. $SO_4^{2-}(aq) + 4H^+(aq) + 2e^- \rightarrow SO_2(g) + H_2O(l)$ $\qquad E^\circ_{red} = 0.155\ V$

$MnO_4^-(aq) + 8H^+(aq) + 5e^- \rightarrow Mn^{2+}(aq) + H_2O(l)$ $\qquad E^\circ_{red} = 1.512\ V$

MnO_4^- is the better oxidizing agent.

c. $AgBr(s) + e^- \rightarrow Ag(s) + Br^-(aq)$ $\qquad E^\circ_{red} = 0.073\ V$

$Ag^+(aq) + e^- \rightarrow Ag(s)$ $\qquad E^\circ_{red} = 0.799\ V$

Ag^+ is the better oxidizing agent.

d. $O_2(g) + 4H^+(aq) + 4e^- \rightarrow 2H_2O(l)$ $\qquad E^\circ_{red} = 1.229\ V$

$O_2(g) + 2H_2O(l) + 4e^- \rightarrow 4OH^-(aq)$ $\qquad E^\circ_{red} = 0.401\ V$

$O_2(g)$ acidic is the better oxidizing agent.

10. *Refer to Section 18.2 and Example 18.2.*

	Br⁻	Zn	Co	PbSO₄	H₂S(acidic)
E_{ox}° (V)	-1.077	0.762	0.282	-1.687	-0.144

The species with the highest E_{ox}° is the strongest reducing agent, thus:

$PbSO_4 < Br^- < H_2S(acidic) < Co < Zn$

---

12. *Refer to Section 18.2.*

Cu^+ can be both an oxidizing and a reducing agent (E_{red}° = 0.518 V, E_{ox}° = -0.161 V).

Zn is a reducing agent (E_{ox}° = 0.762 V).

Ni^{2+} is a oxidizing agent (E_{red}° = -0.236 V).

Fe^{2+} can be both an oxidizing and a reducing agent (E_{red}° = -0.409 V, E_{ox}° = -0.769 V).

H^+ is a oxidizing agent (E_{red}° = 0.00 V).

The species with the highest E_{red}° is the strongest oxidizing agent, thus:

$Fe^{2+} < Ni^{2+} < H^+ < Cu^+$

The species with the highest E_{ox}° is the stronger reducing agent, thus:

$Fe^{2+} < Cu^+ < Zn$

Cu^+ and Fe^{2+} can be both oxidizing and reducing agents.

---

14. *Refer to Section 18.2.*

Write the half-reactions for the conversions listed, with the corresponding E° value. A species that will effect one transformation without the other will have an E° value between that of the two half-reactions.

a. $Pb^{2+}(aq) + 2e^- \rightarrow Pb(s)$ E_{red}° = -0.127 V

    $Tl^+(aq) + e^- \rightarrow Tl(s)$ E_{red}° = -0.336 V

    reducing agent: Sn(s) (anything between Sn(s) and Co(s))

        $Sn^{2+}(aq) + 2e^- \rightarrow Sn(s)$ E_{red}° = -0.141 V

b. $Fe^{2+}(aq) + 2e^- \rightarrow Fe(s)$ E_{red}° = -0.409 V

    $Co^{2+}(aq) + 2e^- \rightarrow Co(s)$ E_{red}° = -0.282 V

    oxidizing agent: $Tl^+(aq)$ (anything between $Tl^+(aq)$ and $Cr^{3+}(aq)$)

        $Tl^+(aq) + e^- \rightarrow Tl(s)$ E_{red}° = -0.336 V

c. $Au^{3+}(aq) + 3e^- \rightarrow Au(s)$ $\qquad\qquad\qquad\qquad\qquad E^\circ_{red} = 1.498$ V

$AuCl_4^-(aq) + 3e^- \rightarrow Au(s) + 4Cl^-(aq)$ $\qquad\qquad E^\circ_{red} = 1.001$ V

reducing agent: $Cl_2(g)$ (anything between $Cl_2(g)$ and $Br^-(aq)$)

$\qquad ClO_3^-(aq) + 6H^+(aq) + 5e^- \rightarrow \frac{1}{2}Cl_2(g) \quad + 3H_2O(l)$ $\qquad E^\circ_{red} = 1.458$ V

---

16. Refer to Section 18.2 and Example 18.3.

Write the half-reactions and the associated redox potentials. When combining the half-reactions, add the potentials to calculate E°. Bear in mind that **multiplying a half-reaction by a factor does not change that E° value.**

a. $Pb(s) \rightarrow Pb^{2+}(aq) + 2e^-$ $\qquad\qquad\qquad\qquad\qquad E^\circ_{ox} = 0.127$ V

$\underline{2[\, Ag^+(aq) + e^- \rightarrow Ag(s)\,]}$ $\qquad\qquad\qquad\qquad E^\circ_{red} = 0.799$ V

$Pb(s) + 2Ag^+(aq) + 2e^- \rightarrow 2Ag(s) + Pb^{2+}(aq) + 2e^-$

$Pb(s) + 2Ag^+(aq) \rightarrow 2Ag(s) + Pb^{2+}(aq)$ $\qquad\qquad E^\circ = 0.926$ V

b. $4[\, Fe^{2+}(aq) \rightarrow Fe^{3+}(aq) + e^-\,]$ $\qquad\qquad\qquad\quad E^\circ_{ox} = -0.769$ V

$\underline{O_2(g) + 4H^+(aq) + 4e^- \rightarrow 2H_2O(l)}$ $\qquad\qquad\quad E^\circ_{red} = 1.229$ V

$4Fe^{2+}(aq) + O_2(g) + 4H^+(aq) + 4e^- \rightarrow 2H_2O(l) + 4Fe^{3+}(aq) + 4e^-$

$4Fe^{2+}(aq) + O_2(g) + 4H^+(aq) \rightarrow 2H_2O(l) + 4Fe^{3+}(aq)$ $\qquad E^\circ = 0.460$ V

c. Reverse the Zn-Zn^{2+} reaction and change the sign of E° to obtain a positive E°(cell).

$Zn(s) \rightarrow Zn^{2+}(aq) + 2e^-$ $\qquad\qquad\qquad\qquad\qquad E^\circ_{ox} = 0.762$ V

$\underline{Cd^{2+}(aq) + 2e^- \rightarrow Cd(s)}$ $\qquad\qquad\qquad\qquad\quad E^\circ_{red} = -0.402$ V

$Zn(s) + Cd^{2+}(aq) + 2e^- \rightarrow Cd(s) + Zn^{2+}(aq) + 2e^-$

$Zn(s) + Cd^{2+}(aq) \rightarrow Cd(s) + Zn^{2+}(aq)$ $\qquad\qquad E^\circ = 0.360$ V

---

18. Refer to Section 18.2 and Example 18.4.

Write the half-reactions and the associated redox potentials. When combining the half-reactions, add the potentials to calculate E°. Bear in mind that **multiplying a half-reaction by a factor does not change that E° value.**

a. $2[Cr^{2+}(aq) \rightarrow Cr^{3+}(aq) + e^-]$ $\qquad\qquad\qquad\quad E^\circ_{ox} = 0.408$ V

$\underline{Sn^{4+}(aq) + 2e^- \rightarrow Sn^{2+}(aq)}$ $\qquad\qquad\qquad\quad E^\circ_{red} = 0.154$ V

$Sn^{4+}(aq) + 2e^- + 2Cr^{2+}(aq) \rightarrow 2Cr^{3+}(aq) + 2e^- + Sn^{2+}(aq)$

$Sn^{4+}(aq) + 2Cr^{2+}(aq) \rightarrow 2Cr^{3+}(aq) + Sn^{2+}(aq)$ $\qquad E^\circ = 0.562$ V

b. $Mn^{2+}(aq) + 2H_2O(l) \rightarrow MnO_2(s) + 4H^+(aq) + 2e^-$ $E^\circ_{ox} = -1.229$ V

$H_2O_2(aq) + 2H^+(aq) + 2e^- \rightarrow 2H_2O(l)$ $E^\circ_{red} = 1.763$ V

$Mn^{2+}(aq) + 2H_2O(l) + H_2O_2(aq) + 2H^+(aq) + 2e^-$
$\rightarrow 2H_2O(l) + MnO_2(s) + 4H^+(aq) + 2e^-$

$Mn^{2+}(aq) + H_2O_2(aq) \rightarrow MnO_2(s) + 2H^+(aq)$ $E^\circ = 0.534$ V

c. $2[Fe(s) + 2OH^-(aq) \rightarrow Fe(OH)_2(s) + 2e^-]$ $E^\circ_{ox} = 0.891$ V

$O_2(g) + 2H_2O(l) + 4e^- \rightarrow 4OH^-(aq)$ $E^\circ_{red} = 0.401$ V

$2Fe(s) + 4OH^-(aq) + O_2(g) + 2H_2O(l) + 4e^- \rightarrow 4OH^-(aq) + 2Fe(OH)_2(s) + 4e^-$

$2Fe(s) + O_2(g) + 2H_2O(l) \rightarrow 2Fe(OH)_2(s)$ $E^\circ = 1.292$ V

---

20. Refer to Section 18.2.

Write the half-reactions and the associated redox potentials. When combining the half-reactions, add the potentials to calculate E°. Bear in mind that **multiplying a half-reaction by a factor does not change that E° value.**

a. $Al(s) \rightarrow Al^{3+}(aq) + 3e^-$ $E^\circ_{ox} = 1.68$ V

$NO_3^-(aq) + 4H^+(aq) + 3e^- \rightarrow NO(g) + 2H_2O(l)$ $E^\circ_{red} = 0.964$ V

$Al(s) + NO_3^-(aq) + 4H^+(aq) + 3e^- \rightarrow NO(g) + 2H_2O(l) + Al^{3+}(aq) + 3e^-$

$Al(s) + NO_3^-(aq) + 4H^+(aq) \rightarrow NO(g) + 2H_2O(l) + Al^{3+}(aq)$ $E^\circ = 2.64$ V

b. $4[Cr^{2+}(aq) \rightarrow Cr^{3+}(aq) + e^-]$ $E^\circ_{ox} = 0.408$ V

$O_2(g) + 4H^+(aq) + 4e^- \rightarrow 2H_2O(l)$ $E^\circ_{red} = 1.229$ V

$4Cr^{2+}(aq) + O_2(g) + 4H^+(aq) + 4e^- \rightarrow 2H_2O(l) + 4Cr^{3+}(aq) + 4e^-$

$4Cr^{2+}(aq) + O_2(g) + 4H^+(aq) \rightarrow 2H_2O(l) + 4Cr^{3+}(aq)$ $E^\circ = 1.637$ V

c. $Cu(aq) \rightarrow Cu^{2+}(aq) + 2e^-$ $E^\circ_{ox} = -0.339$ V

$I_2(s) + 2e^- \rightarrow 2I^-(aq)$ $E^\circ_{red} = 0.534$ V

$Cu(aq) + I_2(s) + 2e^- \rightarrow 2I^-(aq) + Cu^{2+}(aq) + 2e^-$

$Cu(aq) + I_2(s) \rightarrow 2I^-(aq) + Cu^{2+}(aq)$ $E^\circ = 0.195$ V

a. Calculate $E°$ for the reaction of Ag^+ with H_2. Then set $E°_{red}$ (Ag^+) to zero and use the calculated $E°$ to determine $E°_{ox}$ (H_2) and $E°_{red}$ (H^+).

$2[\ Ag^+(aq) + e^- \rightarrow Ag(s)\ ]$	$E°_{red} = 0.799$ V
$H_2(g) \rightarrow 2H^+(aq) + 2e^-$	$E°_{ox} = 0.0$ V

$2Ag^+(aq) + 2e^- + H_2(g) \rightarrow 2H^+(aq) + 2e^- + 2Ag(s)$

$2Ag^+(aq) + H_2(g) \rightarrow 2H^+(aq) + 2Ag(s)$	$E° = 0.799$ V

$2Ag^+(aq) + H_2(g) \rightarrow 2H^+(aq) + 2Ag(s)$	$E° = 0.799$ V
$2[\ Ag^+(aq) + e^- \rightarrow Ag(s)\ ]$	$E°_{red} = 0.0$ V

$H_2(g) \rightarrow 2H^+(aq) + 2e^-$	$E°_{ox} = 0.799$ V

$2H^+(aq) + 2e^- \rightarrow H_2(g)$	$E°_{red} = -0.799$ V

b. Calculate $E°$ for the reaction of Ag^+ with Ca. Then set $E°_{red}$ (Ag^+) to zero and use the calculated $E°$ to determine $E°_{ox}$ (Ca).

$2[\ Ag^+(aq) + e^- \rightarrow Ag(s)\ ]$	$E°_{red} = 0.799$ V
$Ca(s) \rightarrow Ca^{2+}(aq) + 2e^-$	$E°_{ox} = 2.869$ V

$2Ag^+(aq) + 2e^- + Ca(s) \rightarrow Ca^{2+}(aq) + 2e^- + 2Ag(s)$

$2Ag^+(aq) + Ca(s) \rightarrow Ca^{2+}(aq) + 2Ag(s)$	$E° = 3.668$ V

$2Ag^+(aq) + Ca(s) \rightarrow Ca^{2+}(aq) + 2Ag(s)$	$E° = 3.668$ V
$2[\ Ag^+(aq) + e^- \rightarrow Ag(s)\ ]$	$E°_{red} = 0.0$ V

$Ca(s) \rightarrow Ca^{2+}(aq) + 2e^-$	$E°_{ox} = 3.668$ V

c. As is seen in parts a and b above, all the reduction potentials have been decreased by 0.799 V (and oxidation potentials have been increased by an equal amount).

$Cu(aq) \rightarrow Cu^{2+}(aq) + 2e^-$	$E°_{ox} = -0.339$ V + 0.799 V = 0.460 V
$I_2(s) + 2e^- \rightarrow 2I^-(aq)$	$E°_{red} = 0.534$ V - 0.799 V = -0.265

$Cu(aq) + I_2(s) + 2e^- \rightarrow 2I^-(aq) + Cu^{2+}(aq) + 2e^-$

$Cu(aq) + I_2(s) \rightarrow 2I^-(aq) + Cu^{2+}(aq)$	$E° = 0.195$ V

$E°$ does not change. This makes sense logically since any change in $E°_{red}$ would be offset by an equal and opposite change in $E°_{ox}$.

A reaction is spontaneous when $\Delta G°$ is negative. Since $\Delta G° = -nFE°$, and n and F are positive constants, $\Delta G°$ is negative (and the reaction spontaneous) when $E°$ is positive.

a. $Zn(s) \rightarrow Zn^{2+}(aq) + 2e^{-}$ $E_{ox}° = 0.762$ V

$\underline{2[\ Fe^{3+}(aq) + e^{-} \rightarrow Fe^{2+}(aq)\ ]}$ $E_{red}° = 0.799$ V

$Zn(s) + 2Fe^{3+}(aq) + 2e^{-} \rightarrow 2Fe^{2+}(aq) + Zn^{2+}(aq) + 2e^{-}$

$Zn(s) + 2Fe^{3+}(aq) \rightarrow 2Fe^{2+}(aq) + Zn^{2+}(aq)$ $E° = 1.561$ V

$E°$ is positive, so the reaction is **spontaneous**.

b. $Cu(s) \rightarrow Cu^{2+}(aq) + 2e^{-}$ $E_{ox}° = -0.339$ V

$\underline{2H^{+}(aq) + 2e^{-} \rightarrow H_2(g)}$ $E_{red}° = 0.0$ V

$Cu(s) + 2H^{+}(aq) + 2e^{-} \rightarrow Cu^{2+}(aq) + H_2(g) + 2e^{-}$

$Cu(s) + 2H^{+}(aq) \rightarrow Cu^{2+}(aq) + H_2(g)$ $E° = -0.339$ V

$E°$ is negative, so the reaction is **not** spontaneous.

c. $2Br^{-}(aq) \rightarrow Br_2(l) + 2e^{-}$ $E_{ox}° = -1.077$ V

$\underline{I_2(s) + 2e^{-} \rightarrow 2I^{-}(aq)}$ $E_{red}° = 0.534$ V

$2Br^{-}(aq) + 2e^{-} + I_2(s) \rightarrow 2I^{-}(aq) + 2e^{-} + Br_2(l)$

$2Br^{-}(aq) + I_2(s) \rightarrow 2I^{-}(aq) + Br_2(l)$ $E° = -0.543$ V

$E°$ is negative, so the reaction is **not** spontaneous.

Write the half-reactions and the associated redox potentials. When combining the half-reactions, add the potentials to calculate $E°$. Bear in mind that multiplying a half-reaction by a factor does not change that $E°$ value. A reaction is spontaneous when $E°$ is positive (see problem 24 above).

a.   See problem 2(c) for step-by-step solution to balancing this reaction.

$5[\ 2Cl^{-}(aq) \rightarrow Cl_2(g) + 2e^{-}\ ]$ $E_{ox}° = -1.360$ V

$\underline{2[\ MnO_4^{-}(aq) + 5e^{-} + 8H^{+}(aq) \rightarrow Mn^{2+}(aq) + 4H_2O(l)\ ]}$ $E_{red}° = 1.512$ V

$2MnO_4^{-}(aq) + 16H^{+}(aq) + 10Cl^{-}(aq) \rightarrow 5Cl_2(g) + 2Mn^{2+}(ag) + 8H_2O(l)$ $E° = 0.152$ V

Since $E°$ is positive, this reaction is spontaneous and **MnO_4^{-} will oxidize Cl^{-}**.

b. $3[\ 2Cl^-(aq)\ \rightarrow\ Cl_2(g) + 2e^-\ ]$ $\qquad\qquad\qquad\qquad\qquad\qquad\qquad\qquad\quad E^\circ_{ox} = -1.360\ V$

$\underline{Cr_2O_7^{2-}(aq) + 14H^+(aq) + 6e^- \rightarrow 2Cr^{3+}(aq) + 7H_2O(l) \qquad\qquad E^\circ_{red} = 1.33\ V}$

$Cr_2O_7^{2-}(aq) + 14H^+(aq) + 6e^- + 6Cl^-(aq) \rightarrow 3Cl_2(g) + 6e^- + 2Cr^{3+}(aq) + 7H_2O(l)$

$Cr_2O_7^{2-}(aq) + 14H^+(aq) + 6Cl^-(aq) \rightarrow 3Cl_2(g) + 2Cr^{3+}(aq) + 7H_2O(l) \quad E^\circ = -0.03\ V$

Since E° is negative, this reaction is spontaneous and $Cr_2O_7^{2-}$ **will not oxidize** Cl^-.

c. $3[\ 2Cl^-(aq)\ \rightarrow\ Cl_2(g) + 2e^-\ ]$ $\qquad\qquad\qquad\qquad\qquad\qquad\qquad\qquad\quad E^\circ_{ox} = -1.360\ V$

$\underline{2[\ NO_3^-(aq) + 4H^+(aq) + 3e^- \rightarrow NO(g) + 2H_2O(l)\ ] \qquad\qquad E^\circ_{red} = 0.964\ V}$

$2NO_3^-(aq) + 8H^+(aq) + 6e^- + 6Cl^-(aq) \rightarrow 3Cl_2(g) + 6e^- + 2NO(g) + 4H_2O(l)$

$2NO_3^-(aq) + 8H^+(aq) + 6Cl^-(aq) \rightarrow 3Cl_2(g) + 2NO(g) + 4H_2O(l) \qquad E^\circ = -0.396\ V$

Since E° is negative, this reaction is spontaneous and $NO_3^-(aq)$ **will not oxidize** Cl^-.

---

28. Refer to Sections 18.2 and 18.3 and Example 18.4.

Consider the possible half-reactions and associated potentials. If the potentials of a pair of redox half-reactions give a positive E°, then a reaction will occur.

a. $I_2(s) + KBr(aq) \rightarrow\ ?$

$I_2(g) + 2e^- \rightarrow 2I^-(aq)$ $\qquad\qquad\qquad\qquad\qquad\qquad E^\circ_{red} = 0.534\ V$

$K^+(aq) + e^- \rightarrow K(s)$ $\qquad\qquad\qquad\qquad\qquad\qquad E^\circ_{red} = -2.936\ V$

$2Br^-(aq) \rightarrow Br_2(l) + 2e^-$ $\qquad\qquad\qquad\qquad\qquad E^\circ_{ox} = -1.077\ V$

Adding E°_{ox} to either the E°_{red} gives a negative E°, thus no reaction will occur.

b. $Br_2(l) + NaCl(aq) \rightarrow\ ?$

$Br_2(l) + 2e^- \rightarrow 2Br^-(aq)$ $\qquad\qquad\qquad\qquad\qquad\quad E^\circ_{red} = 1.077\ V$

$Na^+(aq) + e^- \rightarrow Na(s)$ $\qquad\qquad\qquad\qquad\qquad\qquad E^\circ_{red} = -2.714\ V$

$2Cl^-(aq) \rightarrow Cl_2(g) + 2e^-$ $\qquad\qquad\qquad\qquad\qquad E^\circ_{ox} = -1.360\ V$

Adding E°_{ox} to either the E°_{red} gives a negative E°, thus no reaction will occur.

c. $Cr(s) + NiCl_2(aq) \rightarrow\ ?$

$Cr(s) \rightarrow Cr^{3+}(aq) + 3e^-$ $\qquad\qquad\qquad\qquad\qquad\quad E^\circ_{ox} = 0.744\ V$

$Ni^{2+}(aq) + 2e^- \rightarrow Ni(s)$ $\qquad\qquad\qquad\qquad\qquad\quad E^\circ_{red} = -0.236\ V$

$2Cl^-(aq) \rightarrow Cl_2(g) + 2e^-$ $\qquad\qquad\qquad\qquad\qquad E^\circ_{ox} = -1.360\ V$

Adding E°_{ox} (Cr) to E°_{red} (Ni^{2+}) gives a positive E°, thus a reaction will occur.

$$2[\ Cr(s) \rightarrow Cr^{3+}(aq) + 3e^-]$$
$$3[\ Ni^{2+}(aq) + 2e^- \rightarrow Ni(s)]$$
$$2Cr(s) + 3Ni^{2+}(aq) + 6e^- \rightarrow 3Ni(s) + 2Cr^{3+}(aq) + 6e^-$$
$$2Cr(s) + 3Ni^{2+}(aq) \rightarrow 3Ni(s) + 2Cr^{3+}(aq) \qquad E° = 0.508\ V$$

---

30. Refer to Sections 18.2 and 18.3 and Example 18.4.

If HCl is oxidizing one of the listed species, then the HCl itself must be reduced. Since the Cl⁻ is already in a reduced state, the likely species to be reduced is H^+.

a. $2H^+(aq) + 2e^- \rightarrow 2H_2(g)$ $E^°_{red} = 0.0\ V$

     $Au(s) \rightarrow Au^{3+}(aq) + 3e^-$ $E^°_{ox} = -1.498\ V$

     $Au(s) + 4Cl^- \rightarrow AuCl_4^-(aq) + 3e^-$ $E^°_{ox} = -1.001\ V$

     Adding $E^°_{red}$ to either $E^°_{ox}$ gives a negative $E°$, thus HCl will **not** oxidize Au.

b. $2H^+(aq) + 2e^- \rightarrow 2H_2(g)$ $E^°_{red} = 0.0\ V$

     $Mg(s) \rightarrow Mg^{2+}(aq) + 2e^-$ $E^°_{ox} = 2.357\ V$

     Adding $E^°_{red}$ to $E^°_{ox}$ gives a positive $E°$, thus HCl **will** oxidize Mg.

c. $2H^+(aq) + 2e^- \rightarrow 2H_2(g)$ $E^°_{red} = 0.0\ V$

     $Cu(s) \rightarrow Cu^{2+}(aq) + 2e^-$ $E^°_{ox} = -0.339\ V$

     $Cu(s) \rightarrow Cu^+(aq) + e^-$ $E^°_{ox} = -0.518\ V$

     Adding $E^°_{red}$ to either $E^°_{ox}$ gives a negative $E°$, thus HCl will **not** oxidize Cu.

d. $2H^+(aq) + 2e^- \rightarrow 2H_2(g)$ $E^°_{red} = 0.0\ V$

     $2F^-(aq) \rightarrow F_2(g) + 2e^-$ $E^°_{ox} = -2.889\ V$

     Adding $E^°_{red}$ to either $E^°_{ox}$ gives a negative $E°$, thus HCl will **not** oxidize F⁻.

---

32. Refer to Sections 18.2 and 18.3.

Consider the possible half-reactions and associated potentials. If the potentials of a pair of redox half-reactions give a positive $E°$, then a reaction will occur.

a. $Fe^{2+}(aq) + 2e^- \rightarrow Fe(s)$ $E^°_{red} = -0.409\ V$

     $Co^{2+}(aq) + 2e^- \rightarrow Co(s)$ $E^°_{red} = -0.282\ V$

     $Fe^{2+}(aq) \rightarrow Fe^{3+}(aq) + e^-$ $E^°_{ox} = -0.769\ V$

$$Co(s) \rightarrow Co^{2+}(aq) + 2e^- \qquad\qquad E^\circ_{ox} = 0.282 \text{ V}$$
$$Co^{2+}(aq) \rightarrow Co^{3+}(aq) + e^- \qquad\qquad E^\circ_{ox} = -1.953 \text{ V}$$

Adding any E°_{ox} to any E°_{red} gives a negative E°, thus no reaction will occur.

b.
$$Fe^{2+}(aq) + 2e^- \rightarrow Fe(s) \qquad\qquad E^\circ_{red} = -0.409 \text{ V}$$
$$Fe^{3+}(aq) + e^- \rightarrow Fe^{2+}(aq) \qquad\qquad E^\circ_{red} = 0.769 \text{ V}$$
$$Ag^+(aq) + e^- \rightarrow Ag(s) \qquad\qquad E^\circ_{red} = 0.799 \text{ V}$$
$$Fe^{2+}(aq) \rightarrow Fe^{3+}(aq) + e^- \qquad\qquad E^\circ_{ox} = -0.769 \text{ V}$$

The only $E^\circ_{ox} + E^\circ_{red}$ that gives a positive E° is:

$$Fe^{2+}(aq) \rightarrow Fe^{3+}(aq) + e^-$$
$$\underline{Ag^+(aq) + e^- \rightarrow Ag(s)}$$
$$Ag^+(aq) + Fe^{2+}(aq) \rightarrow Fe^{3+}(aq) + Ag(s) \qquad\qquad E^\circ = 0.030 \text{ V}$$

c.
$$2Hg(l) \rightarrow Hg_2^{2+}(aq) + 2e^- \qquad\qquad E^\circ_{ox} = -0.796 \text{ V}$$
$$ClO_3^-(aq) + 6H^+(aq) + 5e^- \rightarrow \tfrac{1}{2}Cl_2(g) + 3H_2O(l) \qquad\qquad E^\circ_{red} = 1.458 \text{ V}$$
$$2H^+(aq) + 2e^- \rightarrow H_2(g) \qquad\qquad E^\circ_{red} = 0.0 \text{ V}$$

The only $E^\circ_{ox} + E^\circ_{red}$ that gives a positive E° is:

$$5[\ 2Hg(l) \rightarrow Hg_2^{2+}(aq) + 2e^-\ ]$$
$$\underline{2[\ ClO_3^-(aq) + 6H^+(aq) + 5e^- \rightarrow \tfrac{1}{2}Cl_2(g) + 3H_2O(l)\ ]}$$
$$10Hg(l) + 2ClO_3^-(aq) + 12H^+(aq) \rightarrow Cl_2(g) + 6H_2O(l) + 5Hg_2^{2+}(aq) \quad E^\circ = 0.662 \text{ V}$$

---

34. Refer to Section 18.3.

	ΔG°	E°	K
a.	19 kJ	**-0.098 V**	**5.0×10^{-4}**
b.	**-6.8 kJ**	0.035 V	15
c.	**5.8 kJ**	**-0.030 V**	0.095

a. $\Delta G^\circ = -nFE^\circ$

$19000 \text{ J} = -(2 \text{ mol.})(9.648 \times 10^4 \text{ J/mol.}\cdot\text{V})(E^\circ)$

$\Delta E^\circ = -0.098 \text{ V}$

$$E^\circ = \frac{0.0257 \text{ V}}{n} \cdot \ln K \quad\Rightarrow\quad -0.098 \text{ V} = \frac{0.0257 \text{ V}}{2} \cdot \ln K$$

$\ln K = -7.6$

$K = e^{-7.6} = 5.0 \times 10^{-4}$

b. $\Delta G^\circ = -nFE^\circ$
$\Delta G^\circ = -(2 \text{ mol.})(9.648 \times 10^4 \text{ J/mol.} \cdot \text{V})(0.035 \text{ V})$
$\Delta G^\circ = -6.8 \times 10^3 \text{ J} = -6.8 \text{ kJ}$

$$E^\circ = \frac{0.0257 \text{ V}}{n} \cdot \ln K \quad \Rightarrow \quad 0.035 \text{ V} = \frac{0.0257 \text{ V}}{2} \cdot \ln K$$

$\ln K = 2.7$
$K = e^{2.7} = 15$

c. $$E^\circ = \frac{0.0257 \text{ V}}{n} \cdot \ln K = \frac{0.0257 \text{ V}}{2} \cdot \ln (0.095)$$

$E^\circ = -0.030 \text{ V}$

$\Delta G^\circ = -nFE^\circ$
$\Delta G^\circ = -(2 \text{ mol.})(9.648 \times 10^4 \text{ J/mol.} \cdot \text{V})(-0.030 \text{ V})$
$\Delta G^\circ = 5.8 \times 10^3 \text{ J} = 5.8 \text{ kJ}$

---

36. Refer to Section 18.3 and Example 18.5.

---

a. $\Delta G^\circ = -nFE^\circ$
$\Delta G^\circ = -(1 \text{ mol.})(9.648 \times 10^4 \text{ J/mol.} \cdot \text{V})(1.20 \text{ V})$
$\Delta G^\circ = -1.16 \times 10^5 \text{ J} = -1.16 \times 10^2 \text{ kJ}$

b. $\Delta G^\circ = -nFE^\circ$
$\Delta G^\circ = -(2 \text{ mol.})(9.648 \times 10^4 \text{ J/mol.} \cdot \text{V})(1.20 \text{ V})$
$\Delta G^\circ = -2.32 \times 10^5 \text{ J} = -2.32 \times 10^2 \text{ kJ}$

c. $\Delta G^\circ = -nFE^\circ$
$\Delta G^\circ = -(3 \text{ mol.})(9.648 \times 10^4 \text{ J/mol.} \cdot \text{V})(1.20 \text{ V})$
$\Delta G^\circ = -3.47 \times 10^5 \text{ J} = -3.47 \times 10^2 \text{ kJ}$

The number of electrons exchanged has no effect on the spontaneity of the reaction since changes in the number of electrons does not change the sign of ΔG°.

There are two ways to approach this problem. One can calculate $\Delta G°$ (from $\Delta G_f°$ or from $\Delta H°$ and $\Delta S°$) and from that, $E°$ and K, or one can calculate $E°$ first, and $\Delta G°$ from that value. Given the lack of a $\Delta G_f°$ value for $Fe(OH)_2$, however, the first method is not available for this problem.

Method 1

$\Delta G° = \Sigma \Delta G_{f\ products}° - \Sigma \Delta G_{f\ reactants}°$

Method 2

$2Fe(OH)_3(s) + 2e^- \rightarrow 2Fe(OH)_2(s) + 2OH^-(aq)$	$E_{red}° = -0.547$ V
$2OH^-(aq) + NO_2^-(aq) \rightarrow NO_3^-(aq) + H_2O(l) + 2e^-$	$E_{ox}° = -0.004$ V
$2Fe(OH)_3(s) + NO_2^-(aq) \rightarrow NO_3^-(aq) + H_2O(l) + 2Fe(OH)_2(s)$	$E° = -0.551$ V

$\Delta G° = -nFE°$
$\Delta G° = -(2\text{ mol.})(9.648 \times 10^4\text{ J/mol.·V})(-0.551\text{ V})$
$\Delta G° = 1.06 \times 10^5\text{ J} = 1.06 \times 10^2\text{ kJ}$

$E° = \dfrac{0.0257\text{ V}}{n}\ln K \Rightarrow -0.551\text{ V} = \dfrac{0.0257\text{ V}}{2}\ln K$

$\ln K = -42.9$
$K = e^{-42.9} = 2.34 \times 10^{-19}$

a. $E° = 0.926$ V, $n = 2e^-$
   $\Delta G° = -nFE°$
   $\Delta G° = -(2\text{ mol.})(9.648 \times 10^4\text{ J/mol.·V})(0.926\text{ V})$
   $\Delta G° = -1.79 \times 10^5\text{ J} = -1.79 \times 10^2\text{ kJ}$

b. $E° = 0.460$ V, $n = 4e^-$
   $\Delta G° = -nFE°$
   $\Delta G° = -(4\text{ mol.})(9.648 \times 10^4\text{ J/mol.·V})(0.460\text{ V})$
   $\Delta G° = -1.78 \times 10^5\text{ J} = -1.78 \times 10^2\text{ kJ}$

c. $E° = 0.360$ V, $n = 2e^-$
   $\Delta G° = -nFE°$
   $\Delta G° = -(2\text{ mol.})(9.648 \times 10^4\text{ J/mol.·V})(0.360\text{ V})$
   $\Delta G° = -6.95 \times 10^4\text{ J} = -69.5\text{ kJ}$

42. Refer to Section 18.3, Example 18.6 and Problem 18 above.

$$Sn^{4+}(aq) + 2Cr^{2+}(aq) \rightarrow 2Cr^{3+}(aq) + Sn^{2+}(aq) \qquad n = 2 \quad E° = 0.562 \text{ V}$$
$$Mn^{2+}(aq) + H_2O_2(aq) \rightarrow MnO_2(s) + 2H^+(aq) \qquad n = 2 \quad E° = 0.534 \text{ V}$$
$$2Fe(s) + O_2(g) + 2H_2O(l) \rightarrow 2Fe(OH)_2(s) \qquad n = 4 \quad E° = 1.292 \text{ V}$$

a. $E° = \dfrac{RT}{nF} \cdot \ln K = \dfrac{0.0257}{n} \cdot \ln K \Rightarrow 0.562 \text{ V} = \dfrac{0.0257 \text{ V}}{2} \cdot \ln K$

$\ln K = 43.7$

$K = e^{43.7} = 9.5 \times 10^{18}$

b. $E° = \dfrac{RT}{nF} \cdot \ln K = \dfrac{0.0257}{n} \cdot \ln K \Rightarrow 0.534 \text{ V} = \dfrac{0.0257 \text{ V}}{2} \cdot \ln K$

$\ln K = 41.6$

$K = e^{41.6} = 1.1 \times 10^{18}$

c. $E° = \dfrac{RT}{nF} \cdot \ln K = \dfrac{0.0257}{n} \cdot \ln K \Rightarrow 1.292 \text{ V} = \dfrac{0.0257 \text{ V}}{4} \cdot \ln K$

$\ln K = 201$

$K = e^{201} = 2.0 \times 10^{87}$

44. Refer to Sections 18.2 and 18.4 and Example 18.7.

a. $2[\ Cr(s) \rightarrow Cr^{3+}(aq) + 3e^-\ ]$ $\qquad\qquad E°_{ox} = 0.744 \text{ V}$

$3[\ 2H^+(aq) + 2e^- \rightarrow H_2(g)\ ]$ $\qquad\qquad E°_{red} = 0.0 \text{ V}$
_____
$2Cr(s) + 6H^+(aq) + 6e^- \rightarrow 2Cr^{3+}(aq) + 3H_2(g) + 6e^-$

$2Cr(s) + 6H^+(aq) \rightarrow 2Cr^{3+}(aq) + 3H_2(g)$ $\qquad E° = 0.744 \text{ V}$

b. $E = E° - \dfrac{RT}{nF} \cdot \ln Q = 0.744 \text{ V} - \dfrac{0.0257 \text{ V}}{6} \cdot \ln \dfrac{(P_{H_2})^3[Cr^{3+}]^2}{[H^+]^6}$

c. $E = 0.744 \text{ V} - \dfrac{0.0257 \text{ V}}{6} \cdot \ln \dfrac{(0.929)^3[1.21]^2}{[0.00931]^6}$

$E = 0.744 \text{ V} - (0.00428 \text{ V}) \cdot \ln(1.80 \times 10^{12}) = 0.623 \text{ V}$

46. *Refer to Sections 18.2 and 18.4, Example 18.7 and Problem 44 above.*

a. $2[\ Fe^{2+}(aq)\ \rightarrow\ Fe^{3+}(aq) + e^-\ ]$ $E^{\circ}_{ox} = -0.769\ V$

  $\underline{H_2O_2(aq) + 2H^+(aq) + 2e^- \rightarrow H_2O(l)}$ $E^{\circ}_{red} = 1.763\ V$

  $H_2O_2(aq) + 2H^+(aq) + 2e^- + 2Fe^{2+}(aq) \rightarrow 2Fe^{3+}(aq) + 2e^- + H_2O(l)$

  $H_2O_2(aq) + 2H^+(aq) + 2Fe^{2+}(aq) \rightarrow 2Fe^{3+}(aq) + H_2O(l)$ $E^{\circ} = 0.994\ V$

b. $E = E^{\circ} - \dfrac{RT}{nF}\cdot \ln Q = 0.994\ V - \dfrac{0.0257\ V}{2}\cdot \ln \dfrac{[Fe^{3+}]^2}{[Fe^{2+}]^2[H_2O_2][H^+]^2}$

c. $E = 0.994\ V - \dfrac{0.0257\ V}{2}\cdot \ln \dfrac{(0.199)^2}{(0.00813)^2(0.914)(1.3\times 10^{-3})^2}$

  $E = 0.994\ V - (0.0129\ V)\cdot \ln (3.9\times 10^8) = 0.739\ V$

---

48. *Refer to Sections 18.1, 18.2 and 18.4.*

Write the half-reactions, calculate E° and then calculate E using the Nernst equation.

a. $Zn(s) \rightarrow Zn^{2+}(aq) + 2e^-$ $E^{\circ}_{ox} = 0.762\ V$

  $\underline{Cd^{2+}(aq) + 2e^- \rightarrow Cd(s)}$ $E^{\circ}_{red} = -0.402\ V$

  $Cd^{2+}(aq) + Zn(s) \rightarrow Zn^{2+}(aq) + Cd(s)$ $E^{\circ} = 0.360\ V$

  $E = E^{\circ} - \dfrac{RT}{nF}\cdot \ln Q = 0.360\ V - \dfrac{0.0257\ V}{2}\cdot \ln \dfrac{[Zn^{2+}]}{[Cd^{2+}]}$

  $E = 0.360\ V - 0.0129\cdot \ln \dfrac{0.50}{0.020} = 0.318\ V$

b. $Cu(s) \rightarrow Cu^{2+}(aq) + 2e^-$ $E^{\circ}_{ox} = -0.339\ V$

  $\underline{2H^+(aq) + 2e^- \rightarrow H_2(g)}$ $E^{\circ}_{red} = 0.00\ V$

  $2H^+(aq) + Cu(s) \rightarrow Cu^{2+}(aq) + H_2(g)$ $E^{\circ} = -0.339\ V$

  $E = E^{\circ} - \dfrac{RT}{nF}\cdot \ln Q = -0.339\ V - \dfrac{0.0257\ V}{2}\cdot \ln \dfrac{[Cu^{2+}](P_{H_2})}{[H^+]^2}$

  $E = -0.339\ V - 0.0129\cdot \ln \dfrac{(0.0010)(1.00)}{(0.010)^2} = -0.369\ V$

353

Write the half-reactions and calculate $E°$. Then solve the Nernst equation for [Cl⁻].

$$2[\ Au(s) + 4Cl^-(aq) \rightarrow AuCl_4^-(aq) + 3e^-\ ] \qquad E^°_{ox} = -1.001 \text{ V}$$

$$3[\ Br_2(l) + 2e^- \rightarrow 2Br^-(aq)\ ] \qquad E^°_{red} = 1.077 \text{ V}$$

$$2Au(s) + 8Cl^-(aq) + 3Br_2(l) + 6e^- \rightarrow 6Br^-(aq) + 2AuCl_4^-(aq) + 6e^-$$

$$2Au(s) + 8Cl^-(aq) + 3Br_2(l) \rightarrow 6Br^-(aq) + 2AuCl_4^-(aq) \qquad E° = 0.076 \text{ V}$$

$$E = E° - \frac{RT}{nF} \cdot \ln Q = 0.076 \text{ V} - \frac{0.0257 \text{ V}}{6} \cdot \ln \frac{[AuCl_4^-]^2[Br^-]^6}{[Cl^-]^8}$$

$$0 \text{ V} = 0.076 \text{ V} - \frac{0.0257 \text{ V}}{6} \cdot \ln \frac{(0.200)^2(0.200)^6}{[Cl^-]^8} \Rightarrow -0.076 \text{ V} = -0.00428 \cdot \ln \frac{(0.200)^8}{[Cl^-]^8}$$

$$17.7 \text{ V} = 8 \cdot \ln \frac{(0.200)}{[Cl^-]} = 8(\ln (0.200) - \ln [Cl^-])$$

$$2.2 = \ln (0.200) - \ln [Cl^-]$$

$$3.8 = -\ln [Cl^-]$$

$$[Cl^-] = e^{-3.8} = 0.022 \ M$$

Write the half-reactions and calculate $E°$. Then solve the Nernst equation for [H⁺] and calculate pH.

$$2[\ Ag(s) + Br^-(aq) \rightarrow AgBr(s) + e^-\ ] \qquad E^°_{ox} = -0.073 \text{ V}$$

$$2H^+(aq) + 2e^- \rightarrow H_2(g) \qquad E^°_{red} = 0.00 \text{ V}$$

$$2Ag(s) + 2Br^-(aq) + 2H^+(aq) + 2e^- \rightarrow H_2(g) + 2AgBr(s) + 2e^-$$

$$2Ag(s) + 2Br^-(aq) + 2H^+(aq) \rightarrow H_2(g) + 2AgBr(s) \qquad E° = -0.073 \text{ V}$$

$$E = E° - \frac{RT}{nF} \cdot \ln Q \Rightarrow -0.030 \text{ V} = -0.073 \text{ V} - \frac{0.0257 \text{ V}}{2} \cdot \ln \frac{(P_{H_2})}{[H^+]^2[Br^-]^2}$$

$$0.043 \text{ V} = -0.0129 \text{ V} \cdot \ln \frac{1}{[H^+]^2 (3.73)^2} = -3.33 = \ln (1) - \ln [H^+]^2 - \ln (3.73)^2$$

$$-3.33 \text{ V} = 0 - 2 \cdot \ln [H^+] - 2.63 \Rightarrow 2 \cdot \ln [H^+] = 0.70$$

$$\ln [H^+] = 0.35$$
$$[H^+] = e^{0.35} = 1.4 \ M$$

Write the balanced half-reactions and calculate $E°$. From pH, find $[H^+]$. Substitute these values into the Nernst equation and calculate E.

$$2H_2O(l) \rightarrow O_2(g) + 4H^+(aq) + 4e^- \qquad\qquad E°_{ox} = -1.229 \text{ V}$$

$$\underline{2[\ Br_2(l) + 2e^- \rightarrow 2Br^-(aq)\ ] \qquad\qquad\qquad E°_{red} = 1.077 \text{ V}}$$

$$2Br_2(l) + 4e^- + 2H_2O(l) \rightarrow O_2(g) + 4H^+(aq) + 4e^- + 4Br^-(aq)$$

$$2Br_2(l) + 2H_2O(l) \rightarrow O_2(g) + 4H^+(aq) + 4Br^-(aq) \qquad E° = -0.152 \text{ V}$$

$$[H^+] = 10^{-pH} = 10^{-4.50} = 3.16 \times 10^{-5}\ M$$

$$E = E° - \frac{RT}{nF} \cdot \ln Q \ \Rightarrow\ E = -0.152 \text{ V} - \frac{0.0257 \text{ V}}{4} \cdot \ln (P_{O_2})[H^+]^4[Br^-]^4$$

$$E = -0.152 \text{ V} - (0.00643 \text{ V}) \cdot \ln (1)(3.16 \times 10^{-5})^4(1)^4$$

$$E = 0.115 \text{ V}$$

E is positive, so the reaction is spontaneous (it is not non-spontaneous).

---

a. $2[\ Ag(s) \rightarrow Ag^+(aq) + e^-\ ]$ $E°_{ox} = -0.799 \text{ V}$

$$\underline{Cu^{2+}(aq) + 2e^- \rightarrow Cu(s) \qquad\qquad\qquad\quad E°_{red} = 0.339 \text{ V}}$$

$$Cu^{2+}(aq) + 2Ag(s) \rightarrow 2Ag^+(aq) + Cu(s) \qquad\quad E° = -0.460 \text{ V}$$

b. $E = E° - \dfrac{RT}{nF} \cdot \ln Q \ \Rightarrow\ 0.060 \text{ V} = -0.460 \text{ V} - \dfrac{0.0257 \text{ V}}{2} \cdot \ln \dfrac{[Ag^+]^2}{[Cu^{2+}]}$

$$0.520 \text{ V} = -0.0129 \text{ V} \cdot \ln \frac{[Ag^+]^2}{1.0\,M} \ \Rightarrow\ 40.3 = 2 \ln [Ag^+]$$

$$\ln [Ag^+] = -20.2$$

$$[Ag^+] = e^{-20.2} = 1.7 \times 10^{-9}\ M$$

c. $K_{sp} = [Ag^+][Cl^-] = (1.7 \times 10^{-9})(0.10)$

$$K_{sp} = 1.7 \times 10^{-10}$$

a. $Al_2O_3(s) \rightarrow 2Al(s) + \frac{3}{2}O_2(g)$
   $Al^{3+}(aq) + 3e^- \rightarrow Al(s)$

   $10 \text{ kg Al} \times \dfrac{1000 \text{ g}}{1 \text{ kg}} \times \dfrac{1 \text{ mol. Al}}{26.98 \text{ g}} \times \dfrac{3 \text{ mol. } e^-}{1 \text{ mol Al}} = 1112 \text{ mol. } e^-$

b. $1 \text{ day} \times \dfrac{24 \text{ hr}}{1 \text{ day}} \times \dfrac{60 \text{ min}}{1 \text{ hr}} \times \dfrac{60 \text{ s}}{1 \text{ min}} = 86400 \text{ s}$

   $1112 \text{ mol. } e^- \times \dfrac{9.648 \times 10^4 \text{ C}}{1 \text{ mol. } e^-} \times \dfrac{1 \text{ A} \cdot \text{s}}{1 \text{ C}} \times \dfrac{1}{86400 \text{ s}} = 1.242 \times 10^3 \text{ A}$

c. $10 \text{ kg Al} \times \dfrac{1000 \text{ g}}{1 \text{ kg}} \times \dfrac{1 \text{ mol. Al}}{26.98 \text{ g}} \times \dfrac{\frac{3}{2} \text{ mol. O}_2}{2 \text{ mol Al}} = 278.0 \text{ mol. O}_2$

a. Calculate the volume and mass of gold to be plated out. Bear in mind that the gold will be plated on both sides of the thin sheet.

   $2(1.5 \text{ in} \times 8.5 \text{ in} \times 0.0020 \text{ in}) = 0.051 \text{ in}^3$

   $0.051 \text{ in}^3 \times \dfrac{(2.54 \text{ cm})^3}{(1 \text{ in})^3} \times \dfrac{19.3 \text{ g Au}}{1 \text{ cm}^3 \text{ Au}} = 16 \text{ g Au}$

b. $AuCN(s) \rightarrow Au(s) + CN^-(aq) + e^-$

   $16 \text{ g Au} \times \dfrac{1 \text{ mol. Au}}{197.0 \text{ g Au}} \times \dfrac{1 \text{ mol. } e^-}{1 \text{ mol. Au}} \times \dfrac{9.648 \times 10^4 \text{ C}}{1 \text{ mol. } e^-} = 7.8 \times 10^3 \text{ C}$

   $7.8 \times 10^3 \text{ C} \times \dfrac{1 \text{ A} \cdot \text{s}}{1 \text{ C}} \times \dfrac{1}{7.00 \text{ A}} = 1.1 \times 10^3 \text{ s}$

   $1.1 \times 10^3 \text{ s} \times \dfrac{1 \text{ min}}{60 \text{ s}} = 19 \text{ min}$

Use the equation on page 538 for the reaction of a lead storage battery to determine the moles of e^- per mole of Pb.

a. $95 \text{ min} \times \dfrac{60 \text{ s}}{1 \text{ min}} = 5700 \text{ s}$

$5.00 \text{ A} \times 5700 \text{ s} = 29000 \text{ A} \cdot \text{s} = 29000 \text{ C}$

$29000 \text{ C} \times \dfrac{1 \text{ mol. } e^-}{9.648 \times 10^4 \text{ C}} \times \dfrac{1 \text{ mol. Pb}}{2 \text{ mol. } e^-} \times \dfrac{207.2 \text{ g}}{1 \text{ mol. Pb}} = 31 \text{ g Pb}$

b. $29000 \text{ C} \times 12.0 \text{ V} = 3.5 \times 10^5 \text{ C} \cdot \text{V} = 3.5 \times 10^5 \text{ J}$

$3.5 \times 10^5 \text{ J} \times \dfrac{1 \text{ kWh}}{3.600 \times 10^6 \text{ J}} = 9.7 \times 10^{-2} \text{ kWh}$

Use the ideal gas equation to calculate the moles of $H_2(g)$ needed. Write the half-reactions needed to derive the equation for the electrolysis of water. Then use the moles of $H_2(g)$ and moles of e^- to determine the current and then the time needed for the electrolysis.

$n = \dfrac{PV}{RT} = \dfrac{(0.924 \text{ atm})(10.00 \text{ L})}{(0.0821 \text{ L} \cdot \text{atm/mol.} \cdot \text{K})(295 \text{ K})} = 0.382 \text{ mol. } H_2$

$2[\ 2H_2O(l) + 2e^- \rightarrow H_2(g) + 2OH^-(aq)\ ]$
$2H_2O(l) \rightarrow O_2(g) + 4H^+(aq) + 4e^-$
$\underline{4[\ H^+(aq) + OH^-(aq) \rightarrow H_2O(l)\ ]}$
$2H_2O(l) \rightarrow O_2(g) + 2H_2(g)$ (thus 4 e^- are passed for each 2 moles of $H_2(g)$ produced)

$0.382 \text{ mol. } H_2 \times \dfrac{4 \text{ mol. } e^-}{2 \text{ mol. } H_2} \times \dfrac{9.648 \times 10^4 \text{ C}}{1 \text{ mol. } e^-} = 7.36 \times 10^4 \text{ C}$

$7.36 \times 10^4 \text{ C} \times \dfrac{1 \text{ A} \cdot \text{s}}{1 \text{ C}} \times \dfrac{1}{12.0 \text{ A}} = 6.13 \times 10^3 \text{ s}$

$6.13 \times 10^3 \text{ s} \times \dfrac{1 \text{ min}}{60 \text{ s}} \times \dfrac{1 \text{ hr}}{60 \text{ min}} = 1.70 \text{ hr}$

$$2[\ 2Br^-(aq) \rightarrow Br_2(l) + 2e^-] \qquad\qquad E^\circ_{ox} = -1.077\ V$$

$$\underline{O_2(g) + 4H^+(aq) + 4e^- \rightarrow 2H_2O(l)} \qquad\qquad E^\circ_{red} = 1.229\ V$$

$$4Br^-(aq) + O_2(g) + 4H^+(aq) + 4e^- \rightarrow 2H_2O(l) + 2Br_2(l) + 4e^-$$

$$4Br^-(aq) + O_2(g) + 4H^+(aq) \rightarrow 2H_2O(l) + 2Br_2(l) \qquad E^\circ = 0.152\ V$$

$$[H^+] = K_a \times \frac{[HB]}{[B^-]} = 1.8 \times 10^{-5} \times \frac{0.1}{0.1} = 1.8 \times 10^{-5}\ M$$

$$E = E^\circ - \frac{RT}{nF} \cdot \ln Q = 0.152\ V - \frac{0.0257\ V}{4} \cdot \ln \frac{1}{(P_{O_2})[H^+]^4[Br^-]^4}$$

$$E = 0.152\ V - 0.00643\ V \cdot \ln \frac{1}{(1)(1.8 \times 10^{-5})^4(0.100)^4} = -0.188\ V$$

Calculate ΔH°, ΔS° and ΔG° using the thermodynamic data from Appendix 1. Then calculate E° from ΔG°.

$\Delta H^\circ = \Sigma\ \Delta H_f^\circ{}_{(products)} - \Sigma\ \Delta H_f^\circ{}_{(reactants)}$
$\Delta H^\circ = [(1\ mol.)(-393.5\ kJ/mol.)] - [(1\ mol.)(-110.5\ kJ/mol.) + (\frac{1}{2}\ mol.)(0.0\ kJ/mol.)]$
$\Delta H^\circ = -283.0\ kJ$

$\Delta S^\circ = \Sigma\ S^\circ{}_{(products)} - \Sigma\ S^\circ{}_{(reactants)}$

$\Delta S^\circ = [(1\ mol.)(0.2136\ kJ/mol.\cdot K)]$
$\qquad - [(1\ mol.)(0.1976\ kJ/mol.\cdot K) + (\frac{1}{2}\ mol.)(0.2050\ kJ/mol.\cdot K)]$
$\Delta S^\circ = -0.0865\ kJ/K$

$\Delta G^\circ = \Delta H^\circ - T\Delta S^\circ$
$\Delta G^\circ = -283.0\ kJ - (1275\ K)(-0.0865\ kJ/K)$
$\Delta G^\circ = -172.9\ kJ$

$$CO(g) + \tfrac{1}{2}O_2(g) \rightarrow CO_2(g)$$
$$O^0 + 2e^- \rightarrow O^{2-}$$
$$C^{2+} \rightarrow C^{4+} + 2e^-$$

$\Delta G^\circ = -nFE^\circ$
$(-172.9\ kJ)(1000\ J/kJ) = -(2\ mol.)(9.648 \times 10^4\ J/mol.\cdot V)E^\circ$
$E^\circ = 0.896\ V$

70. *Refer to Sections 18.1 and 18.6.*

a. A salt bridge allows the movement of ions. Without such flow, one part of the cell would develop a build-up of positive charge and the other, negative charge. This build-up of charge would soon become large enough that it would stop further flow of charge.

b. Reduction occurs in the cathodic department; this is accompanied by a reduction in charge. This results in a build-up of negative charge, which is offset by the migration of the cations from the salt bridge to the cathodic compartment.

$$[Cu^{2+}(aq) + SO_4^{2-}(aq)] + 2e^- \rightarrow [Cu(s) + SO_4^{2-}(aq)]$$

charges balanced   *excess negative charge*

c. $H_2SO_4(aq)$ is one of the reactants in the overall equation representing the lead storage battery (see p. 538). In order for the reaction to occur, all the reactants must be present.

72. *Refer to Section 18.4.*

Write the equation for the cell reaction and evaluate Q. Use the Nernst equation to determine the effect on E.

$$Co(s) \rightarrow Co^{2+}(aq) + 2e^-$$
$$\underline{2H^+(aq) + 2e^- \rightarrow H_2(g)}$$
$$Co(s) + 2H^+(aq) \rightarrow H_2(g) + Co^{2+}(aq)$$

$$E = E^\circ - \frac{RT}{nF} \cdot \ln Q = E^\circ - \frac{0.0257 \text{ V}}{2} \cdot \ln \frac{[Co^{2+}](P_{H_2})}{[H^+]^2}$$

a. The voltage is dependent on the concentration of Co^{2+}, not the volume. So changing the volume will have no effect on the voltage of the cell.

b. The voltage of the cell is dependent on the concentration of H^+. Increasing $[H^+]$ will decrease Q, which will decrease $\ln Q$. Consequently, the value being subtracted from E° will be smaller, and E will be larger.

c. The voltage of the cell is dependent on the pressure of H_2. Increasing $P(H_2)$ will increase Q, which will increase $\ln Q$. Consequently, the value being subtracted from E° will be larger, and E will be smaller.

d. The voltage of the cell does not depend on the amount of $Co(s)$ present. Changing this value will have no effect on the voltage of the cell.

e. The voltage of the cell does not depend on the surface area of the electrode. Changing this value will have no effect on the voltage of the cell.

Thus, only (b) will increase the voltage of the cell (E).

$$Ag(s) \rightarrow Ag^+(aq) + e^- \qquad\qquad E^\circ_{ox} = -0.799 \text{ V}$$

$$AgSCN(s) + e^- \rightarrow Ag(s) + SCN^-(aq) \qquad E^\circ_{red} = 0.0895 \text{ V}$$

$$AgSCN(s) \rightarrow Ag^+(aq) + SCN^-(aq) \qquad E^\circ = -0.710 \text{ V}$$

Note that the equation above is the equation for K_{sp}. Thus one can use the Nernst equation to calculate K_{sp} from the calculated E° for the cell.

$$E^\circ = \frac{RT}{nF} \cdot \ln K = -0.710 \text{ V} = \frac{0.0257}{1} \cdot \ln K$$

$$\ln K = -27.6$$

$$K = K_{sp} = e^{-27.6} = 1.0 \times 10^{-12}$$

76. Refer to Sections 18.4 and 18.6 and Chapter 13.

Calculate E°, $[H^+]$ in the acid solution, and from K_a, the $[SO_4^{2-}]$. Substitute these values into the Nernst equation and calculate E.

$$Pb(s) + SO_4^{2-}(aq) \rightarrow PbSO_4(s) + 2e^- \qquad\qquad E^\circ_{ox} = 0.356 \text{ V}$$

$$PbO_2(s) + SO_4^{2-}(aq) + 4H^+(aq) + 2e^- \rightarrow PbSO_4(s) + 2H_2O(l) \qquad E^\circ_{red} = 1.687 \text{ V}$$

$$Pb(s) + PbO_2(s) + 2SO_4^{2-}(aq) + 4H^+(aq) + 2e^- \rightarrow 2PbSO_4(s) + 2H_2O(l) + 2e^-$$

$$Pb(s) + PbO_2(s) + 2SO_4^{2-}(aq) + 4H^+(aq) \rightarrow 2PbSO_4(s) + 2H_2O(l) \qquad E^\circ = 2.043 \text{ V}$$

$$\frac{1.286 \text{ g solution}}{1 \text{ cm}^3} \times \frac{1 \text{ cm}^3}{1 \text{ mL}} \times \frac{1000 \text{ mL}}{1 \text{ L}} \times \frac{38 \text{ g H}_2\text{SO}_4}{100 \text{ g sol'n}} \times \frac{1 \text{ mol.}}{98.1 \text{ g}} = 4.98 \text{ M H}_2\text{SO}_4$$

$$[H^+] = [HSO_4^-] = [H_2SO_4] = 4.98 \, M$$

$$K_a = \frac{[H^+][SO_4^{2-}]}{[HSO_4^-]} \Rightarrow 1.0 \times 10^{-2} = \frac{(4.98 + x)(x)}{(4.98 - x)}$$

$$K_a/a = 1.0 \times 10^{-2} / 4.98 = 2.0 \times 10^{-3}$$
$$K_a/a = 2.0 \times 10^{-3} < 2.5 \times 10^{-3}$$

(The 5% rule applies. See the explanation accompanying Problem 38 in Chapter 13.)

$$1.0 \times 10^{-2} = \frac{(4.98)(x)}{4.98}$$

$$x = [SO_4^{2-}] = 0.0100 \, M$$

$$E = E^\circ - \frac{0.0257}{n} \cdot \ln \frac{1}{[H^+]^4 [SO_4^{2-}]^2} = 2.043 - \frac{0.0257}{2} \cdot \ln \frac{1}{(4.98)^4 (0.0100)^2} = 2.01 \text{ V}$$

77. Refer to Sections 18.2 and 18.4.

a.

$$Zn(s) \rightarrow Zn^{2+}(aq) + 2e^- \qquad\qquad E°_{ox} = 0.762 \text{ V}$$

$$\underline{Sn^{2+}(aq) + 2e^- \rightarrow Sn(s) \qquad\qquad E°_{red} = -0.141 \text{ V}}$$

$$Sn^{2+}(aq) + Zn(s) \rightarrow Zn^{2+}(aq) + Sn(s) \qquad\qquad E° = 0.621 \text{ V}$$

b. $E°$ is positive, so the reaction is spontaneous in the forward direction. As the cell operates, $[Zn^{2+}]$ increases and $[Sn^{2+}]$ decreases.

c. $$E = E° - \frac{0.0257}{n} \cdot \ln \frac{[Zn^{2+}]}{[Sn^{2+}]} \quad\Rightarrow\quad 0 = 0.621 - \frac{0.0257}{2} \cdot \ln \frac{[Zn^{2+}]}{[Sn^{2+}]}$$

$$\ln \frac{[Zn^{2+}]}{[Sn^{2+}]} = 48.3 \quad\Rightarrow\quad \frac{[Zn^{2+}]}{[Sn^{2+}]} = 1 \times 10^{21}$$

d. $[Zn^{2+}] = [Zn^{2+}]_i + x$ and $[Sn^{2+}] = [Sn^{2+}]_i - x$
$[Zn^{2+}] = 1.0\,M + x$ and $[Sn^{2+}] = 1.0\,M - x$

$$1 \times 10^{21} = \frac{[Zn^{2+}]}{[Sn^{2+}]} = \frac{1.0\,M + x}{1.0\,M - x}$$

$1 \times 10^{21} - 1 \times 10^{21}x = 1 + x$
$1 \times 10^{21} = 1 \times 10^{21}x$
$x = 1.00$

$[Zn^{2+}] = 1.0\,M + x = 2.0\,M$
$[Sn^{2+}] = 1.0\,M - x = 0.0\,M$ **or**
$[Sn^{2+}] = [Zn^{2+}] / 1 \times 10^{21} = 2.0\,M / 1 \times 10^{21} = 2 \times 10^{-21}\,M\ (\cong 0.0\,M)$

Since K is so large, the reaction lies almost completely to the right at equilibrium (when $E = 0$). Thus we can assume that essentially all the Sn^{2+} reacts to form Zn^{2+}. Consequently, $[Zn^{2+}] = 1.0\,M + 1.0\,M = 2.0\,M$. The residual $[Sn^{2+}]$ can be calculated from:

$$K = 1 \times 10^{21} = \frac{[Zn^{2+}]}{[Sn^{2+}]}$$

78. Refer to Section 18.3.

a. For step 1:
$\Delta G°' = -nFE°'$
$\Delta G°' = -(2 \text{ mol.})(9.648 \times 10^4 \text{ J/mol.·V})(-0.581 \text{ V})$
$\Delta G°' = 1.12 \times 10^5 \text{ J}$

For step 2:
$\Delta G°' = -nFE°'$
$\Delta G°' = -(2 \text{ mol.})(9.648 \times 10^4 \text{ J/mol.·V})(-0.197 \text{ V})$
$\Delta G°' = 3.80 \times 10^4 \text{ J}$

Overall:

$$\Delta G^{o\prime} = 1.12 \times 10^5 \text{ J} + 3.80 \times 10^4 \text{ J}$$
$$\Delta G^{o\prime} = 1.50 \times 10^5 \text{ J}$$

b. In the overall process, 4 electrons are transferred.

$$\Delta G^{o\prime} = -nFE^{o\prime}$$
$$1.50 \times 10^5 \text{ J} = -(4 \text{ mol.})(9.648 \times 10^4 \text{ J/mol.} \cdot \text{V})(E^{o\prime})$$
$$E^{o\prime} = -0.389 \text{ V}$$

---

79. Refer to Sections 18.2 and 18.4.

$$H_2(g) \rightarrow 2H^+(aq) + 2e^- \qquad\qquad E^\circ_{ox} = 0.0 \text{ V}$$

$$2H^+(aq) + 2e^- \rightarrow H_2(g) \qquad\qquad E^\circ_{red} = 0.0 \text{ V}$$

$$H_{2\ (anode)}(g) + 2H^+_{\ (cathode)}(aq) \rightarrow 2H^+_{\ (anode)}(aq) + H_{2\ (cathode)}(g) \qquad E^\circ = 0.00 \text{ V}$$

$$[H^+]_{(anode)} = 10^{-pH} = 10^{-7.0} = 1.0 \times 10^{-7} \ M$$
$$[H^+]_{(cathode)} = 10^{-pH} = 10^{0.0} = 1.0 \ M$$

$$E = E^\circ - \frac{0.0257}{n} \cdot \ln \frac{(P_{H_2})_{cathode} [H^+]^2_{anode}}{(P_{H_2})_{anode} [H^+]^2_{cathode}} = 0.00 - \frac{0.0257}{2} \cdot \ln \frac{(1.0)(1.0 \times 10^{-7})^2}{(1.0)(1.0)^2}$$

$$E = 0.414 \text{ V}$$

Chapter 19: Nuclear Chemistry

---

2. *Refer to Section 19.1 and Example 19.1.*

$^{241}_{95}\text{Am} \rightarrow {}^{237}_{93}\text{Np} + {}^{A}_{Z}?$

$A = 241 - 237 = 4$
$Z = 95 - 93 = 2$

$^{241}_{95}\text{Am} \rightarrow {}^{237}_{93}\text{Np} + {}^{4}_{2}\textbf{He}$ ^{241}Am undergoes alpha emission.

---

4. *Refer to Section 19.1 and Example 19.1.*

a. $^{230}_{90}\text{Th} \rightarrow {}^{4}_{2}\text{He} + {}^{A}_{Z}?$

$A = 230 - 4 = 226$
$Z = 90 - 2 = 88$

$^{230}_{90}\text{Th} \rightarrow {}^{4}_{2}\text{He} + {}^{226}_{88}\textbf{Ra}$

b. $^{210}_{82}\text{Pb} \rightarrow {}^{0}_{-1}e + {}^{A}_{Z}?$

$A = 210 - 0 = 210$
$Z = 82 - (-1) = 83$

$^{210}_{82}\text{Pb} \rightarrow {}^{0}_{-1}e + {}^{210}_{83}\textbf{Bi}$

c. Fission indicates that a neutron reacted with the nucleus. To produce an *excess* of 2 neutrons, a total of 3 must be produced (the initial one plus 2 more).

$^{235}_{92}\text{U} + {}^{1}_{0}n \rightarrow {}^{140}_{56}\text{Ba} + 3{}^{1}_{0}n + {}^{A}_{Z}?$

$A = (235 + 1) - (140 + 3(1)) = 93$
$Z = (92 + 0) - (56 + 3(0)) = 36$

$^{235}_{92}\text{U} + {}^{1}_{0}n \rightarrow {}^{140}_{56}\text{Ba} + 3{}^{1}_{0}n + {}^{93}_{36}\textbf{Kr}$

d. $^{37}_{18}\text{Ar} + {}^{0}_{-1}e \rightarrow {}^{A}_{Z}?$

$A = 37 + 0 = 37$
$Z = 18 + (-1) = 17$

$^{37}_{18}\text{Ar} + {}^{0}_{-1}e \rightarrow {}^{37}_{17}\textbf{Cl}$

6. Refer to Section 19.1 and Example 19.1.

The sum of the mass numbers and the atomic numbers on either side of the arrow must be equal to get the missing species. Calculate A and Z for the missing species and determine the element from the atomic number.

a. $^{87}_{37}\text{Rb} \rightarrow \, ^{0}_{-1}e + \, ^{A}_{Z}\text{X}$

$A = 87 - 0 = 87$
$Z = 37 - (-1) = 38$

$^{87}_{37}\text{Rb} \rightarrow \, ^{0}_{-1}e + \, ^{87}_{38}\textbf{Sr}$

b. $^{A}_{Z}\text{X} \rightarrow \, ^{0}_{+1}e + \, ^{87}_{37}\text{Rb}$

$A = 87 - 0 = 87$
$Z = 37 + 1 = 38$

$^{87}_{38}\textbf{Sr} \rightarrow \, ^{0}_{+1}e + \, ^{87}_{37}\text{Rb}$

8. Refer to Section 19.1 and Example 19.1.

$^{282}_{115}\text{X} + \, ^{0}_{-1}e \rightarrow \, ^{A}_{Z}\text{Y}$

$A = 282 + 0 = 282$
$Z = 115 + (-1) = 114$

$^{282}_{115}\text{X} + \, ^{0}_{-1}e \rightarrow \, ^{282}_{114}\text{Y}$

$^{282}_{115}\text{X} \rightarrow \, ^{A}_{Z}\text{Y} + \, ^{0}_{1}e$

$A = 282 - 0 = 87$
$Z = 115 - 1 = 114$

$^{282}_{115}\text{X} \rightarrow \, ^{282}_{114}\text{Y} + \, ^{0}_{1}e$

The product nuclide is the same for both reactions.

10. Refer to Section 19.1 and Example 19.1.

a. $^{54}_{26}\text{Fe} + \, ^{4}_{2}\text{He} \rightarrow 2^{1}_{1}\text{H} + \, ^{A}_{Z}?$

$A = 54 + 4 - 2(1) = 56$
$Z = 26 + 2 - 2(1) = 26$

$^{54}_{26}\text{Fe} + \, ^{4}_{2}\text{He} \rightarrow 2^{1}_{1}\text{H} + \, ^{56}_{26}\textbf{Fe}$

b. $^{96}_{42}\text{Mo} + \, ^{2}_{1}H \rightarrow \, ^{1}_{0}n + \, ^{A}_{Z}?$

$A = 96 + 2 - 1 = 97$
$Z = 42 + 1 - 0 = 43$

$^{96}_{42}\text{Mo} + \, ^{2}_{1}H \rightarrow \, ^{1}_{0}n + \, ^{97}_{43}\textbf{Tc}$

c. $^{40}_{18}\text{Ar} + \, ^{A}_{Z}? \rightarrow \, ^{43}_{19}\text{K} + \, ^{1}_{1}\text{H}$

$A = (43 + 1) - 40 = 4$
$Z = (19 + 1) - 18 = 2$

$^{40}_{18}\text{Ar} + \, ^{4}_{2}\textbf{He} \rightarrow \, ^{43}_{19}\text{K} + \, ^{1}_{1}\text{H}$

d. $^{A}_{Z}? + \, ^{1}_{0}n \rightarrow \, ^{1}_{1}H + \, ^{31}_{15}\text{P}$

$A = 31 + 1 - 1 = 31$
$Z = 15 + 1 - 0 = 16$

$^{31}_{16}\textbf{S} + \, ^{1}_{0}n \rightarrow \, ^{1}_{1}H + \, ^{31}_{15}\text{P}$

12. Refer to Section 19.1 and Example 19.1.

a. $^{121}_{51}Sb + ^{4}_{2}He \rightarrow ^{1}_{1}H + ^{A}_{Z}?$

$A = 121 + 4 - 1 = 124$
$Z = 51 + 2 - 1 = 52$

$^{121}_{51}Sb + ^{4}_{2}He \rightarrow ^{1}_{1}H + ^{124}_{52}Te$

b. $^{238}_{92}U + ^{1}_{0}n \rightarrow ^{0}_{-1}e + ^{A}_{Z}?$

$A = 238 + 1 - 0 = 239$
$Z = 92 + 0 - (-1) = 93$

$^{238}_{92}U + ^{1}_{0}n \rightarrow ^{0}_{-1}e + ^{239}_{93}Np$

c. $^{14}_{7}N + ^{A}_{Z}? \rightarrow ^{1}_{1}H + ^{17}_{8}O$

$A = 17 + 1 - 14 = 4$
$Z = 8 + 1 - 7 = 2$

$^{14}_{7}N + ^{4}_{2}He \rightarrow ^{1}_{1}H + ^{17}_{8}O$

d. $^{A}_{Z}? + ^{4}_{2}He \rightarrow ^{27}_{14}Si + ^{1}_{0}n$

$A = 27 + 1 - 4 = 24$
$Z = 14 + 0 - 2 = 12$

$^{24}_{12}Mg + ^{4}_{2}He \rightarrow ^{27}_{14}Si + ^{1}_{0}n$

14. Refer to Section 19.2.

$1\ Ci = 3.700 \times 10^{10}$ atoms/s

$2793\ Ci \times \dfrac{3.700 \times 10^{10}\ \text{atoms/s}}{1\ Ci} = 1.033 \times 10^{14}$ atoms/s

$\dfrac{1.033 \times 10^{14}\ \text{atoms}}{1\ s} \times \dfrac{60\ s}{1\ \text{min}} = 6.198 \times 10^{15}$ disintegrations / min

16. Refer to Section 19.2.

$\dfrac{3.00 \times 10^{4}\ \text{disintegrations}}{5.00\ \text{min}} \times \dfrac{1\ \text{atom}}{1\ \text{disintegration}} \times \dfrac{1\ \text{min}}{60\ s} \times \dfrac{1\ Ci}{3.700 \times 10^{10}\ \text{atoms/s}} = 2.70 \times 10^{-9}\ Ci$

18. Refer to Section 19.2 and Example 19.2.

Calculate N for the sample, then use $A = k \cdot N$ to calculate the activity.

$2.00\ mg \times \dfrac{1\ g}{1000\ mg} \times \dfrac{1\ \text{mol. Kr - 87}}{87\ \text{g Kr - 87}} \times \dfrac{6.02 \times 10^{23}\ \text{atoms}}{1\ \text{mol.}} = 1.38 \times 10^{19}$ atoms

$A = (1.5 \times 10^{-4}\ /\text{sec})(1.38 \times 10^{19}\ \text{atoms}) = 2.1 \times 10^{15}$ atoms/sec

$2.1 \times 10^{15}\ \text{atoms/sec} \times \dfrac{1\ Ci}{3.700 \times 10^{10}\ \text{atoms/sec}} = 5.6 \times 10^{4}\ Ci$

a. $_{82}^{210}\text{Pb} \rightarrow _{-1}^{0}e + _{83}^{210}\text{Bi}$ (see Problem 4b)

b. Convert $t_{1/2}$ to seconds and then calculate k. Calculate atoms of Pb-210. Then calculate the activity (in atoms/s) and convert that to mCi.

$$20.4 \text{ yr} \times \frac{365 \text{ d}}{1 \text{ yr}} \times \frac{24 \text{ hr}}{1 \text{ d}} \times \frac{3600 \text{ s}}{1 \text{ hr}} = 6.43 \times 10^8 \text{ s}$$

$$k = \frac{0.693}{t_{1/2}} = \frac{0.693}{6.43 \times 10^8 \text{ s}} = 1.08 \times 10^{-9} \text{ /s}$$

$$0.500 \text{ g Pb-210} \times \frac{1 \text{ mol.}}{210.0 \text{ g}} \times \frac{6.02 \times 10^{23} \text{ atoms}}{1 \text{ mol.}} = 1.43 \times 10^{21} \text{ atoms}$$

$$A = kN = (1.08 \times 10^{-9} \text{ /s})(1.43 \times 10^{21} \text{ atoms}) = 1.54 \times 10^{12} \text{ atoms/s}$$

$$1.54 \times 10^{12} \text{ atoms/s} \times \frac{1 \text{ Ci}}{3.700 \times 10^{10} \text{ atoms/s}} \times \frac{1000 \text{ mCi}}{1 \text{ Ci}} = 4.16 \times 10^4 \text{ mCi}$$

---

22. *Refer to Section 19.2.*

Calculate A, then N. Then convert the number of atoms to mass of Pb-210.

$$\frac{1.3 \times 10^4 \text{ disintegrations}}{5.00 \text{ min}} \times \frac{1 \text{ atom}}{1 \text{ disintegration}} \times \frac{1 \text{ min}}{60 \text{ s}} = 43 \text{ atoms/s}$$

$$A = kN \implies 43 \text{ atoms/s} = (1.08 \times 10^{-9} \text{ /s})(N)$$

$$N = 4.0 \times 10^{10} \text{ atoms}$$

$$4.0 \times 10^{10} \text{ atoms} \times \frac{1 \text{ mol.}}{6.022 \times 10^{23} \text{ atoms}} \times \frac{210.0 \text{ g}}{1 \text{ mol.}} = 1.4 \times 10^{-11} \text{ g}$$

---

24. *Refer to Sections 19.1 and 19.2.*

$_{17}^{36}\text{Cl} \rightarrow _{-1}^{0}e + _{18}^{36}\text{A}$ (Note there is 1 β-particle per 1 $_{17}^{36}\text{Cl}$ decay.)

$$\frac{2.3 \times 10^{-6}}{\text{yr}} \times \frac{1 \text{ yr}}{365 \text{ d}} \times \frac{1 \text{ d}}{24 \text{ hr}} \times \frac{1 \text{ hr}}{60 \text{ min}} = 4.4 \times 10^{-12} \text{ /min}$$

$$1.00 \text{ mg Cl-36} \times \frac{1\text{ g}}{1000\text{ mg}} \times \frac{1\text{ mol. Cl-36}}{36.0\text{ g}} = 2.78 \times 10^{-5} \text{ mol. Cl-36}$$

$$2.78 \times 10^{-5} \text{ mol. Cl-36} \times \frac{1\text{ mol. }\beta\text{-particles}}{1\text{ mol. }^{36}\text{Cl}} \times \frac{6.022 \times 10^{23}}{1\text{ mol.}} = 1.67 \times 10^{19} \ \beta\text{-particles}$$

$$A = kN = (4.4 \times 10^{-12} \text{/min})(1.67 \times 10^{19} \text{ β-particles}) = 7.3 \times 10^{7} \text{ particles/min}$$

$$\frac{7.5 \times 10^{7} \text{ particles}}{1\text{ min}} \times \frac{1\text{ min}}{60\text{ s}} \times \frac{1\text{ Ci}}{3.700 \times 10^{10} \text{ particles/s}} = 3.4 \times 10^{-5} \text{ Ci}$$

---

26. *Refer to Section 19.2 and Example 19.3.*

Calculate k from the half-life, then calculate time.

$$k = \frac{0.693}{5730 \text{ yr}} = 1.21 \times 10^{-4} \text{/yr}$$

$$\ln\left(\frac{X_0}{X}\right) = kt \ \Rightarrow \ \ln\left(\frac{1}{0.975}\right) = (1.21 \times 10^{-4} \text{/yr}) \cdot t$$

$t = 209$ yr

This could **not** be a painting from Rembrandt (active in the 17th century, 300 years ago) since it is from the 18th century (200 years ago).

---

28. *Refer to Section 19.2 and Example 19.3.*

Calculate k from the half-life, then calculate time.

$$k = \frac{0.693}{5730 \text{ yr}} = 1.21 \times 10^{-4} \text{/yr}$$

$$\ln\left(\frac{X_0}{X}\right) = kt \ \Rightarrow \ \ln\left(\frac{22.1}{12.0}\right) = (1.21 \times 10^{-4} \text{/yr}) \cdot t$$

$t = 5.05 \times 10^{3}$ yr

---

30. *Refer to Section 19.2.*

Calculate k from the half-life, then calculate time.

$$k = \frac{0.693}{12.3 \text{ yr}} = 0.0563 \text{/yr}$$

If the tritium content is ⅗ the original, then $X = \frac{3}{5}X_0$. If $X_0 = 1$, then $X = 0.600$.

$$\ln\left(\frac{X_0}{X}\right) = kt \implies \ln\left(\frac{1}{0.600}\right) = (0.0563\,/\text{yr}) \cdot t$$

$t = 9.07$ yr

---

32. Refer to Section 19.3 and Example 19.4.

a. $^{230}_{90}\text{Th} \rightarrow {}^{4}_{2}\text{He} + {}^{226}_{88}\text{Ra}$ (See problem 4a)

$\Delta m = 4.00150$ g/mol. $+ 225.9771$ g/mol. $- 229.9837$ g/mol. $= -0.0051$ g/mol.

b. Calculate ΔE for one mole, then convert to kJ/g.

$\Delta E = 9.00 \times 10^{10}$ kJ/g $\times \Delta m = (9.00 \times 10^{10}$ kJ/g$)(-0.0051$ g/mol.$) = -4.6 \times 10^{8}$ kJ/mol.

$$1.00\,\text{g} \times \frac{1\,\text{mol.}}{230\,\text{g}} \times \frac{-4.6 \times 10^{8}\,\text{g}}{1\,\text{mol.}} = -2.0 \times 10^{6}\,\text{kJ}$$

---

34. Refer to Section 19.3 and Example 19.5.

$$^{10}_{4}\text{Be} \rightarrow 6\,^{1}_{0}n + 4\,^{1}_{1}\text{H}$$

a. $\Delta m = $ [mass of neutrons + mass of protons] – mass of nucleus
$\Delta m = [6(1.00867$ g$) + 4(1.00728$ g$)] - 10.01134 = 0.06980$ g

b. $\Delta E = 9.00 \times 10^{10}$ kJ/g $\times \Delta m$
$\Delta E = (9.00 \times 10^{10}$ kJ/g$)(0.06980$ g$) = 6.28 \times 10^{9}$ kJ

---

36. Refer to Section 19.3, Example 19.5 and Table 19.3.

Calculate the mass defect for each isotope. Then calculate and compare the ΔE's.

$$^{26}_{12}\text{Mg} \rightarrow 12\,^{1}_{1}\text{H} + 14\,^{1}_{0}n$$
$$^{26}_{13}\text{Al} \rightarrow 13\,^{1}_{1}\text{H} + 13\,^{1}_{0}n$$

Mg-26: $12(1.00728$ g$) + 14(1.00867$ g$) - 25.97600$ g $= 0.2327$ g
Al-26: $13(1.00728$ g$) + 13(1.00867$ g$) - 25.97977$ g $= 0.2276$ g

Since Mg-26 has the larger mass defect, it will have the greater binding energy.

$\Delta E_{\text{Mg-26}} = 9.00 \times 10^{10}$ kJ/g $\times \Delta m = 9.00 \times 10^{10}$ kJ/g $\times (0.2327$ g$) = 2.09 \times 10^{10}$ kJ
$\Delta E_{\text{Al-26}} = 9.00 \times 10^{10}$ kJ/g $\times \Delta m = 9.00 \times 10^{10}$ kJ/g $\times (0.2276$ g$) = 2.05 \times 10^{10}$ kJ

Calculate the mass defect for the nuclear reaction, adjust for 1.00 g H, then calculate ΔE.

Δm = 4.00150 g + 2(0.00055 g) - 4(1.00728 g) = -0.02652 g (per 4 mole H)

$$1.00 \text{ g H} \times \frac{1 \text{ mol. H}}{1.0073 \text{ g}} \times \frac{-0.02652 \text{ g}}{4 \text{ mol. H}} = -0.006582 \text{ g}$$

ΔE = 9.00 x 10^{10} kJ/g x Δm = 9.00 x 10^{10} kJ/g x (-0.006582 g) = -5.92 x 10^{8} kJ

Write the nuclear reaction for the fission of U-235 and calculate the mass defect. Calculate the enery associated with this Δm. Then convert to kJ per 1 mg. Finally, calculate the mass of NH_4NO_3 required to produce an equal amount of energy.

$$^{235}_{92}U + ^{1}_{0}n \rightarrow ^{144}_{58}Ce + ^{89}_{37}Rb + 3^{0}_{-1}e + 3^{1}_{0}n$$

Δm = [143.8817 + 88.8913 + 3(0.00055) + 3(1.00867)] - [234.9934 + 1.00867]
Δm = -0.2014 g (per mole of U-235)

ΔE = 9.00 x 10^{10} kJ/g x Δm = (9.00 x 10^{10} kJ/g)(-0.2014 g/mol.) = -1.81 x 10^{10} kJ/mol.

Energy from one milligram of U-235:

$$1.00 \text{ mg} \times \frac{1 \text{ g}}{1000 \text{ mg}} \times \frac{1 \text{ mol.}}{234.99 \text{ g}} \times \frac{-1.81 \times 10^{10} \text{ kJ}}{1 \text{ mol.}} = -7.70 \times 10^{4} \text{ kJ}$$

Thus 7.70 x 10^{4} kJ of energy is released.

Mass of NH_4NO_3 need to produce an equivalent amount of energy:

$$7.70 \times 10^{4} \text{ kJ} \times \frac{1 \text{ mol. } NH_4NO_3}{37.0 \text{ kJ}} \times \frac{80.04 \text{ g } NH_4NO_3}{1 \text{ mol. } NH_4NO_3} \times \frac{1 \text{ kg}}{1000 \text{ g}} = 167 \text{ kg}$$

a. $k = \dfrac{0.693}{8.1 \text{ d}} = 0.086 \text{ /d}$

$\ln\left(\dfrac{X_0}{X}\right) = kt = (0.086 \text{ /d})(2.0 \text{ d}) = 0.17$

$\dfrac{X_0}{X} = 1.19 \Rightarrow X_0 = 1.19X$

If X_0 is set to 100%, then $X = 84.4\%$, thus 84.4% remains after 2.0 days, meaning: **15.6% has disintegrated**.

b. The goal is provide, with this partially decayed I-131, the same activity the patient would receive from fresh I-131. This can be expressed mathematically as:

$(15.0 \text{ mg})(100\%) = (x \text{ mg})(84.4\%)$
$x = 17.8 \text{ mg}$

---

44. Refer to Sections 19.3 and 19.5, Example 19.6 and Problem 40 (above).

---

a. $5 \text{ ton} \times \dfrac{2000 \text{ lb}}{1 \text{ ton}} \times \dfrac{453.6 \text{ g}}{1 \text{ lb}} \times \dfrac{1 \text{ mol.}}{80.04 \text{ g}} \times \dfrac{-37.0 \text{ kJ}}{1 \text{ mol.}} = -2.10 \times 10^6 \text{ kJ}$

Thus 2.10×10^6 kJ of energy is released.

b. $-2.10 \times 10^6 \text{ kJ} \times \dfrac{1 \text{ g}}{-2.76 \text{ kJ}} = 7.61 \times 10^5 \text{ g} = 761 \text{ kg}$

c. $\Delta m = [88.8913 + 143.8817 + 3(0.00055) + 3(1.00867)] - [1.00867 + 234.9934]$
$\Delta m = -0.2014 \text{ g}$ (per mole of U-235)

$\Delta E = -2.10 \times 10^6 \text{ kJ} = 9.00 \times 10^{10} \text{ kJ/g} \times \Delta m$
$\Delta m = -2.33 \times 10^{-5} \text{ g}$

$-2.33 \times 10^{-5} \text{ g} \times \dfrac{1 \text{ mol. U-235}}{-0.2014 \text{ g}} \times \dfrac{235.0 \text{ g U-235}}{1 \text{ mol. U-235}} = 0.0272 \text{ g U-235}$

---

46. Refer to Chapter 16.

The activity of the solution is proportional to the concentration of I^-. Use the activity to set up a ratio for the two solutions and solve for the concentration of I^- in the filtrate.

$\dfrac{[I^-]_{0.050 \text{ M}}}{A_{0.050 \text{ M}}} = \dfrac{[I^-]_{\text{filtrate}}}{A_{\text{filtrate}}} \Rightarrow \dfrac{0.050 M}{1.25 \times 10^{10}} = \dfrac{[I^-]_{\text{filtrate}}}{2.50 \times 10^3}$

$[I^-]_{\text{filtrate}} = 1.0 \times 10^{-8} M$

$K_{sp} = [Ag^+][I^-] = [I^-]^2$ (since $[Ag^+] = [I^-]$)

$K_{sp} = (1.0 \times 10^{-8})^2 = 1.0 \times 10^{-16}$

Calculate k (in s^{-1}) from the half-life. Then calculate moles of $^3H_2$ using the ideal gas law. Finally, calculate N and A.

$$k = \frac{0.693}{12.3\,\text{yr}} \times \frac{1\,\text{yr}}{365\,\text{d}} \times \frac{1\,\text{d}}{24\,\text{hr}} \times \frac{1\,\text{hr}}{3600\,\text{s}} = 1.79 \times 10^{-9}/\text{s}$$

$$n = \frac{PV}{RT} = \frac{(1\,\text{atm})(0.00100\,\text{L})}{(0.0821\,\text{L} \cdot \text{atm/mol} \cdot \text{K})(273\,\text{K})} = 4.46 \times 10^{-5}\,\text{mol.}\ ^3H_2$$

$$N = 4.46 \times 10^{-5}\,\text{mol.}\ ^3H_2 \times \frac{6.02 \times 10^{23}\,\text{molecules}\ ^3H_2}{1\,\text{mol.}} \times \frac{2\,\text{atoms}\ ^3H}{1\ ^3H_2} = 5.37 \times 10^{19}\,\text{atoms}\ ^3H$$

$$A = 5.37 \times 10^{19}\,\text{atoms} \times 1.79 \times 10^{-9}/\text{s} \times \frac{1\,\text{Ci}}{3.700 \times 10^{10}\,\text{atoms/s}} = 2.60\,\text{Ci}$$

The A_0/A ratio represents a "dilution factor" that can be used to calculate the total volume (volume of blood).

$$5.0\,\text{mL} \times \frac{1.7 \times 10^5\,\text{cps}}{1.3 \times 10^3\,\text{cps}} = 6.5 \times 10^2\,\text{mL}$$

$$C_8H_{18}(l) + \text{}^{25}/_2 O_2(g) \rightarrow 8CO_2(g) + 9H_2O(g)$$

$\Delta H° = \Sigma\ \Delta H_f°_{(products)} - \Sigma\ \Delta H_f°_{(reactants)}$
$\Delta H° = [(8\,\text{mol.})(-393.5\,\text{kJ/mol.}) + (9\,\text{mol.})(-241.8\,\text{kJ/mol.})]$
$\quad\quad - [(1\,\text{mol.})(-249.9\,\text{kJ/mol.}) + (^{25}/_2\,\text{mol.})(0\,\text{kJ/mol.})]$
$\Delta H° = -5074\,\text{kJ}$

Burning 1 mol. octane produces 5074 kJ

Energy from one gram of U-235 (see problem 40):

$$1.00\,\text{g U-235} \times \frac{1\,\text{mol. U-235}}{234.99\,\text{g U-235}} \times \frac{-1.81 \times 10^{10}\,\text{kJ}}{1\,\text{mol. U-235}} = -7.70 \times 10^7\,\text{kJ}$$

Mass of octane need to produce an equivalent amount of energy.

$$7.70 \times 10^7 \text{ kJ} \times \frac{1 \text{ mol. C}_8\text{H}_{18}}{5074 \text{ kJ}} \times \frac{114.2 \text{ g C}_8\text{H}_{18}}{1 \text{ mol. C}_8\text{H}_{18}} \times \frac{1 \text{ mL}}{0.703 \text{ g}} \times \frac{1 \text{ L}}{1000 \text{ mL}} = 2.47 \times 10^3 \text{ L}$$

(Note: put in more familiar units, 1 oz of U-235 produces as much energy as 18,000 gal of octane!)

---

54. Refer to Sections 19.1 and 19.2 and Chapter 5.

Calculate k using $t_{1/2}$. Then convert the mass of Po-210 to moles to get X_0. Then solve for X to determine the amount of Po-210 remaining. Calculate the amount of Po-210 that decayed (and thus the amount of He that is formed). Finally, apply the ideal gas equation to determine the volume of He produced.

$$^{210}_{84}\text{Po} \rightarrow {}^{4}_{2}\text{He} + {}^{206}_{82}\text{Pb}$$

$$k = \frac{0.693}{138 \text{ d}} \times \frac{1 \text{ d}}{24 \text{ hr}} = 2.09 \times 10^{-4} / \text{hr}$$

$$25.00 \text{ g Po-210} \times \frac{1 \text{ mol. Po-210}}{209.94 \text{ g Po-210}} = 0.1191 \text{ mol Po-210}$$

$$\ln\left(\frac{X_0}{X}\right) = kt \implies \ln\left(\frac{0.1191 \text{ mol.}}{X}\right) = (2.09 \times 10^{-4} / \text{hr})(75 \text{ hr})$$

$\ln (0.1191 \text{ mol.}) - \ln (X) = 0.016$
$\ln (X) = -2.144$
$X = e^{-2.144} = 0.117$ (mol. of Po-210 remaining)

$X_0 - X = 0.1191 - 0.117 = 0.002$ (mol. Po-210 that decayed = mol. He produced)

$$V = \frac{nRT}{P} = \frac{(0.002 \text{ mol.})(0.0821 \text{ L} \cdot \text{atm/mol.} \cdot \text{K})(298 \text{ K})}{1.20 \text{ atm}} = 0.04 \text{ L He} = 40 \text{ mL}$$

---

56. Refer to Sections 19.1, 19.2, 19.4 and 19.5.

---

a. **False.** The mass number is unchanged during β-emission since the β-particle has no mass.

b. **False.** Rate = kX. Therefore a large k corresponds to a rapid rate of decay.

c. **False.** Fusion gives off more energy per gram of fuel than fission.

Calculate the percent Am-241 that remains after one year, setting X_0 to 100%.

$$\ln\left(\frac{X_0}{X}\right) = kt \;\Rightarrow\; \ln\left(\frac{100}{X}\right) = (1.51 \times 10^{-3}\,/\text{yr})(1\,\text{yr})$$

$\ln(100) - \ln(X) = 1.51 \times 10^{-3}$

$\ln(X) = 4.60$

$X = e^{4.60} = 99.8$ (percent Am-241 remaining)

Thus 99.8% of the Am-241 remains. At that rate, the Am-241 will last a lifetime (only 14% decays in 100 years).

60. *Refer to Sections 11.3 and 19.2.*

Plot $\ln(A)$ versus time (in hours). The slope of the line is $-k$, which is used to calculate $t_{1/2}$.

Time (h)	0.00	0.50	1.00	1.50	2.00	2.50
A (disintegrations/h)	14,472	13,095	11,731	10,615	9,605	8,504
ln (A)	9.580	9.480	9.370	9.270	9.170	9.048

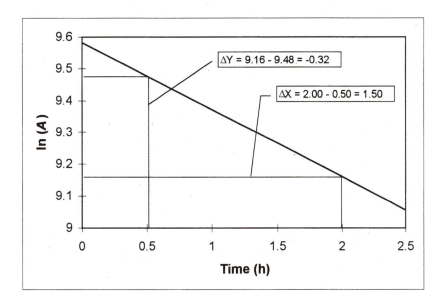

$$-k = \text{slope} = \frac{\Delta Y}{\Delta X} = \frac{-0.32}{1.5} = -0.21$$

$k = 0.21$

$$k = \frac{0.693}{t_{1/2}} \Rightarrow 0.21 = \frac{0.693}{t_{1/2}}$$

$t_{1/2} = 3.3 \text{ hr}$

---

62. *Refer to Section 19.2.*

From $t_{1/2}$, calculate k and, from k and A, calculate N.

$$k = \frac{0.693}{3.82 \text{ d}} \times \frac{1 \text{ d}}{24 \text{ hr}} \times \frac{1 \text{ hr}}{3600 \text{ s}} = 2.10 \times 10^{-6} / \text{s}$$

$$N = \frac{20 \times 10^{-12} \text{ Ci}}{2.10 \times 10^{-6} / \text{s}} \times \frac{3.700 \times 10^{10} \text{ atoms/s}}{1 \text{ Ci}} = 3.5 \times 10^{5} \text{ atoms}$$

This is a concentration of 3.5×10^{-5} atoms/L. Convert to moles per liter.

$$\frac{3.5 \times 10^{-5} \text{ atoms}}{1 \text{ L}} \times \frac{1 \text{ mol.}}{6.02 \times 10^{23} \text{ atoms}} = 5.8 \times 10^{-19} \text{ mol./L}$$

---

63. *Refer to Sections 19.2 and 19.3 and "Chemistry Beyond the Classroom."*

---

a. $1.00 \text{ g} \times 5.5 \times 10^{-11} / \text{min.} \times 45 \text{ min.} = 2.5 \times 10^{-9} \text{ g}$

b. $\Delta m = 234.9934 + 4.00150 - 239.0006 = -0.0057 \text{ g}$

$$\Delta m_{(1g)} = 2.5 \times 10^{-9} \text{ g} \times \frac{1 \text{ mol.}}{239.0 \text{ g}} \times \frac{-0.0057 \text{ g}}{1 \text{ mol.}} = -6.0 \times 10^{-14} \text{ g}$$

$\Delta E = 9.00 \times 10^{10} \text{ kJ/g} \times (-6.0 \times 10^{-14}) \text{ g} = -5.4 \times 10^{-3} \text{ kJ} = -5.4 \text{ J}$

Thus 5.4 J of energy is given off.

c. $1 \text{ rad} = 10^{-2} \text{ J/kg}$
    rems $= n(\text{rads})$ $n = 10$ for α rays

$$\frac{5.4 \text{ J}}{75 \text{ kg}} \times \frac{1 \text{ rad}}{10^{-2} \text{ J/kg}} = 7.2 \text{ rads}$$

rems = 7.2 rads $\times$ 10 rems/rad = 72 rems

64. Refer to the equations stated in the problem.

a. $E = 8.99 \times 10^9 q_1 q_2 / r = (8.99 \times 10^9)(1.60 \times 10^{-19}\ \text{C})^2 / (2 \times 10^{-15}\ \text{m}) = 1 \times 10^{-13}\ \text{J}$

b. Convert the mass of the deuteron to kilograms.

$$2\ \text{deuterons} \times \frac{2.01355\ \text{g}}{1\ \text{mol.}} \times \frac{1\ \text{mol.}}{6.022 \times 10^{23}} \times \frac{1\ \text{kg}}{1000\ \text{g}} = 6.687 \times 10^{-27}\ \text{kg}$$

$E = mv^2/2 \quad = \quad 2E/m = v^2$

$v = [(2)(1 \times 10^{-13}\ \text{J}) / (6.687 \times 10^{-27}\ \text{kg})]^{\frac{1}{2}} = 5 \times 10^6\ \text{m/s}$

65. Refer to Section 19.3 and Example 19.4.

a. $\Delta m = 4.00150 - 2(2.01355) = -0.02560\ \text{g (per 2 mol. H-2)}$

$$\Delta m_{(1g)} = 1.0000\ \text{g} \times \frac{1\ \text{mol.}}{2.01355\ \text{g}} \times \frac{-0.02560\ \text{g}}{2\ \text{mol.}} = -0.006357\ \text{g}$$

$\Delta E = 9.00 \times 10^{10}\ \text{kJ/g} \times (-0.006357\ \text{g}) = -5.72 \times 10^8\ \text{kJ}$ (per gram of deuterium fused)

b. $1.3 \times 10^{24}\ \text{g} \times \dfrac{0.0017\ \text{g}}{100\ \text{g}} = 2.2 \times 10^{19}\ \text{g H-2}$

$2.2 \times 10^{19}\ \text{g} \times -5.72 \times 10^8\ \text{kJ/g} = -1.3 \times 10^{28}\ \text{kJ}$

c. $\dfrac{2.3 \times 10^{17}\ \text{kJ}}{1.3 \times 10^{28}\ \text{kJ}} = 1.8 \times 10^{-11}$

Chapter 20: Chemistry of the Metals

---

2. *Refer to Section 20.1 and Chapter 5.*

$2Al_2O_3(l) \rightarrow 4Al(l) + 3O_2(g)$

$$n = \frac{PV}{RT} = \frac{(0.988\,atm)(2.00\,L)}{(0.0821\,L\cdot atm/mol.\cdot K)(298\,K)} = 0.0808\,mol.\,O_2$$

$$0.0808\,mol.\,O_2 \times \frac{4\,mol.\,Al}{3\,mol.\,O_2} \times \frac{26.98\,g\,Al}{1\,mol.\,Al} = 2.91\,g\,Al$$

---

4. *Refer to Section 20.1.*

$Cu_2S(s) + O_2(g) \rightarrow 2Cu(s) + SO_2(g)$

---

6. *Refer to Example 20.1 and Chapter 17.*

$\Delta H° = \Sigma\,\Delta H_f°_{(products)} - \Sigma\,\Delta H_f°_{(reactants)}$
$\Delta H° = [(2\,mol.)(0.0\,kJ/mol.) + (1\,mol.)(-296.8\,kJ/mol.)]$
$\quad\quad - [(1\,mol.)(-79.5\,kJ/mol.) + (1\,mol.)(0.0\,kJ/mol.)]$
$\Delta H° = -217.3\,kJ$

$\Delta S° = \Sigma\,S°_{(products)} - \Sigma\,S°_{(reactants)}$
$\Delta S° = [(2\,mol.)(0.0332\,kJ/mol.\cdot K) + (1\,mol.)(0.2481\,kJ/mol.\cdot K)]$
$\quad\quad - [(1\,mol.)(0.1209\,kJ/mol.\cdot K) + (1\,mol.)(0.2050\,kJ/mol.\cdot K)]$
$\Delta S° = -0.0114\,kJ/K$

$\Delta G_f° = \Delta H° - T\Delta S°$
$\Delta G° = -217.3\,kJ - (473\,K)(-0.0114\,kJ/K) = -211.9\,kJ$

---

8. *Refer to Section 20.1.*

a. $Fe_2O_3(s) + 3CO(g) \rightarrow 2Fe(l) + 3CO_2(g)$

b. $C(s) + O_2(g) \rightarrow CO_2(g)$

10. Refer to Section 20.1 and Chapter 18.

one metric ton = 1×10^3 kg

$$1.000 \times 10^3 \text{ kg} \times \frac{1000 \text{ g}}{1 \text{ kg}} \times \frac{1 \text{ mol. Zn}}{65.39 \text{ g}} = 1.529 \times 10^4 \text{ mol. Zn}$$

$$1.529 \times 10^4 \text{ mol. Zn} \times \frac{2 \text{ mol. } e^-}{1 \text{ mol. Zn}} \times \frac{9.648 \times 10^4 \text{ C}}{1 \text{ mol. } e^-} = 2.950 \times 10^9 \text{ C}$$

$$J = VC = (3.0 \text{ V})(2.950 \times 10^9 \text{ C}) = 8.9 \times 10^9 \text{ J}$$

$$8.9 \times 10^9 \text{ J} \times \frac{1 \text{ kWh}}{3.600 \times 10^6 \text{ J}} = 2.5 \times 10^3 \text{ kWh}$$

12. Refer to Section 20.1.

Chalcopyrite, $CuFeS_2$ has a ratio of Cu:S of 1:2; thus for each mole of Cu, 2 moles of SO_2 are formed. Chalcopyrite ore is not pure chalcopyrite, but contains $CuFeS_2$ and other minerals. The ore contains 0.75% Cu, meaning 100 g ore has 0.75 g Cu.

$$4.00 \times 10^3 \text{ ft}^3 \times \left(\frac{12 \text{ in}}{1 \text{ ft}}\right)^3 \times \left(\frac{2.54 \text{ cm}}{1 \text{ in}}\right)^3 \times \frac{2.6 \text{ g ore}}{1 \text{ cm}^3} = 2.95 \times 10^8 \text{ g ore}$$

$$2.95 \times 10^8 \text{ g ore} \times \frac{0.75 \text{ g Cu}}{100 \text{ g ore}} \times \frac{1 \text{ mol. Cu}}{63.55 \text{ g Cu}} = 3.47 \times 10^4 \text{ mol. Cu}$$

$$3.47 \times 10^4 \text{ mol. Cu} \times \frac{2 \text{ mol. } SO_2}{1 \text{ mol. Cu}} = 6.94 \times 10^4 \text{ mol. } SO_2$$

$$V = \frac{nRT}{P} = \frac{(6.94 \times 10^4 \text{ mol.})(0.0821 \text{ L} \cdot \text{atm/mol.} \cdot \text{K})(298 \text{ K})}{1 \text{ atm}} = 1.70 \times 10^6 \text{ L } SO_2$$

14. Refer to Section 20.2 and Table 20.1.

a. potassium nitride K_3N

b. potassium iodide KI

c. potassium hydroxide KOH

d. potassium hydride KH

e. potassium sulfide K_2S

a. $Na_2O_2(s) + 2H_2O(l) \rightarrow 2Na^+(aq) + 2OH^-(aq) + H_2O_2(aq)$
   sodium ion, hydroxide ion, hydrogen peroxide.

b. $2Ca(s) + O_2(g) \rightarrow 2CaO(s)$
   calcium oxide.

c. $Rb(s) + O_2(g) \rightarrow RbO_2(s)$
   rubidium superoxide.

d. $SrH_2(s) + 2H_2O(l) \rightarrow Sr^{2+}(aq) + 2OH^-(aq) + 2H_2(g)$
   strontium ion, hydroxide ion, hydrogen gas.

$$n = \frac{PV}{RT} = \frac{(1.00\,atm)(1.00\,L)}{(0.0821\,L \cdot atm/mol. \cdot K)(310\,K)} = 0.0393\,mol.\,air$$

$$0.0393\,mol.\,air \times \frac{5.00\,mol.\,CO_2}{100\,mol.\,air} = 0.00196\,mol.\,CO_2$$

$0.00196\,mol.\,CO_2 \times 0.90 = 0.00177\,mol.\,CO_2$ (to be removed).

$4KO_2(s) + 2H_2O(g) \rightarrow 3O_2(g) + 4KOH(s)$
$KOH(s) + CO_2(g) \rightarrow KHCO_3(g)$

$$0.00177\,mol.\,CO_2 \times \frac{1\,mol.\,KOH}{1\,mol.\,CO_2} \times \frac{4\,mol.\,KO_2}{4\,mol.\,KOH} \times \frac{71.10\,g}{1\,mol.\,KO_2} = 0.126\,g\,KO_2$$

a. $\quad Co^0 \rightarrow Co^{2+} + 2e^-$ oxidation half-reaction
   $\quad \underline{2H^+ + 2e^- \rightarrow H_2}$ reduction half-reaction
   $\quad\quad Co(s) + 2H^+(aq) \rightarrow Co^{2+}(aq) + H_2(g)$

b. $\quad Cu^0 \rightarrow Cu^{2+} + 2e^-$ oxidation half-reaction
   $\quad 4H^+ + NO_3^- + 3e^- \rightarrow NO + 2H_2O$ reduction half-reaction

          balancing the electrons gives:
   $3Cu^0 \rightarrow 3Cu^{2+} + 6e^-$
   $\underline{8H^+ + 2NO_3^- + 6e^- \rightarrow 2NO + 4H_2O}$
   $3Cu(s) + 8H^+(aq) + 2NO_3^-(aq) \rightarrow 3Cu^{2+}(aq) + 2NO(g) + 4H_2O(l)$

c. $14H^+(aq) + Cr_2O_7^{2-}(aq) + 6e^- \rightarrow 2Cr^{3+}(aq) + 7H_2O(l)$
  (see Chapter 4, Problem 54b.)

---

22. Refer to Section 20.3 and Chapter 4.

$Cd(s) + H^+(aq) + Cl^-(aq) + NO_3^-(aq) \rightarrow CdCl_4^{2-}(aq) + NO(g) + H_2O(l)$

$Cd^0 \rightarrow Cd^{2+} + 2e^-$ Oxidation half-reaction
$Cd(s) + 4Cl^-(aq) \rightarrow CdCl_4^{2-}(aq) + 2e^-$

$N^{5+} + 3e^- \rightarrow N^{2+}$ Reduction half-reaction
$NO_3^-(aq) + 3e^- \rightarrow NO(g)$
$4H^+(aq) + NO_3^-(aq) + 3e^- \rightarrow NO(g) + 2H_2O(l)$

    Balancing electrons gives:
$3Cd(s) + 12Cl^-(aq) + 8H^+(aq) + 2NO_3^-(aq) \rightarrow 2NO(g) + 4H_2O(l) + 3CdCl_4^{2-}(aq)$

---

24. Refer to Section 20.3 and Chapter 4.

---

a. $Fe^0 \rightarrow Fe^{3+} + 3e^-$ Oxidation half-reaction

  $N^{5+} + e^- \rightarrow N^{4+}$ Reduction half-reaction
  $2H^+(aq) + NO_3^-(aq) + e^- \rightarrow NO_2(g) + H_2O(l)$

    Balancing electrons gives:
  $6H^+(aq) + Fe(s) + 3NO_3^-(aq) \rightarrow 3NO_2(g) + Fe^{3+}(aq) + 3H_2O(l)$

b. $Cr^{3+} \rightarrow Cr^{6+} + 3e^-$ Oxidation half-reaction
  $Cr(OH)_3(s) \rightarrow CrO_4^{2-}(aq) + 3e^-$
  $5\,OH^-(aq) + Cr(OH)_3(s) \rightarrow CrO_4^{2-}(aq) + 3e^-$
  $5\,OH^-(aq) + Cr(OH)_3(s) \rightarrow CrO_4^{2-}(aq) + 3e^- + 4H_2O(l)$

  $O^0 + 2e^- \rightarrow O^{2-}$ Reduction half-reaction
  $O_2(g) + 4e^- \rightarrow 2H_2O(l)$
  $O_2(g) + 4e^- \rightarrow 2H_2O(l) + 4\,OH^-(aq)$
  $4H_2O(l) + O_2(g) + 4e^- \rightarrow 2H_2O(l) + 4\,OH^-(aq)$
  $2H_2O(l) + O_2(g) + 4e^- \rightarrow 4\,OH^-(aq)$

    Balancing electrons gives:
  $20\,OH^-(aq) + 4Cr(OH)_3(s) + 6H_2O(l) + 3O_2(g) + 12e^-$
                        $\rightarrow 12\,OH^-(aq) + 4CrO_4^{2-}(aq) + 12e^- + 16H_2O(l)$
  $8\,OH^-(aq) + 4Cr(OH)_3(s) + 3O_2(g) \rightarrow 4CrO_4^{2-}(aq) + 10H_2O(l)$

Calculate $E°$ for the reaction. If $E°$ is positive, then the reaction will occur.

$$NO_3^-(aq) \rightarrow NO(g) \qquad\qquad E_{red}° = +0.964 \text{ V}$$

a. $Cd \rightarrow Cd^{2+}$ $\qquad\qquad\qquad\qquad E_{ox}° = 0.402 \text{ V}$

$E° = E_{red}° + E_{ox}° = 0.964 + 0.402 = 1.366 \text{ V}$
$E°$ is positive, so the reaction will occur.

b. $Cr \rightarrow Cr^{2+}$ $\qquad\qquad\qquad\qquad E_{ox}° = 0.912 \text{ V}$

$E° = E_{red}° + E_{ox}° = 0.964 + 0.912 = 1.876 \text{ V}$
$E°$ is positive, so the reaction will occur.

$Cr \rightarrow Cr^{3+}$ $\qquad\qquad\qquad\qquad E_{ox}° = 0.744 \text{ V}$

$E° = E_{red}° + E_{ox}° = 0.964 + 0.744 = 1.708 \text{ V}$
$E°$ is positive, so the reaction will occur.

c. $Co \rightarrow Co^{2+}$ $\qquad\qquad\qquad\qquad E_{ox}° = 0.282 \text{ V}$

$E° = E_{red}° + E_{ox}° = 0.964 + 0.282 = 1.246 \text{ V}$
$E°$ is positive, so the reaction will occur.

d. $Ag \rightarrow Ag^+$ $\qquad\qquad\qquad\qquad E_{ox}° = -0.799 \text{ V}$

$E° = E_{red}° + E_{ox}° = 0.964 - 0.799 = 0.165 \text{ V}$
$E°$ is positive, so the reaction will occur.

e. $Au \rightarrow Au^{3+}$ $\qquad\qquad\qquad\qquad E_{ox}° = -1.498 \text{ V}$

$E° = E_{red}° + E_{ox}° = 0.964 - 1.498 = -0.534 \text{ V}$
$E°$ is negative, so the reaction will **not** occur.

---

28. *Refer to Section 20.3 and Chapter 18.*

a. $2Co^{3+}(aq) + 2e^- \rightarrow 2Co^{2+}(aq)$ $\qquad E_{red}° = 1.953 \text{ V}$

$H_2O \rightarrow \frac{1}{2}O_2(g) + 2H^+(aq) + 2e^-$ $\qquad E_{ox}° = -1.229 \text{ V}$

$E° = E_{red}° + E_{ox}° = 1.953 - 1.229 = 0.724 \text{ V}$

b. $I_2(s) + 2e^- \rightarrow 2I^-(aq)$ $E^\circ_{red} = 0.534$ V

    $2Cr^{2+}(aq) \rightarrow 2Cr^{3+}(aq) + 2e^-$ $E^\circ_{ox} = 0.408$ V

    $E^\circ = E^\circ_{red} + E^\circ_{ox} = 0.534 + 0.408 = 0.942$ V

---

30. Refer to Section 20.3 and Chapter 18.

---

a. $Au^+ + e^- \rightarrow Au^0$ $E^\circ_{red} = 1.695$ V

    $Au^+ \rightarrow Au^{3+} + 2e^-$ $E^\circ_{red} = -1.400$ V

           Balancing electrons:

    $3Au^+ \rightarrow Au^{3+} + 2Au^0$ $E^\circ = 0.295$ V

$$E^\circ = \frac{0.0257}{n} \ln K \quad \Rightarrow \quad 0.295 = \frac{0.0257}{2} \ln K$$

$\ln K = 23.0$

$K = 9.74 \times 10^9$

b. $K = \dfrac{[Au^{3+}]}{[Au^+]^3} \quad \Rightarrow \quad 9.74 \times 10^9 = \dfrac{0.10\,M}{[Au^+]^3}$

$[Au^+]^3 = 1.03 \times 10^{-11}$

$[Au^+] = 2.2 \times 10^{-4}$

---

32. Refer to Section 20.2, Problem 18 above and Chapter 5.

     Calculate the volume of water vapor in the exhaled air, calculate the moles of water and then the amount of KO_2 that reacts, and finally the amount of KO_2 remaining.

$$116 \text{ L air} \times \frac{6.2 \text{ L H}_2\text{O}}{100 \text{ L air}} = 7.19 \text{ L H}_2\text{O}(g)$$

$$n = \frac{PV}{RT} = \frac{(0.984 \text{ atm})(7.19 \text{ L})}{(0.0821 \text{ L} \cdot \text{atm/mol} \cdot \text{K})(310 \text{ K})} = 0.278 \text{ mol. H}_2\text{O}$$

$$0.278 \text{ mol. H}_2\text{O} \times \frac{4 \text{ mol. KO}_2}{2 \text{ mol. H}_2\text{O}} \times \frac{71.10 \text{ g KO}_2}{1 \text{ mol. KO}_2} = 39.5 \text{ g KO}_2 \text{ (reacts)}$$

248 g - 39.5 g = 208 g (remains)

Calculate the [H$^+$] for the given concentrations of chromate and dichromate, then calculate pH from that.

$$K_a = \frac{[Cr_2O_7^{2-}]}{[CrO_4^{2-}]^2[H^+]^2} \quad \Rightarrow \quad 3 \times 10^{14} = \frac{0.10\,M}{(0.10\,M)^2[H^+]^2}$$

$[H^+]^2 = 3 \times 10^{-14}$

$[H^+] = 2 \times 10^{-7}\,M$

pH = -log[H$^+$] = - log(2 × 10^{-7}) = 6.7

Calculate the total pressure of H$_2$(g) and then the moles of H$_2$. From that calculate the mass of Zn and the mass percent of zinc in the alloy.

$$P_{H_2} = P_{total} - P_{H_2O} = 755 - 26.74 = 728\,mm\,Hg = 0.958\,atm$$

$$n = \frac{PV}{RT} = \frac{(0.958\,atm)(0.1057\,L)}{(0.0821\,L \cdot atm/mol. \cdot K)(300\,K)} = 0.00411\,mol.\,H_2$$

$$Zn(s) + 2H^+(aq) \rightarrow Zn^{2+}(aq) + H_2(g)$$

$$0.00411\,mol.\,H_2 \times \frac{1\,mol.\,Zn}{1\,mol.\,H_2} \times \frac{65.39\,g\,Zn}{1\,mol.\,Zn} = 0.269\,g\,Zn$$

$$mass\,\%\,Zn = \frac{0.269\,g\,Zn}{0.500\,g\,alloy} \times 100\% = 53.8\,\%\,Zn$$

100% - 53.8% Zn = 46.2% Cu

Write a balanced redox equation for the process, assuming basic conditions (since gold is extracted similarly, under basic conditions). To simplify the process, consider the oxygen to be reduced from O$_2$(g) to hydroxide ion, and ignore the spectator ions (silver and cyanide).

$Ag_2S(s) + CN^-(aq) + O_2(g) \rightarrow SO_2(g) + Ag(CN)_2^-(aq)$

$O^0 + 2e^- \rightarrow O^{2-}$

$O_2(g) + 4e^- \rightarrow 2OH^-(aq)$

$O_2(g) + 4e^- \rightarrow 2OH^-(aq) + 2OH^-(aq)$

$2H_2O(l) + O_2(g) + 4e^- \rightarrow 4OH^-(aq)$

$S^{2-} \rightarrow S^{4+} + 6e^-$

$S^{2-}(aq) \rightarrow SO_2(g) + 6e^-$

$4OH^-(aq) + S^{2-}(aq) \rightarrow SO_2(g) + 6e^-$

$4OH^-(aq) + S^{2-}(aq) \rightarrow SO_2(g) + 6e^- + 2H_2O(l)$

Balancing electrons gives:

$6H_2O(l) + 3O_2(g) + 12e^- + 8OH^-(aq) + 2S^{2-}(aq) \rightarrow 2SO_2(g) + 12e^- + 4H_2O(l) + 12OH^-(aq)$

$2H_2O(l) + 3O_2(g) + 2S^{2-}(aq) \rightarrow 2SO_2(g) + 4OH^-(aq)$

Now add in the spectator ions, keeping the proper ratio.

$2H_2O(l) + 3O_2(g) + 2Ag_2S(s) \rightarrow 2SO_2(g) + 4OH^-(aq) + 4Ag^+(aq)$

$2H_2O(l) + 3O_2(g) + 2Ag_2S(s) + 8CN^-(aq) \rightarrow 2SO_2(g) + 4OH^-(aq) + 4Ag(CN)_2^-(aq)$

This equation describes the conversion of insoluble Ag_2S to the soluble $Ag(CN)_2^-$ complex. The final step is the reduction of the silver cation to silver metal.

$Zn^0 \rightarrow Zn^{2+} + 2e^-$

$Ag^+ + e^- \rightarrow Ag^0$

Balancing electrons gives:

$Zn(s) + 2Ag^+(aq) + 2e^- \rightarrow 2Ag(s) + Zn^{2+}(aq) + 2e^-$

$Zn(s) + 2Ag^+(aq) \rightarrow 2Ag(s) + Zn^{2+}(aq)$

Adding back the cyanide spectator ion:

$Zn(s) + 2Ag(CN)_2^-(aq) \rightarrow 2Ag(s) + Zn(CN)_4^{2-}(aq)$

---

40. Refer to Section 20.3 and Chapter 18.

---

a. The strongest reducing agent is the one most easily oxidized. Of the choices, **Cr^{2+}** is the only one with a positive E°_{ox} (+0.408 V) and is thus the strongest reducing agent.

b. The strongest oxidizing agent is the one most easily reduced. Of the choices, **Au^+** has the highest E°_{red} (+1.695 V), and is thus the strongest oxidizing agent.

c. The weakest reducing agent is the one least readily oxidized. Of the choices, **Co^{2+}** has the lowest E°_{ox} (-1.953 V), and is thus the weakest reducing agent.

d. The weakest oxidizing agent is the one least readily reduced. Of the choices, **Mn^{2+}** has the lowest E°_{red} (-1.182 V), and is thus the weakest oxidizing agent.

The only reactants in the two reactions are barium and oxygen. Calculate the amount of oxygen that reacts and then the mole ratio of barium to oxygen. The excess moles of oxygen must be used to form BaO_2, and thus indicate the amount of BaO_2 formed.

22.38 g total - 20.00 g Ba = 2.38 g oxygen.

$$20.00 \, g \times \frac{1 \, mol. \, Ba}{137.3 \, g} = 0.1457 \, mol. \, Ba$$

$$2.38 \, g \times \frac{1 \, mol. \, O}{16.00 \, g} = 0.149 \, mol. \, O$$

$$mole \, ratio = \frac{0.149 \, mol. \, O}{0.1457 \, mol. \, Ba} = 1.02 \, mol. \, O \, per \, mol. \, Ba \quad (in \, the \, product \, mixture)$$

This indicates that there are 0.02 moles 'excess' oxygen. Since barium oxide (BaO) has a 1:1 Ba:O ratio, the 'excess' oxygens must belong to barium peroxide (BaO_2); giving 0.02 mol. BaO_2, with the remainder (0.98 mol.) BaO.

This can be shown mathematically by setting x = mol. Ba in BaO, and y = mol. Ba in BaO_2. Consequently, x = mol. O in Ba, and $2y$ = mol. O in BaO_2. Then:

Ba:	$x + y = 1$	total moles Ba
O:	$x + 2y = 1.02$	total moles O
	$y = 0.02$	This is moles of BaO_2 (and x = 0.98 mol. BaO)

$$\frac{0.98 \, mol. \, BaO}{1 \, mol. \, total} \times 100\% = 98\% \, BaO$$

$$\frac{0.02 \, mol. \, BaO_2}{1 \, mol. \, total} \times 100\% = 2\% \, BaO_2$$

a. The oxalic acid is a chelating ligand, and Fe^{3+} has 6 coordination sites, suggesting that a complex with 3 oxalates forms. Recall that oxalic acid is a diprotic acid (has 2 H^+).

$$Fe(OH)_3(s) + 3H_2C_2O_4(aq) \rightarrow [Fe(C_2O_4)_3]^{3-}(aq) + 3H_2O(l) + 3H^+(aq)$$

b. $1.0 \, g \, Fe(OH)_3 \times \dfrac{1 \, mol. \, Fe(OH)_3}{106.9 \, g} \times \dfrac{3 \, mol. \, H_2C_2O_4}{1 \, mol. \, Fe(OH)_3} \times \dfrac{1 \, L \, solution}{0.10 \, mol. \, H_2C_2O_4} = 0.28 \, L$

Write a balanced redox equation for the dichromate-ferric ion couple. Then calculate the initial amount of Fe^{2+} used, and the amount in excess. The difference gives the amount that reacted with MnO_4^-. Write a balanced redox equation for the reduction of permanganate with ferrous ion to get a mole ratio of iron to manganese. From the mole ratio, calculate the amount of permanganate and then the percent manganese in the sample.

$$Fe^{2+} \rightarrow Fe^{3+} + e^- \qquad\qquad \text{oxidation half-reaction}$$

$$Cr^{6+} + 3e^- \rightarrow Cr^{3+} \qquad\qquad \text{reduction half-reaction}$$
$$14H^+(aq) + Cr_2O_7^{2-}(aq) + 6e^- \rightarrow 2Cr^{3+}(aq) + 7H_2O(l) \quad \textit{(see problem 20c)}$$

balancing electrons and adding the equations gives:
$$14H^+(aq) + Cr_2O_7^{2-}(aq) + 6Fe^{2+}(aq) \rightarrow 6Fe^{3+}(aq) + 2Cr^{3+}(aq) + 7H_2O(l)$$

$$0.07500\,\text{L sol'n} \times \frac{0.125\,\text{mol. FeSO}_4}{1\,\text{L sol'n}} \times \frac{1\,\text{mol. Fe}^{2+}}{1\,\text{mol. FeSO}_4} = 0.00938\,\text{mol. Fe}^{2+} \quad \text{(initial)}$$

$$0.01350\,\text{L sol'n} \times \frac{0.100\,\text{mol. Cr}_2\text{O}_7^{2-}}{1\,\text{L sol'n}} \times \frac{6\,\text{mol. Fe}^{2+}}{1\,\text{mol. Cr}_2\text{O}_7^{2-}} = 0.00810\,\text{mol. Fe}^{2+} \quad \text{(excess)}$$

0.00938 mol. $- 0.00810 = 0.00128$ mol. Fe^{2+} (reacted with MnO_4^-)

$$Fe^{2+} \rightarrow Fe^{3+} + e^- \qquad\qquad \text{oxidation half-reaction}$$

$$Mn^{7+} + 5e^- \rightarrow Mn^{2+} \qquad\qquad \text{reduction half-reaction}$$
$$8H^+(aq) + MnO_4^-(aq) + 5e^- \rightarrow Mn^{2+}(aq) + 4H_2O(l) \quad \textit{(see Chapter 4, problem 50a)}$$

balancing electrons and adding the equations gives:
$$8H^+(aq) + MnO_4^-(aq) + 5Fe^{2+}(aq) \rightarrow 5Fe^{3+}(aq) + Mn^{2+}(aq) + 4H_2O(l)$$

$$0.00128\,\text{mol. Fe}^{2+} \times \frac{1\,\text{mol. MnO}_4^-}{5\,\text{mol. Fe}^{2+}} \times \frac{1\,\text{mol. Mn}}{1\,\text{mol. MnO}_4^-} \times \frac{54.94\,\text{g}}{1\,\text{mol. Mn}} = 0.0141\,\text{g Mn}$$

$$\text{mass \% Mn} = \frac{\text{mass Mn}}{\text{mass sample}} \times 100\% = \frac{0.0141\,\text{g}}{0.500\,\text{g}} \times 100\% = 2.82\,\%$$

$\Delta G^\circ = \Delta H^\circ - T\Delta S^\circ$ and,
$\Delta G^\circ = -RT \ln K$, but $\ln(1.00) = 0$, thus,
$\Delta G^\circ = 0 = \Delta H^\circ - T\Delta S^\circ$

Calculate ΔH° and ΔS°, and then solve for T.

$\Delta H_{(reaction)} = \Sigma\ \Delta H_f^\circ\ _{(products)} - \Sigma\ \Delta H_f^\circ\ _{(reactants)}$

$\Delta H_{(reaction)} = [(1\ mol.)(0\ kJ/mol.) + (1\ mol.)(0\ kJ/mol.)] - [(1\ mol.)(-520.0\ kJ/mol.)]$

$\Delta H_{(reaction)} = +520.0\ kJ$

$\Delta S_{(reaction)} = \Sigma\ S_f^\circ\ _{(products)} - \Sigma\ S_f^\circ\ _{(reactants)}$

$\Delta S_{(reaction)} = [(1\ mol.)(0.0320\ kJ/mol.\cdot K) + (1\ mol.)(0.2050\ kJ/mol.\cdot K)]$
$\qquad\qquad - [(1\ mol.)(0.0530\ kJ/mol.\cdot K)]$

$\Delta S_{(reaction)} = 0.184\ kJ/K$

$\Delta G^\circ = 0 = 520.0\ kJ - (T)(0.184\ kJ/K)$
$T = 2.83 \times 10^3\ K = 2.56 \times 10^3\ ^\circ C$

---

45. Refer to Section 20.3, Figure 20.11 and Chapters 4 and 15.

$Cr_2O_7^{2-}(aq) + 2OH^-(aq) \rightarrow CrO_4^{2-}(aq) + H_2O(l)$

$CrO_4^{2-}(aq) + 2Ag^+(aq) \rightarrow Ag_2CrO_4(s)$

$Ag_2CrO_4(s) + 4NH_3(aq) \rightarrow 2[Ag(NH_3)_2]^+(aq) + CrO_4^{2-}(aq)$

$2[Ag(NH_3)_2]^+(aq) + CrO_4^{2-}(aq) + 4H^+(aq) \rightarrow Ag_2CrO_4(s) + 4NH_4^+(aq)$

Chapter 21: Chemistry of the Nonmetals

---

2. Refer to Chapter 2.

    a. bromic acid   b. potassium hypoiodate

    c. sodium chlorite   d. sodium perbromate

---

4. Refer to Chapter 2.

    a. $KBrO_2$ b. $CaBr_2$

    c. $NaIO_4$ d. $Mg(ClO)_2$

---

6. Refer to Section 21.4.

To act as a reducing agent, the anion itself must be oxidized. Thus an oxoanion in which the nonmetal is in its highest oxidation state cannot be further oxidized and cannot act as a reducing agent.

    a. NO_3^- b. SO_4^{2-} c. ClO_4^-

---

8. Refer to Section 21.2 and Chapter 2.

    a. $SO_2 + H_2O \rightarrow H_2SO_3$

    b. $Cl_2O + H_2O \rightarrow 2HClO$

    c. $P_4O_6 + 6H_2O \rightarrow 4H_3PO_3$

---

10. Refer to Section 21.2 and Chapter 2.

    a. NaN_3 b. H_2SO_3

    c. N_2H_4 d. NaH_2PO_4

12. Refer to Section 21.2 and Chapter 2.

    a. H_2S b. N_2H_4 c. PH_3

14. Refer to Sections 21.2 and 21.4.

    a. NH_3, N_2H_4 b. HNO_3

    c. HNO_2 d. HNO_3

16. Refer to Section 21.1 and Chapter 4.

First write the half-reactions, then combine them and cancel common elements.

    a. $I^- \rightarrow I^0 + e^-$

        $2I^-(aq) \rightarrow I_2(s) + 2e^-$ oxidation half-reaction

        $S^{6+} + 2e^- \rightarrow S^{4+}$

        $SO_4^{2-}(aq) + 2e^- \rightarrow SO_2(g)$

        $4H^+(aq) + SO_4^{2-}(aq) + 2e^- \rightarrow SO_2(g)$

        $4H^+(aq) + SO_4^{2-}(aq) + 2e^- \rightarrow SO_2(g) + 2H_2O(l)$ reduction half-reaction

           $4H^+(aq) + SO_4^{2-}(aq) + 2e^- + 2I^-(aq) \rightarrow I_2(s) + 2e^- + SO_2(g) + 2H_2O(l)$

           $4H^+(aq) + SO_4^{2-}(aq) + 2I^-(aq) \rightarrow I_2(s) + SO_2(g) + 2H_2O(l)$

    b. $I^- \rightarrow I^0 + e^-$

        $2I^-(aq) \rightarrow I_2(s) + 2e^-$ oxidation half-reaction

        $Cl^0 + e^- \rightarrow Cl^-$

        $Cl_2(g) + 2e^- \rightarrow 2Cl^-(aq)$ reduction half-reaction

        $Cl_2(g) + 2e^- + 2I^-(aq) \rightarrow I_2(s) + 2e^- + 2Cl^-(aq)$

        $Cl_2(g) + 2I^-(aq) \rightarrow I_2(s) + 2Cl^-(aq)$

18. Refer to Sections 21.1 and 21.4 and Chapter 4.

First write the half-reactions, then combine them and cancel common elements.

    a. In this reaction, one HClO molecule oxidizes a second HClO molecule.

        $Cl^+ \rightarrow Cl^{3+} + 2e^-$

        $HClO(aq) \rightarrow HClO_2(aq) + 2e^-$

        $HClO(aq) \rightarrow HClO_2(aq) + 2e^- + 2H^+(aq)$

        $H_2O(l) + HClO(aq) \rightarrow HClO_2(aq) + 2e^- + 2H^+(aq)$ oxidation half-reaction

$Cl^+ + e^- \rightarrow Cl^0$

$2HClO(aq) + 2e^- \rightarrow Cl_2(g)$

$2H^+(aq) + 2HClO(aq) + 2e^- \rightarrow Cl_2(g)$

$2H^+(aq) + 2HClO(aq) + 2e^- \rightarrow Cl_2(g) + 2H_2O(l)$ reduction half-reaction

$2H^+(aq) + 2HClO(aq) + 2e^- + H_2O(l) + HClO(aq)$
$$\rightarrow HClO_2(aq) + 2e^- + 2H^+(aq) + Cl_2(g) + 2H_2O(l)$$

$3HClO(aq) \rightarrow HClO_2(aq) + Cl_2(g) + 2H_2O(l)$

(Note: this equation could also be written with the anions ClO^- and ClO_2^- instead of the acids. However, since these are weak acids, they exist in solution mainly as the acid.)

b. $Cl^{5+} \rightarrow Cl^{7+} + 2e^-$

$ClO_3^-(aq) \rightarrow ClO_4^-(aq) + 2e^-$

$ClO_3^-(aq) \rightarrow ClO_4^-(aq) + 2e^- + 2H^+(aq)$

$H_2O(l) + ClO_3^-(aq) \rightarrow ClO_4^-(aq) + 2e^- + 2H^+(aq)$ oxidation half-reaction

$Cl^{5+} + 2e^- \rightarrow Cl^{3+}$

$ClO_3^-(aq) + 2e^- \rightarrow ClO_2^-(aq)$

$2H^+(aq) + ClO_3^-(aq) + 2e^- \rightarrow ClO_2^-(aq)$

$2H^+(aq) + ClO_3^-(aq) + 2e^- \rightarrow ClO_2^-(aq) + H_2O(l)$ reduction half-reaction

$2H^+(aq) + ClO_3^-(aq) + 2e^- + H_2O(l) + ClO_3^-(aq)$
$$\rightarrow ClO_4^-(aq) + 2e^- + 2H^+(aq) + ClO_2^-(aq) + H_2O(l)$$

$2ClO_3^-(aq) \rightarrow ClO_4^-(aq) + ClO_2^-(aq)$

---

20. Refer to Section 21.1 and Table 18.1.

---

a. $2Br^-(aq) \rightarrow Br_2(l) + 2e^-$ $E^\circ_{ox} = -1.077$ V

$Cl_2(g) + 2e^- \rightarrow 2Cl^-(aq)$ $E^\circ_{red} = +1.360$ V

$Cl_2(g) + 2Br^-(aq) \rightarrow 2Cl^-(aq) + Br_2(l)$ E° is (+) thus reaction occurs.

b. $2Cl^-(aq) \rightarrow Cl_2(g) + 2e^-$ $E^\circ_{ox} = -1.360$ V

$I_2(s) + 2e^- \rightarrow 2I^-(aq)$ $E^\circ_{red} = +0.534$ V

$I_2(s) + Cl^-(aq) \rightarrow$ **N.R.** E° is (-) thus no reaction occurs.

c. $2Br^-(aq) \rightarrow Br_2(l) + 2e^-$ $E^\circ_{ox} = -1.077$ V

$I_2(s) + 2e^- \rightarrow 2I^-(aq)$ $E^\circ_{red} = +0.534$ V

$I_2(s) + Br^-(aq) \rightarrow$ **N.R.** E° is (-) thus no reaction occurs.

d. $2Cl^-(aq) \rightarrow Cl_2(g) + 2e^-$ $E_{ox}^{\circ} = -1.360$ V

$Br_2(l) + 2e^- \rightarrow 2Br^-(aq)$ $E_{red}^{\circ} = +1.077$ V

$Br_2(l) + Cl^-(aq) \rightarrow$ **N.R.** E° is (-) thus no reaction occurs.

---

22. Refer to Section 21.1.

a. $Pb(N_3)_2(s) \rightarrow Pb(s) + 3N_2(g)$

b. $2O_3(g) \rightarrow 3O_2(g)$

c. $2H_2S(g) + O_2(g) \rightarrow 2H_2O(l) + 2S(s)$

---

24. Refer to Section 21.2.

a. $Cd^{2+}(aq) + H_2S(aq) \rightarrow 2H^+(aq) + CdS(s)$ (metathesis reaction)

b. $OH^-(aq) + H_2S(aq) \rightarrow H_2O(l) + HS^-(aq)$ (acid-base reaction)

c. $O_2(g) + 2H_2S(aq) \rightarrow 2H_2O(l) + 2S(s)$ (redox reaction)

---

26. Refer to Section 21.4 and Chapter 4.

a. $CaCO_3(s) + H_2SO_4(aq) \rightarrow CaSO_4(aq) + H_2O(l) + CO_2(g)$
Writing the soluble species as ions gives:
$CaCO_3(s) + 2H^+(aq) + SO_4^{2-}(aq) \rightarrow Ca^{2+}(aq) + SO_4^{2-}(aq) + H_2O(l) + CO_2(g)$
Net ionic: $CaCO_3(s) + 2H^+(aq) \rightarrow Ca^{2+}(aq) + H_2O(l) + CO_2(g)$

b. $2NaOH(aq) + H_2SO_4(aq) \rightarrow Na_2SO_4(aq) + 2H_2O(l)$
Writing the soluble species as ions gives:
$2Na^+(aq) + 2OH^-(aq) + 2H^+(aq) + SO_4^{2-}(aq) \rightarrow 2Na^+(aq) + SO_4^{2-}(aq) + 2H_2O(l)$
Net ionic: $H^+(aq) + OH^-(aq) \rightarrow H_2O(l)$

c. This is a redox reacton and best approached from that standpoint.

$Cu^0 \rightarrow Cu^{2+}(aq) + 2e^-$ oxidation half-reaction

$SO_4^{2-}(aq) \rightarrow SO_2(g)$
$S^{6+} + 2e^- \rightarrow S^{4+}$
$SO_4^{2-}(aq) + 2e^- \rightarrow SO_2(g)$
$4H^+(aq) + SO_4^{2-}(aq) + 2e^- \rightarrow SO_2(g)$
$4H^+(aq) + SO_4^{2-}(aq) + 2e^- \rightarrow SO_2(g) + 2H_2O(l)$ reduction half-reaction

$4H^+(aq) + 2e^- + SO_4^{2-}(aq) + Cu(s) \rightarrow Cu^{2+}(aq) + 2e^- + SO_2(g) + 2H_2O(l)$
$4H^+(aq) + SO_4^{2-}(aq) + Cu(s) \rightarrow Cu^{2+}(aq) + SO_2(g) + 2H_2O(l)$

Of the elements listed, only **P** forms allotropes.

30. *Refer to Section 21.1.*

    a. Ozone is prepared commercially by passing O_2 gas through a high voltage (10^4 V) electric discharge.

    b. Heating rhombic sulfur to 96°C converts it to monoclinic sulfur (slow process). Freezing liquid sulfur at 119°C, then quickly cooling to room temperature (faster) also converts it to monoclinic sulfur.

    c. Red phosphorus can be formed by heating white phosphorus to 300°C in the absence of air.

32. *Refer to Section 21.3 and Chapter 7.*

    a. 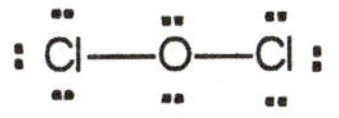 $: \overset{..}{\underset{..}{Cl}} — \overset{..}{\underset{..}{O}} — \overset{..}{\underset{..}{Cl}} :$

    b. 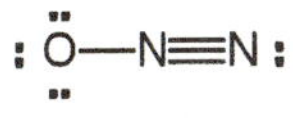 $: \overset{..}{\underset{..}{O}} — N \equiv N :$

    c. 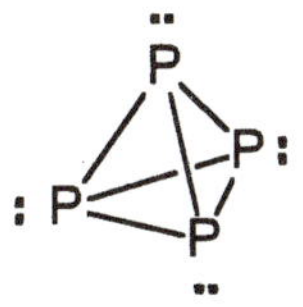 (P₄ tetrahedron structure)

    d.  $: N \equiv N :$

34. *Refer to Chapter 7.*

Both **(a)** Cl_2O and **(b)** N_2O are polar (Since Cl_2O is bent (AX_2E_2), the dipoles do not cancel.)

P_4 and N_2 both lack dipoles and are non-polar.

36. *Refer to Section 21.4 and Chapters 7 and 13.*

    a. NO_3^-

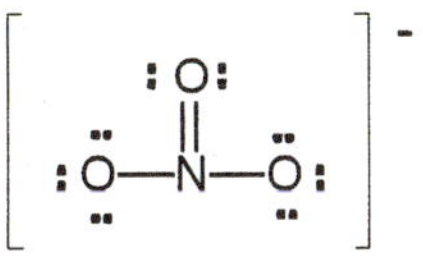

$$\left[\begin{array}{c} : \overset{..}{O} : \\ \| \\ : \overset{..}{\underset{..}{O}} — N — \overset{..}{\underset{..}{O}} : \end{array} \right]^-$$

b. HSO_4^-

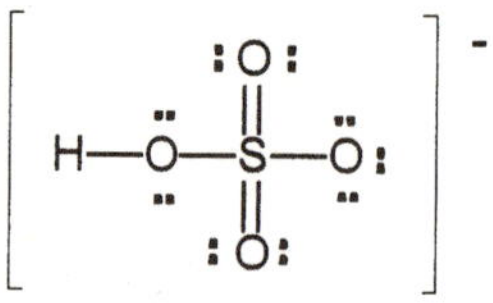

This Lewis structure has a -1 formal charge on the most electronegative atom (O).

c. $H_2PO_4^-$

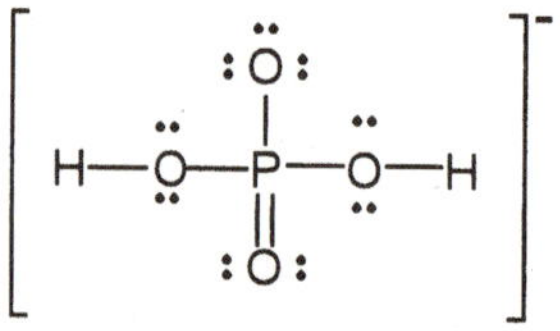

This Lewis structure has a -1 formal charge on the most electronegative atom (O).

---

38. Refer to Section 21.4 and Chapter 7.

a. N_2O_5

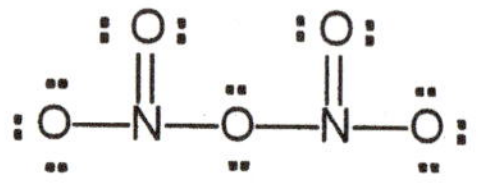

b. HNO_3

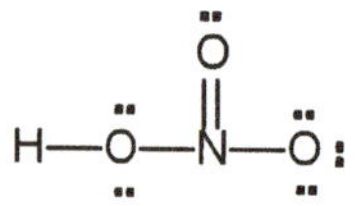

c. SO_4^{2-}

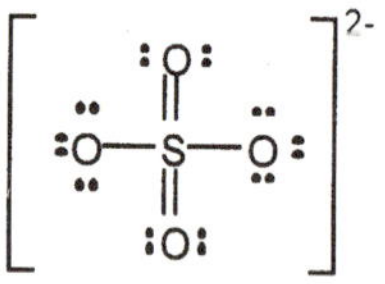

This Lewis structure has a -1 formal charge on two of the oxygens, which are the most electronegative atoms.

---

40. Refer to Section 21.1, Problem 16b above and Chapter 5.

Write the balanced equation and calculate the moles of Cl_2 needed for the oxidation. Then convert the moles to liters using the ideal gas law.

$$2NaI(aq) + Cl_2(g) \rightarrow I_2(s) + 2NaCl(aq)$$

$$175 \text{ g NaI} \times \frac{1 \text{ mol. NaI}}{149.89 \text{ g NaI}} \times \frac{1 \text{ mol. Cl}_2}{2 \text{ mol. NaI}} = 0.584 \text{ mol. Cl}_2$$

$$758 \text{ mm Hg} \times \frac{1 \text{ atm}}{760 \text{ mmHg}} = 0.997 \text{ atm}$$

$$V = \frac{nRT}{P} = \frac{(0.584 \text{ mol.})(0.0821 \text{ L} \cdot \text{atm/mol.} \cdot \text{K})(298 \text{ K})}{0.997 \text{ atm}} = 14.3 \text{ L}$$

$$n = \frac{PV}{RT} = \frac{(0.974\,\text{atm})(1.283\,\text{L})}{(0.0821\,\text{L} \cdot \text{atm/mol.} \cdot \text{K})(298\,\text{K})} = 0.0511\,\text{mol. HBr}$$

$$M = \frac{\text{mol.}}{\text{L}} = \frac{0.0511\,\text{mol.}}{0.250\,\text{L}} = 0.204\,M\,\text{HBr}$$

Calculate the amount of SO_2 produced, and from that, the moles of H_2S needed. Then calculate the volume of H_2S and the mass of S produced.

$$S(s) + O_2(g) \;\rightarrow\; SO_2(g)$$

$$1.00 \times 10^3\,\text{kg} \times \frac{1000\,\text{g}}{1\,\text{kg}} \times \frac{5.0\,\text{g S}}{100\,\text{g coal}} \times \frac{1\,\text{mol. S}}{32.066\,\text{g}} \times \frac{1\,\text{mol. } SO_2}{1\,\text{mol. S}} = 1.6 \times 10^3\,\text{mol. } SO_2$$

$$755\,\text{mm Hg} \times \frac{1\,\text{atm}}{760\,\text{mmHg}} = 0.993\,\text{atm}$$

$$2H_2S(g) + SO_2(g) \;\rightarrow\; 3S(s) + 2H_2O(l)$$

$$1.6 \times 10^3\,\text{mol. } SO_2 \times \frac{2\,\text{mol.} H_2S}{1\,\text{mol. } SO_2} = 3.2 \times 10^3\,\text{mol. } H_2S$$

$$V = \frac{(3.2 \times 10^3\,\text{mol.})(0.0821\,\text{L} \cdot \text{atm/mol.} \cdot \text{K})(300\,\text{K})}{0.993\,\text{atm}} = 7.9 \times 10^4\,\text{L}$$

$$1.6 \times 10^3\,\text{mol. } SO_2 \times \frac{3\,\text{mol. S}}{1\,\text{mol. } SO_2} \times \frac{32.066\,\text{g S}}{1\,\text{mol. S}} = 1.5 \times 10^5\,\text{g} = 0.15\,\text{metric tons S}$$

Convert gal. water to grams H_2S. Then calculate moles of Cl_2 required to react with that mass of H_2S.

$$1.00 \times 10^3\,\text{gal.} \times \frac{4\,\text{qt.}}{1\,\text{gal.}} \times \frac{1\,\text{L}}{1.057\,\text{qt.}} \times \frac{1000\,\text{mL}}{1\,\text{L}} = 3.78 \times 10^6\,\text{mL } H_2O$$

$$3.78 \times 10^6\,\text{mL } H_2O \times \frac{1.00\,\text{g}}{1\,\text{mL}} \times \frac{5\,\text{g } H_2S}{10^6\,\text{g } H_2O} = 19\,\text{g } H_2S$$

$$19 \text{ g H}_2\text{S} \times \frac{1 \text{ mol. H}_2\text{S}}{34.08 \text{ g H}_2\text{S}} \times \frac{1 \text{ mol. Cl}_2}{1 \text{ mol. H}_2\text{S}} = 0.56 \text{ mol. Cl}_2$$

$$V = \frac{(0.56 \text{ mol.})(0.0821 \text{ L} \cdot \text{atm/mol.} \cdot \text{K})(273 \text{ K})}{1 \text{ atm}} = 13 \text{ L}$$

To calculate the pH, calculate the moles of H^+, concentration of H^+, and the pH.

$$19 \text{ g H}_2\text{S} \times \frac{1 \text{ mol. H}_2\text{S}}{34.08 \text{ g H}_2\text{S}} \times \frac{2 \text{ mol. H}^+}{1 \text{ mol. H}_2\text{S}} = 1.1 \text{ mol. H}^+$$

$$1.1 \text{ mol. H}^+ \times \frac{1}{3.78 \times 10^6 \text{ mL H}_2\text{O}} \times \frac{1000 \text{ mL}}{1 \text{ L}} = 2.9 \times 10^{-4} \ M$$

$$\text{pH} = -\log [\text{H}^+] = -\log (2.9 \times 10^{-4}) = 3.54$$

---

48. Refer to Chapter 13.

$$\text{HClO} \rightleftharpoons \text{H}^+ + \text{ClO}^-$$

$$K_a = 2.8 \times 10^{-8} = \frac{[\text{H}^+][\text{ClO}^-]}{[\text{HClO}]}$$

If $[\text{H}^+] = x$, then $[\text{ClO}^-] = x$ and $[\text{HClO}] = 0.10 - x$.

$$K_a = 2.8 \times 10^{-8} = \frac{[x][x]}{[0.10 - x]} \implies 2.8 \times 10^{-9} - (2.8 \times 10^{-8})x = x^2$$

Using successive approximations: $x \cong \sqrt{2.8 \times 10^{-9}}$
$x = [\text{H}^+] = [\text{ClO}^-] = 5.3 \times 10^{-5} \ M$

Equilibrium concentration of HClO = $0.10 - (5.3 \times 10^{-5}) = 0.10 \ M$

$$\text{pH} = -\log [\text{H}^+] = -\log (5.3 \times 10^{-5}) = 4.28$$

---

50. Refer to Chapter 13.

The equilibrium equation for the final reaction is:

$$K_{\text{overall}} = \frac{[\text{H}^+][\text{HF}_2^-]}{[\text{HF}]^2} \qquad\qquad \text{eq. 1}$$

Write the equilibrium equation for the second reaction, and solve for HF_2^-.

$$K_1 = 2.7 = \frac{[\text{HF}_2^-]}{[\text{HF}][\text{F}^-]} \implies [\text{HF}_2^-] = (2.7)[\text{HF}][\text{F}^-]$$

Substituting this equation into equation 1 gives:

$$K_{overall} = \frac{[H^+](2.7)[HF][F^-]}{[HF]^2} = \frac{[H^+](2.7)[F^-]}{[HF]} \qquad \text{eq. 2}$$

Write the equilibrium equation for the first reaction, and solve for HF.

$$K_a = 6.9 \times 10^{-4} = \frac{[H^+][F^-]}{[HF]} \implies [HF] = \frac{[H^+][F^-]}{6.9 \times 10^{-4}}$$

Substituting this equation into equation 2 gives:

$$K_{overall} = \frac{[H^+](2.7)[F^-]}{\dfrac{[H^+][F^-]}{6.9 \times 10^{-4}}} = (2.7)(6.9 \times 10^{-4}) = 1.9 \times 10^{-3}$$

Thus we see that the equilibrium constant for the overall process is the product of the individual constants.

---

52. Refer to Chapter 16.

$K_{sp} = [Ba^{2+}][F^-]^2$

If x = mol. BaF_2, then $[F^-] = 2x$, and $[Ba^{2+}] = x$

$K_{sp} = 1.8 \times 10^{-7} = (0.10 - x)(2x)^2$

$K_{sp} = 1.8 \times 10^{-7} = 0.40x^2 - 4x^3$

Using successive approximations: $\quad 1.8 \times 10^{-7} = 0.40x^2$

$$x = 6.7 \times 10^{-4}\,M$$

Thus 6.7×10^{-4} moles of BaF_2 would dissolve in 1 L of 0.10 M $BaCl_2$

$$100\,mL \times \frac{1\,L}{1000\,mL} \times \frac{6.7 \times 10^{-4}\,mol.}{1\,L} \times \frac{175.33\,g\,BaF_2}{1\,mol.} = 0.012\,g\,BaF_2 \ \ (per\,100\,mL)$$

---

54. Refer to Chapter 17.

Calculate the enthalpy and entropy and then calculate the Gibbs free energy.

$\Delta S^\circ = \Sigma\, S^\circ_{(products)} - \Sigma\, S^\circ_{(reactants)}$
$\Delta S^\circ = [(2\,mol.)(0.1431\,kJ/mol.\cdot K) + (1\,mol.)(0.1522\,kJ/mol.\cdot K)]$
$\qquad - [(2\,mol.)(0.1492\,kJ/mol.\cdot K) + (1\,mol.)(0.2230\,kJ/mol.\cdot K)]$
$\Delta S^\circ = -0.0830\,kJ/K$

$$\Delta H^\circ = \Sigma\ \Delta H_f^\circ{}_{(products)} - \Sigma\ \Delta H_f^\circ{}_{(reactants)}$$
$$\Delta H^\circ = [(2\ mol.)(-397.7\ kJ/mol.) + (1\ mol.)(0.0\ kJ/mol.)]$$
$$- [(2\ mol.)(-360.2\ kJ/mol.) + (1\ mol.)(0.0\ kJ/mol.)]$$
$$\Delta H^\circ = -75.0\ kJ$$

$$\Delta G^\circ = \Delta H^\circ - T\ \Delta S^\circ$$
$$\Delta G^\circ = -75.0\ kJ - (298\ K)(-0.0830\ kJ/K)$$
$$\Delta G^\circ = -50.2\ kJ$$

ΔG is negative, therefore **the reaction is spontaneous**.

The temperature at which the reaction is no longer spontaneous is that at which $\Delta G^\circ = 0$

$$0 = -75.0\ kJ/mol. - (T)(0.0832\ kJ/mol.\cdot K)$$

$T = 901\ K.$ Thus the reaction is spontaneous **up** to 901 K, and the

lowest temperature at which the reaction is spontaneous is 0 K.

---

56. *Refer to Chapter 17.*

---

a.   Calculate the enthalpy and entropy and then calculate the Gibbs free energy.

$$\Delta S^\circ = \Sigma\ S^\circ{}_{(products)} - \Sigma\ S^\circ{}_{(reactants)}$$
$$\Delta S^\circ = [(2\ mol.)(0.2230\ kJ/mol.\cdot K) + (2\ mol.)(0.0699\ kJ/mol.\cdot K)]$$
$$- [(4\ mol.)(0.1868\ kJ/mol.\cdot K) + (1\ mol.)(0.2050\ kJ/mol.\cdot K)]$$
$$\Delta S^\circ = -0.3664\ kJ/K$$

$$\Delta H^\circ = \Sigma\ \Delta H_f^\circ{}_{(products)} - \Sigma\ \Delta H_f^\circ{}_{(reactants)}$$
$$\Delta H^\circ = [(2\ mol.)(0.0\ kJ/mol.) + (2\ mol.)(-285.8\ kJ/mol.)]$$
$$- [(4\ mol.)(-92.3\ kJ/mol.) + (1\ mol.)(0.0\ kJ/mol.)]$$
$$\Delta H^\circ = -202.4\ kJ$$

$$\Delta G^\circ = \Delta H^\circ - T\ \Delta S^\circ$$
$$\Delta G^\circ = -202.4\ kJ - (298\ K)(-0.3664\ kJ/K)$$
$$\Delta G^\circ = -93.2\ kJ$$

ΔG is negative, therefore **the reaction is spontaneous**.

b. $\Delta G^\circ = -RT \ln K$

$$\ln K = \frac{-\Delta G}{RT} = \frac{93200\ J}{(8.314\ J/mol.\cdot K)(298\ K)} = 37.6$$

$$K = e^{37.6} = 2.14 \times 10^{16}$$

Sublimation is the phase change from solid to gas. Calculate T using $\Delta G° = 0$, the point at which the phase change is becoming spontaneous.

$P_4(s) = P_4(g)$

$\Delta S° = (1 \text{ mol.})(0.2800 \text{ kJ/mol.·K}) - (1 \text{ mol.})(0.1644 \text{ kJ/mol.·K})$
$\Delta S° = 0.1156 \text{ kJ/·K}$

$\Delta H° = (1 \text{ mol.})(58.9 \text{ kJ/mol.}) - (1 \text{ mol.})(0.0 \text{ kJ/mol.})$
$\Delta H° = 58.9 \text{ kJ}$

$\Delta G° = \Delta H° - T \Delta S°$
$0 = 58.9 \text{ kJ} - (T)(0.1156 \text{ kJ/K})$
$58.9 \text{ kJ} = (T)(0.1156 \text{ kJ/K})$
$T = 510 \text{ K} = 237°\text{C}$

Calculate the moles of electrons produced by the current. Then calculate the mass of F_2 produced with the given efficiency.

$$7.00 \times 10^3 \text{ C/s} \times \frac{3600 \text{ s}}{1 \text{ hr}} \times \frac{24 \text{ hr}}{1 \text{ d}} \times 2 \text{ d} \times \frac{1 \text{ mol.}}{9.648 \times 10^4 \text{ C}} = 1.25 \times 10^4 \text{ mol. } e^-$$

$2F^- \rightarrow F_2 + 2e^-$

$$1.25 \times 10^4 \text{ mol. } e^- \times 0.95 \times \frac{1 \text{ mol. } F_2}{2 \text{ mol. } e^-} \times \frac{37.996 \text{ g } F_2}{1 \text{ mol. } F_2} = 2.26 \times 10^5 \text{ g } F_2$$

Calculate the moles of electrons produced by the current. Then calculate the mass of $NaClO_4$ produced.

$Cl^{5+} \rightarrow Cl^{7+} + 2e^-$
$ClO_3^- \rightarrow ClO_4^- + 2e^-$

$$1.50 \times 10^3 \text{ C/s} \times \frac{3600 \text{ s}}{1 \text{ hr}} \times 8 \text{ hr} \times \frac{1 \text{ mol.}}{9.648 \times 10^4 \text{ C}} = 448 \text{ mol. } e^-$$

$$448 \text{ mol. } e^- \times \frac{1 \text{ mol. } NaClO_4}{2 \text{ mol. } e^-} \times \frac{122.44 \text{ g}}{1 \text{ mol. } NaClO_4} = 2.74 \times 10^4 \text{ g} = 27.4 \text{ kg } NaClO_4$$

To determine if the species will be oxidized by H_2O_2, add E°_{ox} and E°_{red} (H_2O_2). If the sum is negative, then the species will not be oxidized.

(Note: The values listed in table 18.1 are *reduction* potentials.)

a. Co^{2+} $1.763 \text{ V} + (-1.953 \text{ V}) = -0.190 \text{ V}$ Will **not** be oxidized.

b. Cl^- $1.763 \text{ V} + (-1.360 \text{ V}) = 0.403 \text{ V}$ Will be oxidized.

c. Fe^{2+} $1.763 \text{ V} + (-0.769 \text{ V}) = 0.994 \text{ V}$ Will be oxidized.

d. Sn^{2+} $1.763 \text{ V} + (-0.154 \text{ V}) = 1.609 \text{ V}$ Will be oxidized.

The first steps are to calculate E° and write the balanced redox reaction. Then apply the Nernst equation, solving for $[H^+]$. With $[H^+]$ in hand, calculate pH.

$SO_2 \rightarrow SO_4^{2-}$	-0.115 V
$NO_3^- \rightarrow NO$	0.964 V
$NO_3^- + SO_2 \rightarrow NO + SO_4^{2-}$	0.809 V

$S^{4+} \rightarrow S^{6+} + 2e^-$

$SO_2(g) \rightarrow SO_4^{2-}(aq) + 2e^-$

$SO_2(g) \rightarrow SO_4^{2-}(aq) + 2e^- + 4H^+(aq)$

$2H_2O(l) + SO_2(g) \rightarrow SO_4^{2-}(aq) + 2e^- + 4H^+(aq)$ Oxidation half-reaction

$N^{5+} + 3e^- \rightarrow N^{2+}$

$NO_3^-(aq) + 3e^- \rightarrow NO(g)$

$4H^+(aq) + NO_3^-(aq) + 3e^- \rightarrow NO(g)$

$4H^+(aq) + NO_3^-(aq) + 3e^- \rightarrow NO(g) + 2H_2O(l)$ Reduction half-reaction

To balance the electrons, multiply the oxidation half-reaction by 3 and the reduction half-reaction by 2, and then add the results and simplify.

$6H_2O(l) + 3SO_2(g) + 8H^+(aq) + 2NO_3^-(aq) + 6e^-$
$\qquad\qquad \rightarrow 2NO(g) + 4H_2O(l) + 3SO_4^{2-}(aq) + 6e^- + 12H^+(aq)$

$2H_2O(l) + 3SO_2(g) + 2NO_3^-(aq) \rightarrow 2NO(g) + 3SO_4^{2-}(aq) + 4H^+(aq)$

$$E = E^\circ - \frac{0.0257}{n} \ln Q \Rightarrow 1.000 \text{ V} = 0.809 \text{ V} - \frac{0.0257}{6} \ln \left(\frac{(P_{NO})^2 [SO_4^{2-}]^3 [H^+]^4}{(P_{SO_2})^3 [NO_3^-]^2} \right)$$

$$0.191\,V = -0.00428\ln\left(\frac{(1)^2[0.100]^3[H^+]^4}{(1)^3[0.100]^2}\right)$$

$$-44.6 = \ln(0.100)[H^+]^4$$

$$(0.100)[H^+]^4 = e^{-44.6} = 4.3 \times 10^{-20}$$

$$[H^+]^4 = 4.3 \times 10^{-19}$$

$$[H^+] = 2.6 \times 10^{-5}\,M$$

$$pH = -\log(2.6 \times 10^{-5}) = 4.59$$

---

68. Refer to Chapter 9.

a. dispersion forces

b. dispersion and dipole forces

c. dispersion forces, dipole forces and hydrogen bonding

d. dispersion forces, dipole forces and hydrogen bonding

e. none, this is an ionic solid

---

70. Refer to Chapter 4.

a. NO_2^-
      O: -2
      N: +3 $(N + 2(-2) = -1)$

b. NO_2
      O: -2
      N: +4 $(N + 2(-2) = 0)$

c. HNO_3
      O: -2
      H: +1
      N: +5 $(N + 1 + 3(-2) = 0)$

d. NH_4^+
      H: +1
      N: -3 $(N + 4(+1) = +1)$

---

72. Refer to Sections 21.1 and 21.2.

a. HClO

b. S, $KClO_3$

c. NH_3, NaClO

d. HF

401

a. Increasing oxidation number corresponds to an increase in oxygens atoms around the central atom. These oxygen atoms stabilize the negative charge of the resulting conjugate base by i) delocalizing the negative charge (consider resonance structures) and ii) by the electronegative oxygens pulling electron density away from the atom bearing the charge.

b. NO_2 has an odd number of electrons and will thus have an unpaired electron.

c. In general, the reduction reactions of the oxoanions (oxoanions acting as oxidizing agents) involve H^+ as a reactant. Therefore higher $[H^+]$ (lower pH) causes the reaction to be more spontaneous.

d. The sugar is oxidized to carbon, which is black.

To determine the amount of sulfuric acid that could be produced, one would first need to know the mass of sulfur present. This would require that one know the **depth of the deposit** to calculate the volume of the deposit, the **density** to calculate the mass of the deposit, and the **percent by mass (or purity) of the sulfur**, in the deposit to calculate the mass of sulfur.

Use a given mass of quartz (i.e. 1000 g) to calculate the amount of gold present and the amount of HF solution needed. Then calculate the cost of each.

$$1000 \, g \, SiO_2 \times \frac{1.0 \times 10^{-3} \, g \, Au}{100 \, g \, SiO_2} = 0.010 \, g \, Au$$

$$1000 \, g \, SiO_2 \times \frac{1 \, mol. \, SiO_2}{60.08 \, g \, SiO_2} \times \frac{4 \, mol. \, HF}{1 \, mol. \, SiO_2} \times \frac{20.01 \, g}{1 \, mol. \, HF} = 1330 \, g \, HF$$

$$1330 \, g \, HF \times \frac{100 \, g \, sol'n}{50 \, g \, HF} \times \frac{1 \, mL}{1.17 \, g \, sol'n} = 2270 \, mL \, solution$$

$$2270 \, mL \times \frac{1 \, L}{1000 \, mL} \times \frac{\$ 0.75}{1 \, L} = \$1.70$$

$$0.010 \, \text{g Au} \times \frac{1 \, \text{troy oz}}{31.1 \, \text{g}} \times \frac{\$425}{1 \, \text{troy oz}} = \$0.14$$

Thus it would cost $1.70 of HF solution to recover $0.14 of gold. Definitely not economical.

---

77. *Refer to Chapters 3 and 4.*

Determine the mole ratios of the reductant and oxidant for each redox couple. Then use mass relationships of thiosulfate to iodine and iodine to NaClO to determine the mass of NaClO and from that, the mass percent of NaClO in the bleach solution.

$$S^{2+} \rightarrow S^{2.5+} + \tfrac{1}{2}e^-$$
$$2S_2O_3^{2-} \rightarrow S_4O_6^{2-} + 2e^-$$

$$I^0 + e^- \rightarrow I^-$$
$$I_2 + 2e^- \rightarrow 2I^-$$

$$2S_2O_3^{2-} + I_2 \rightarrow 2I^- + S_4O_6^{2-}$$

This gives the mole ratio of $S_2O_3^{2-}$ to I_2 for the calculation of moles of I_2.

$$25.00 \, \text{mL} \times \frac{1 \, \text{L}}{1000 \, \text{mL}} \times \frac{0.0700 \, \text{mol. S}_2\text{O}_3^{2-}}{1 \, \text{L}} \times \frac{1 \, \text{mol. I}_2}{2 \, \text{mol. S}_2\text{O}_3^{2-}} = 8.75 \times 10^{-4} \, \text{mol. I}_2$$

$$Cl^+(aq) + 2e^- \rightarrow Cl^-(aq)$$
$$2H^+(aq) + ClO^-(aq) + 2e^- \rightarrow Cl^-(aq) + H_2O(l)$$

$$2I^- \rightarrow I_2 + 2e^-$$

$$2H^+(aq) + ClO^-(aq) + 2I^-(aq) \rightarrow I_2(aq) + Cl^-(aq) + H_2O(l)$$

This gives the mole ratio of $ClO^-(aq)$ (or NaClO) to I_2 for the calculation of mass of NaClO.

$$8.75 \times 10^{-4} \, \text{mol. I}_2 \times \frac{1 \, \text{mol. NaClO}}{1 \, \text{mol. I}_2} \times \frac{74.44 \, \text{g NaClO}}{1 \, \text{mol. NaClO}} = 0.0651 \, \text{g NaClO}$$

$$\text{mass \%} = \frac{0.0651 \, \text{g NaClO}}{5.00 \, \text{g sol'n}} \times 100\% = 1.30\%$$

78. *Refer to Section 21.1, Chapter 5 and "Chemistry Beyond the Classroom: Airbags"* *(p. 135).*

Assume that none of the N_2 escapes from the bag. Then use the ideal gas law to calculate the moles of N_2 needed to fill 20.0 L, and the amount of NaN_3 to produce that N_2.

$$n = \frac{PV}{RT} = \frac{(1\,\text{atm})(20.0\,\text{L})}{(0.0821\,\text{L} \cdot \text{atm/mol} \cdot \text{K})(298\,\text{K})} = 0.817\,\text{mol. N}_2$$

$$2NaN_3(s) \rightarrow 2Na(s) + 3N_2(g)$$

$$0.817\,\text{mol. N}_2 \times \frac{2\,\text{mol. NaN}_3}{3\,\text{mol. N}_2} \times \frac{65.01\,\text{g NaN}_3}{1\,\text{mol. NaN}_3} = 35.4\,\text{g NaN}_3$$

Chapter 22: Organic Chemistry

---

2. *Refer to Sections 22.1 and 22.2 and Example 22.2.*

For a hydrocarbon with n carbons, if the number of hydrogens equals $2n+2$, then it is an alkane, if it equals $2n$, then it is an alkene, if it equals $2n-2$, then it is an alkyne. This assumes the hydrocarbons are **not** cyclic and contain at most, one double bond.

a. $C_{12}H_{24}$ $n=12$, number of hydrogens $= 24 = 2n$, thus it is an **alkene**.

b. C_7H_{12} $n=7$, number of hydrogens $= 12 = 2n-2$, thus it is an **alkyne**.

c. $C_{13}H_{28}$ $n=13$, number of hydrogens $= 28 = 2n+2$, thus it is an **alkane**.

---

4. *Refer to Sections 22.1 and 22.2 and Example 22.2.*

For a hydrocarbon with n carbons, the number of hydrogens in an alkane equals $2n+2$, in alkene it equals $2n$, in an alkyne it equals $2n-2$. This assumes the hydrocarbons are **not** cyclic and contain at most, one double bond.

a. For alkynes, the number of hydrogens $= 2n-2 = 16$, thus $n = 9$. C_9H_{16}

b. For alkenes, the number of hydrogens $= 2n = 44$, thus $n = 22$. $C_{22}H_{44}$

c. For alkanes, the number of hydrogens $= 2n+2 = 2(10) + 2 = 22$. $C_{10}H_{22}$

---

6. *Refer to Section 22.4 and Example 22.4.*

a. **Alcohol**. The -OH group is not attached to a C=O.

b. **Ester**. The C=O is attached to a -O-, but not an -OH.

c. **Carboxylic acid and Ester**. The central C=O (the third C from the left) is attached to an -O- which is not an -OH, thus that group is an ester. The terminal COOH is a C=O with and -OH attached, thus it is a carboxylic acid.

a.
$$CH_3-\underset{\underset{OH}{|}}{CH}-CH_3$$

b.
$$\underset{H_3}{\overset{CH_3}{\diagdown}}CH-COOH$$

c.
$$\left.\begin{array}{c}CH_3-\underset{\underset{C=O}{|}}{CH}-CH_3\end{array}\right\}\ \text{From the acid}$$
$$\left.\begin{array}{c}\overset{|}{O}\\CH_3-\overset{|}{CH}-CH_3\end{array}\right\}\ \text{From the alcohol}$$

Frequency	Bond Type
5732 nm	C=O
3305 nm	O-H (acid) or C-H

The frequency at 5732 nm indicates C=O. The frequency at 3305 nm could be an acid O-H or a C-H. The combination of C=O and O-H generally indicates a **carboxylic acid**. That the O-H is in the acid region, confirms this assignment. The combination of C=O and C-H could be either a **ketone** or an **aldehyde**.

Thus the compound could be either an **aldehyde, carboxylic acid or a ketone**.

Recall that placing the double bond between the first two carbons is identical to placing the double bond between the last two (as shown in the 1st structure). Thus there are only three unique isomers.

$$CH_3-CH_2-CH=CH_2 \qquad CH_3-CH=CH-CH_3 \qquad CH_3-\underset{\underset{CH_3}{|}}{C}=CH_2$$

Note that the last structure is identical to the last structure given in the answer in Appendix 6.

This type of problem is best approached by systematically moving the Cl atoms, first keeping two to a carbon, then moving them one at a time. Be aware that the isomer with both Cl's on the 1st carbon is identical to that with both on the last carbon.

$$Cl_2CH-CH_2-CH_3 \qquad CH_3-CCl_2-CH_3 \qquad ClCH_2-\underset{\underset{Cl}{|}}{CH}-CH_3 \qquad ClCH_2-CH_2-CH_2Cl$$

This type of problem is best approached by systematically moving the Cl atoms. Be aware that the isomer with Cl's on carbons 1, 2 and 4 is identical to that with Cl's on carbons 1, 2 and 5.

$$\text{(1,2,3-trichlorobenzene)} \qquad \text{(1,2,4-trichlorobenzene)} \qquad \text{(1,3,5-trichlorobenzene)}$$

Refer to Problem 12 (above) for the three butene isomers, then systematically replace 2 H's of the double-bonded carbon(s), with Cl and Br.

$$\underset{\text{Cl Br}}{CH_3-CH_2-C=CH} \qquad \underset{\text{Br}}{CH_3-CH_2-CH=C-Cl} \qquad \underset{\text{Br Cl}}{CH_3-CH_2-C=CH}$$

$$\underset{\text{Br Cl}}{CH_3-C=C-CH_3} \qquad \underset{\text{CH}_3\ \text{Cl}}{CH_3-C=C{\diagdown}_{Br}}$$

Since the carboxylic acid group can only be on an end of the hydrocarbon chain, the number of isomers is severely limited.

$$CH_3-CH_2-CH_2-COOH \qquad \underset{CH_3}{\overset{CH_3}{\diagdown}}CH-COOH$$

All the isomers, except the last one (($CH_3)_2C=CClBr$), have geometric isomers. The last does not have geometric isomers since it has two identical groups ($-CH_3$) on the same carbon. In assigning *cis* / *trans* to isomers, the terms refer to the relationship between the larger groups attached to each carbon. For example, in the first pair of isomers, Cl is larger than C, and Br is larger than H; thus *cis* refers to the isomer in which the Cl and Br are on the same side.

cis   trans   cis   trans

a) Cl   Br   Cl   H   b) H   Cl   H   Br
     \\ /   \\ /   \\ /   \\ /
      C=C   C=C   C=C   C=C
     /   \\   /   \\   /   \\   /   \\
 CH₃CH₂   H   CH₃CH₂   Br   CH₃CH₂   Br CH₃CH₂   Cl

c) Br   Cl   Br   H   d) Br   Cl   Br   CH₃
     \\ /   \\ /   \\ /   \\ /
      C=C   C=C   C=C   C=C
     /   \\   /   \\   /   \\   /   \\
 CH₃CH₂   H   CH₃CH₂   Cl   CH₃   CH₃   CH₃   Cl

---

24. Refer to Section 22.5.

a. No. The two groups attached to any one of the double-bonded carbons are identical.

b. Yes. Refer to the pair of structures in problem 22(d) and replace the bromines with Cl, note that the two isomers are still unique after the substitution.

c. Yes. Refer to the pair of structures in problem 22(d).

---

26.   Refer to Section 22.5.

To show optical isomerism, the molecule must have a carbon atom with 4 different groups attached.

a. H_2CCl_2. No, the carbon atom has two identical groups attached.

b. $ClCH_2$-CH_2Cl. No, each of the carbons has two identical groups (H's).

c. CBrClFH. Yes, the carbon has 4 different groups (H, F, Cl and Br).

d. CH_3-$CH(Br)OH$. Yes, the alcoholic carbon has 4 different groups (OH, H, Br and CH_3).

---

28. Refer to Section 22.5.

The chiral carbons (denoted with an *) are those that have four different groups attached.

a.   H   H   b.   H-C=C-CH₂-OH   c.   Cl F
         | | | | | |
    CH₃-*C - C-H   H H   no chiral carbons CH₃-C - C*-Cl
         | | | |
         OH OH   Cl H

408

30. Refer to Section 22.3.

The circle represents the six delocalized electrons of an aromatic ring.

32. Refer to Section 22.4.

The first two reactions are given in the text. The text shows the reaction to form methyl acetate. The formation of ethyl acetate is identical except that ethanol is used instead of methanol.

a. $CO(g) + 2H_2(g) \rightarrow CH_3OH(g)$

b. $CH_3CH_2OH(aq) + O_2(g) \rightarrow CH_3CO_2H(aq) + H_2O(l)$

c. $CH_3CO_2H(l) + CH_3CH_2OH(l) \rightarrow CH_3CO_2CH_2CH_3(l) + H_2O(l)$

34. Refer to Section 22.4 and "Chemistry Beyond the Classroom: Cholesterol."

a. A saturated fat is one that has no multiple bonds. Each carbon has the maximum number of H's, thus each carbon is said to be saturated with hydrogens. This compound is an alkane.

b. Soaps are sodium salts of "fatty acids" (long chain carboxylic acids).

c. Proof refers to the concentration of an aqueous solution of ethanol. The proof is twice the concentration, expressed as a volume percent.

d. Ethanol that has been poisoned to render it undrinkable. This is done so that alcohol intended for industrial (non-consumption) purposes, and thus exempt from federal taxes, cannot be bootlegged.

35. *Refer to "Chemistry Beyond the Classroom: Cholesterol."*

In the structure, each vertex not explicitly labeled represents a carbon atom. Each carbon atom forms 4 bonds. Bonds not shown are understood to be hydrogens. Thus:

counting the C's, H's and O gives: $C_{27}H_{46}O$.

36. *Refer to Section 22.5.*

This type of problem is best approached systematically. Start with the linear molecule and move the OH group. Then repeat with a branched molecule, then with a second branched molecule and so on, until all possibilities have been exhausted. Note that the solution below lists the same molecules in the same order as that in Appendix 6; many of the molecules have been "flipped," but they are identical to those in the appendix.

Start by calculating the amount of heat need to raise the temperature of the water to boiling.

$q = mc\Delta t$

$$m = 1\,\text{qt} \times \frac{1\,\text{L}}{1.057\,\text{qt}} \times \frac{1000\,\text{mL}}{1\,\text{L}} \times \frac{1\,\text{g}}{1\,\text{mL}} = 946\,\text{g}$$

$\Delta t = 100°C - 25°C = 75°C$

$q = (946\,\text{g})(4.184\,\text{J/}°\text{C})(75°\text{C}) = 297000\,\text{J} = 297\,\text{kJ}$

Now calculate the amount of heat given off by the combustion of one mole of propane.

$C_3H_8(g) + 5O_2(g) \rightarrow 3CO_2(g) + 4H_2O(l)$

$\Delta H° = \Sigma\,\Delta H_f°\,_{\text{products}} - \Delta H_f°\,_{\text{reactants}}$

$\Delta H° = [3\Delta H_f°CO_2 + 4\Delta H_f°H_2O] - [\Delta H_f°C_3H_8 + 5\Delta H_f°O_2]$

$\Delta H° = [(3\,\text{mol.})(-393.2\,\text{kJ/mol.}) + (4\,\text{mol.})(-285.8\,\text{kJ/mol.})]$
$\quad\quad - [(1\,\text{mol.})(-103.8\,\text{kJ/mol.}) + (5\,\text{mol.})(0\,\text{kJ/mol.})]$

$\Delta H° = -2220\,\text{kJ/mol.}$

Finally, calculate the moles and grams of propane needed to raise the temperature of the water.

$$297\,\text{kJ} \times \frac{1\,\text{mol. C}_3\text{H}_8}{2220\,\text{kJ}} \times \frac{44.10\,\text{g C}_3\text{H}_8}{1\,\text{mol. C}_3\text{H}_8} = 5.9\,\text{g}$$

About 5.9 g propane is needed to the temperature of one quart of water from 25°C (roughly room temperature) to boiling. This neglects the specific heat of the pan and also assumes that all the heat given off is actually transferred to the water.